The City & Guilds textbook

Site Carpentry and Architectural Joinery

LEVEL 3 TECHNICAL CERTIFICATE (7906)
LEVEL 3 DIPLOMA (6706)
LEVEL 3 APPRENTICESHIP (6571)

Stephen Redfern
Stephen Jones
Martin Burdfield
Colin Fearn

Orders: please contact Hachette UK Distribution, Hely Hutchinson Centre, Milton Road, Didcot, Oxfordshire, OX11 7HH. Telephone: +44 (0)1235 827827. Email education@hachette.co.uk Lines are open from 9 a.m. to 5 p.m., Monday to Friday. You can also order through our website: www.hoddereducation.co.uk

ISBN: 978 1 5104 5815 4

© Stephen Redfern, Stephen Jones, Martin Burdfield and Colin Fearn, 2020

First published in 2020 by
Hodder Education,
An Hachette UK Company
Carmelite House
50 Victoria Embankment
London EC4Y 0DZ

www.hoddereducation.co.uk

Impression number   10 9 8 7 6 5 4 3

Year   2024 2023

Cover photo © Shutterstock/stuar

Illustrations by Integra Software Services Pvt, Ltd.

Typeset by Integra Software Services Pvt. Ltd., Pondicherry, India

Printed by CPI Group (UK) Ltd, Croydon CR0 4YY

A catalogue record for this title is available from the British Library.

MIX
Paper | Supporting responsible forestry
FSC™ C104740

Contents

About your qualification

Introduction to the carpentry and joinery qualifications

You are completing one of the following qualifications:
- Level 3 Advanced Technical Diploma in Site Carpentry (7906–30)
- Level 3 Advanced Technical Diploma in Architectural Joinery (7906–31)
- Level 3 Diploma in Site Carpentry (6706–33)
- Level 3 Diploma in Bench Joinery (6706–36)
- Level 3 NVQ Diploma in Wood Occupations – Site Carpentry (6571–07; 6571–27)
- Level 3 NVQ Diploma in Wood Occupations – Architectural Joinery (6571–08; 6571–28)

The Level 3 Advanced Technical Diplomas and Level 3 Diplomas are for learners who are interested in developing the specific technical and professional skills that can lead to meaningful employment in the construction industry as a site carpenter or architectural joiner.

The Level 3 NVQ Diplomas in Wood Occupations are the on-programme qualifications for the Advanced Carpentry and Joinery Apprenticeship (6571-08 or 6571-28) and are designed to provide the apprentice with the opportunity to develop the knowledge, skills and core behaviours that are expected of an advanced site carpenter or architectural joiner. An advanced carpenter or joiner can undertake complex tasks that require high levels of practical skills and knowledge. They also need to be able to manage their own work and lead small teams.

The 7906 Advanced Technical Diplomas and 6706 Level 3 Diplomas provide the knowledge and practical skills to prepare you for an apprenticeship. The apprenticeship will give you an understanding of suitable advanced on-site skills and further knowledge required to work in the construction industry.

How to achieve your qualification

The requirements for successfully obtaining your qualification depend on which programme you are enrolled on.

7906

The Level 3 7906 qualification is assessed using one written exam (externally marked) and one practical synoptic assignment (internally marked and externally moderated).

For the synoptic assignment, a typical brief for site carpenters could be to fit a door into a frame on site, and to adapt the door to fit the frame. You will need to draw on the knowledge, understanding and skills developed across the qualification to prepare a cutting list, machine the timber to size, hang the door and fit a lock, prepare for painting and check correct operation. Architectural joiners may be asked to design and build a bespoke staircase, drawing on knowledge and skills from across the specification to prepare and interpret drawings and client specifications, set out quarter turn of tapered steps, cut joints and fit components accurately and assemble. You will need to demonstrate that you are following Health and Safety regulations at all times, which will draw upon your knowledge of legislation and regulations.

The Level 3 Knowledge test will be made up of ten multiple-choice questions which test 'Recall of Knowledge'. In addition, a number of short written answer style questions will test 'Understanding Knowledge'. The final question will be an 'Extended Response' question, which allows the candidate to show the depth of knowledge they have and will be worth 12 marks.

6706

The Level 3 6706 qualification is assessed through a series of multiple-choice tests (one for each unit within the qualification) and practical assignments.

The practical assignments allow you to demonstrate your practical skills. Prior to carrying out each task, you will be provided with task instructions – for example, you may be asked to set out a shaped door. You will be told how the task will be assessed and how long you have to complete the task. Your assessor will watch you carry out the tasks and check your final pieces of work. You can ask your assessor for help in understanding the task instructions, but all of the work must be your own. You must use safe working practices at all times.

The multiple-choice tests will cover all of the knowledge outcomes for each unit.

6571

The Level 3 NVQ Diplomas in Wood Occupations are assessed by a portfolio of evidence completed for each unit within the qualifications. Evidence produced through performance in the workplace is the main source for meeting the requirements, and may include naturally occurring documentary evidence (hard copy and electronic), direct observation of activities and witness testimony. Workplace evidence is supported by evidence of knowledge and understanding, which may be demonstrated by:

- questioning
- recognised industry education and training programme assessment or professional interview assessment that has been matched to National Occupational Standards (NOS) requirements
- performance evidence.

The apprenticeship is assessed separately to the on-programme qualification and is assessed by an end-point assessment (EPA [9079]). In order to progress through the end-test gateway to end-point assessment, you must complete the following:

- Level 3 NVQ Diploma in Wood Occupations – Site Carpentry (6571–27) or Level 3 NVQ Diploma in Wood Occupations – Architectural Joinery (6571–28) qualification
- Level 2 Maths
- Level 2 English.

The graded EPA will be comprised of the following assessment methods:

- multiple-choice test
- one-day practical test, which includes a practical activity and oral questioning.

Acknowledgements

This book draws on several earlier books that were published by City & Guilds, and we acknowledge and thank the writers of those books:

- Colin Fearn
- Stuart Raine
- Tim Taylor
- Martin Burdfield.

We would also like to thank everyone who has contributed to City & Guilds photoshoots. In particular, thanks to: Andrew Buckle (photographer), Tony Manktelow, Victoria Lockwood, Lindsay Cotte, David Hartsilver and all the staff at Burton and South Derbyshire College and at Central Sussex College, models Charlie Barber, Kieran Kelly, Jake North, Joe Smith and Martin Standbridge, and Collin Fearn and Steve Redfern.

Contains public sector information licensed under the Open Government Licence v3.0.

From the authors

I'd like to thank my wife, Rebecca, for her patience. I love you. I'd also like to thank my son, Daniel, and daughters, Rachel and Jess, for their support whilst working on this book.

Stephen Jones

I would like to say a big thank you to my wife, Sharon. Without her love, support and patience these chapters would not be completed. Also, a big thank you to my two children, Katie and Shaun, and my four grandchildren, Lewis, Charlotte, Evelyn and Emilie, for all their love and support. I would also like to thank Brian, a former tutor of mine and now dear friend, who I have had the pleasure of knowing for around 35 years. He has continually encouraged and pushed me to achieve the success I have.

Stephen Redfern

I wish to thank my darling wife Clare and our grown offspring, Matthew and Eleanor, for their encouragement, and Julia Sandford-Cooke for her editorial support during the writing process.

Martin Burdfield

About the authors

Stephen Jones

I was born and grew up in Newport, South Wales. After leaving school at 16, I started an apprenticeship with a local Joinery business, where I completed my training. I then met my wife and later moved to Torquay in Devon, where we got married and raised three children.

During my career in the South West, I have worked as a shopfitter and joiner and self-employed site carpenter, travelling all over the country on various contracts. I had my own shopfitting and joinery business for a number of years, working on retail, leisure and domestic projects, and employed a small workforce. In 2004, I began teaching at South Devon College. During my time in further education, I have continued to develop my knowledge and skills by completing the Postgraduate Certificate in Education (PGCE), Higher National Certificate (HNC) in Construction and the National Examination Board in Occupational Safety and Health (NEBOSH) Certificate. I am still employed as a full-time lecturer, assessor and internal verifier of Site Carpentry and Architectural Joinery courses from Level 1 to 3.

When I first left school, I never dreamt that I would do or see many of the things that I have during my time in the construction industry. I have been able to travel the length and breadth of the country with work and to visit other countries. This year I have been fortunate enough to have the opportunity to write my third and fourth Carpentry and Joinery textbooks, a feat I never thought possible when I was at school.

If you have a passion and enjoy what you do, 'the world is your oyster'.

Stephen Redfern

I was born and grew up in the Midlands, where I continue to live. I am married with two children and four grandchildren.

On leaving school at 16, I managed to get an indentured apprenticeship with a joinery manufacturer. I have spent the better part of 40 years working in joinery and the construction industry, 26 of which were at a further education college from which I have now retired as a course leader for Joinery. During my time as a course leader, I delivered courses from Level 1 through to Level 3 in Wood Machining, Carpentry and Joinery to apprentices, full-time students and adult learners.

In my spare time, other than working on construction projects, I like fishing, working my spaniels and clay shooting.

Martin Burdfield

Following a long line of Burdfields who have worked in Carpentry, Joinery and Construction contracting I strongly believe that there is rarely a dull moment.

As an apprentice I enjoyed learning from the experience of others and from all the available textbooks. I was fortunate to win the City and Guilds Silver medal for the highest marks in the Advanced Craft Certificate and won The UK's first Gold medal at the Worldskills competition. My career took me from Joinery works supervisor to setting out, estimating and then Works Manager of two Architectural Joinery Works. Concurrently with this, I was invited to start part-time evening teaching at college, which led to a new career in Further Education, culminating in construction department head. For over 40 years I have watched learners' skills improve and am proud to have formed part of their development.

For ten years I ran the Skillbuild Joinery competitions and was the UK's Training Manager for nine years, retiring as Chief Expert elect at the Worldskills competition, having trained the UK's second Gold Medallist in Joinery in 2003.

I have worked for City and Guilds since 1987 in various roles and currently as Chief Examiner, Principal Moderator and External Quality Assurer.

I believe that if you work and study hard you will find a career in the construction industry very rewarding. Be proud of what you produce; it will stand as testament to your skill and will be there for others to admire for many years to come.

Picture credits

Figure 1.8 © BuildPix/Avalon/Construction Photography/Alamy Stock Photo; Figure 1.10 © Alexander Erdbeer/Shutterstock; Figure 1.11 © Avalon/Construction Photography/Alamy Stock Photo; Figure 1.12 © Goodluz - Fotolia; Table 1.4 2nd © ipm/Alamy Stock Photo; 3rd © mike.irwin/Shutterstock; 4th © Mint Images Limited/Alamy Stock Photo; 5th © 29september/Shutterstock; Figure 1.14 © George Dolgikh - Fotolia; Figure 1.15 © A/stock.adobe.com; Figure 1.16 © Virynja/stock.adobe.com; Table 1.5 1st © Ricochet64/stock.adobe.com; 2nd © Seetwo.adobe.com; 3rd © alona_s/stock.adobe.com; 4th © Warning signs/stock.adobe.com; 5th © HSNKRT/stock.adobe.com; 6th luca pb/stock.adobe.com; 7th © T. Michel/stock.adobe.com; Figure 1.17 © John Williams/123RF; Table 1.6 1st © DenisNata/Shutterstock; 2nd © Stockbyte/Getty Images/Entertainment & Leisure CD35; 3rd © George Dolgikh - Fotolia; 4th © Virynja/stock.adobe.com; 5th © James Hughes/Alamy Stock Photo; 6th © Kitten/stock.adobe.com; 7th © photomelon/stock.adobe.com; 8th © modustollens/stock.adobe.com; 9th © IRC/stock.adobe.com; 10th © nd700/stock.adobe.com; 10th © Valentin/stock.adobe.com; 12th © Petrik/stock.adobe.com; 13th © pixelrobot/stock.adobe.com; 14th © Kraska/stock.adobe.com; 15th © Mediscan/Alamy Stock Photo; Figure 1.18 © Rob Kints/Shutterstock; Figure 1.19 © AKP Photos/Alamy Stock Photo; Figure 1.20 © Steve Jones; Figure 1.21 © Israel Hervas Bengochea/Shutterstock; Figure 1.23 © 2019 Screwfix Direct Limited; Figure 1.34 © SergeBertasiusPhotography/Shutterstock; Figure 1.35 © Steroplast Healthcare Limited; Figure 1.37 © Mark Richardson/Alamy Stock Photo; Figure 1.38 © Mr.Zach/Shutterstock; Figure 1.39 © Mark Sykes/Alamy Stock Photo; Figure 1.40 © MARK SYKES/ Science Photo Library/Alamy Stock Photo; Figure 1.41 © Colin Underhill/Alamy Stock Photo; Figure 1.42 © Technicsorn Stocker/Shutterstock.com; Figure 1.43 © NESRUDHEEN/stock.adobe.com; page 37 © Seetwo/stock.adobe.com; Figure 2.1 © Daisy Daisy/stock.adobe.com; Figure 2.2 © BREEAM / BRE Global Ltd; Figure 2.3 © Passivehouseplus; Figure 2.4 Crown copyright; Figure 2.7 © Dario Sabljak/stock.adobe.com; Table 2.4 1st © BanksPhotos/E+/Getty Images; 2nd © Alena Brozova/Shutterstock; 3rd © Alena Brozova/ Shutterstock; 6th ©

CORDELIA MOLLOY/Science Photo Library/Alamy Stock Photo; 7th © SueC/ Shutterstock; 9th © Alterfalter/Shutterstock.com; 11th © eric/stock.adobe.com; 12th © Coprid/stock.adobe.com; Figure 2.11 © Forest Stewardship Council; Figure 2.12 © Matthew siddons/Shutterstock.com; Figure 2.13 © Freebird/stock.adobe.com; Figure 2.14 © Oleg Shipov/stock.adobe.com; Figure 2.15 © goodluz/stock.adobe.com; Figure 2.27 © John Williams/123RF; Figure 2.31 © ivvv1975/Shutterstock.com; Figure 2.36 © tanasan/stock.adobe.com; p.75 © kzww/Shutterstock; Figure 2.40 © Auremar/stock.adobe.com; Figure 2.43 © Murattellioglu/stock.adobe.com; Figure 2.44 © Bokic Bojan/Shutterstock.com; Figure 2.50 © Gorodenkoff/stock.adobe.com; Figure 2.51 © SketchUp and Trimble, Inc; Figure 2.56 © Claudio Divizia/stock.adobe.com; Figure 2.58 © Bozena Fulawka/stock.adobe.com; Figure 3.13 © 2020 Screwfix Direct Limited; Figure 3.14 by kind permission of Conway Saw & Supply Co Ltd; Figure 3.32 © City & Guilds; Figures 3.37–3.45 © City & Guilds; Figure 3.46 © The Family Handyman; Figure 3.49 © HSE; Figure 3.54 (bottom) © City & Guilds; Figure 3.61 © HSE; Figure 3.62 © City & Guilds; Figure 3.63 © City & Guilds; Figure 3.64 courtesy of Axminster Tool Centre; Table 3.4 1st © Toolstation Ltd; 2nd–10th © City & Guilds; Figure 3.76 © SoloStock Industrial/Alamy Stock Photo; Figure 3.77 © Infinity Tools; Figure 3.84–3.87 © City & Guilds; p.141 © City & Guilds; Figure 3.113 © 'The Design and Technology Association www.data.org.uk'; Table 3.9 © City & Guilds; Table 3.10 © City & Guilds; Figure 3.118 © City & Guilds; Figure 3.120 © City & Guilds; Figure 3.122 © City & Guilds; Figure 3.123 © City & Guilds; Figure 3.129–3.134 © City & Guilds; Figure 3.137 courtesy of Axminster Tool Centre; Figure 3.138 © City & Guilds; Figure 3.143 © City & Guilds; Figure 3.144 © City & Guilds; Figure 3.145 © HSE; Figure 3.148 © 2020 Screwfix Direct Limited; Figure 3.157 © Valdas/stock.adobe.com; Figure 3.158 Martin Burdfield/Building Crafts College; Figure 3.163 Martin Burdfield/Building Crafts College; Table 4.1 p.174 © City & Guilds; Table 4.1 p.178 © Traditional Oak Carpentry; Figure 4.18 © Paul D. Smith/Shutterstock; Figure 4.19 © Rob Kints/Shutterstock; Figure 4.26 © City & Guilds; Figure 4.38

How to use this book

Throughout this book you will see the following features:

Industry tips are particularly useful pieces of advice that can assist you in your workplace or help you to remember something important.

> ### INDUSTRY TIP
>
> 'Bearers' are usually short lengths of strong timber placed on the ground, or between packs of materials, to store materials on. This prevents the materials potentially being damaged on the surface of the ground and it also makes it easier to lift them if mechanical lifting aids are used, e.g. pallet truck, forklift truck or telehandler.

Key terms in bold purple in the text are explained in the margin to aid your understanding. (They are also explained in the Glossary at the back of the book.)

> ### KEY TERM
>
> **Compound cut:** a cut that consists of two angles – the bevelled angle from the canted saw blade and the mitre angle (or crosscut angle) from the fence.

Health and safety boxes flag important points to keep yourself, colleagues and clients safe in the workplace. They also link to sections in the health and safety chapter for you to recap learning.

> ### HEALTH AND SAFETY
>
> Not all construction sites permit smoking or the use of e-cigarettes. It is not a duty an employer has to fulfill, and is often discouraged because of the associated health, safety and welfare risks.

Activities help to test your understanding and enable you to learn from your colleagues' experiences.

> ### ACTIVITY
>
> What would motivate you to improve your work? Make a note and discuss with your team to see what motivates them.

Improve your maths items provide opportunities to practise or improve your maths skills.

Improve your English items provide opportunities to practise or improve your English skills.

At the beginning of each chapter, there is a table that shows how the main headings in the chapter cover the learning outcomes for each qualification specification.

At the end of each chapter there are some **Test your knowledge questions** and **Practical tasks**. These are designed to identify any areas where you might need further training or revision.

Apprenticeship only flagging identifies content that is relevant to apprenticeship learners only.

CHAPTER 1

HEALTH, SAFETY AND WELFARE IN CONSTRUCTION AND ASSOCIATED INDUSTRIES

INTRODUCTION

A career in the building industry can be a very rewarding one, both personally and financially. However, building sites and workshops are potentially dangerous places and there are many hazards in the construction industry. Many construction operatives (workers) are injured or even killed each year; they may suffer life-changing injuries or have to live with the long-term effects of illness and diseases caused by poor health, safety and welfare practices.

Over the past 40 years, regulations have been brought in to reduce accidents and improve working conditions. Statistics show that the industry is now much safer to work in than it once was. However, there is still a worrying number of accidents and incidents in construction compared with other industries.

This chapter explains the main regulations that help to make the sites on which you will work safer, and to improve general health and welfare in the workplace. It is not just your employer's responsibility to keep you safe – you also have a personal responsibility to prevent accidents and to report any dangerous practices. This requires you to be able to identify hazards in the workplace and to obey safety signs, notices and processes. You should be aware of your site's accident and emergency reporting procedures and documentation, including what to do if there is a fire. You also need to know how to carry out your work safely, including how to use different techniques and methods to move, handle and store materials, how to use access equipment and basic working platforms when working at height, how to work with electrical equipment in the workplace and, of course, how to use and maintain personal protective equipment (PPE).

LEARNING OBJECTIVES

By reading this chapter you will learn about:
1 health, safety and welfare regulations, roles and responsibilities
2 site inductions and other on-site training
3 first aid and first-aid kits
4 fire.

1 HEALTH, SAFETY AND WELFARE REGULATIONS, ROLES AND RESPONSIBILITIES

Sources of health and safety information

Sources of health and safety information are listed in Table 1.1.

▼ Table 1.1 Sources of health and safety information

Source	How they can help
Health and Safety Executive (HSE)	A government body that oversees health and safety in the workplace. It produces health and safety literature such as **Approved Codes of Practice (ACoPs)** and guidance notes.
Construction Industry Training Board (CITB)	This body produces health and safety publications and is directly involved with construction training.
Royal Society for the Prevention of Accidents (RoSPA)	A charity that produces literature and gives advice about health and safety in many contexts.
Royal Society for Public Health	An independent, multi-disciplinary charity that is dedicated to the promotion and protection of collective health and well-being.
Institution of Occupational Safety and Health (IOSH)	A chartered body for health and safety practitioners. The world's largest health and safety professional membership organisation.
British Safety Council	Helps businesses with their health, safety and environmental management.
Regulations	Provide sources of information on specific issues to enable legal compliance.
Material safety data sheets	Provide information on the safe use, storage, transportation and disposal etc. of hazardous materials (e.g. control of substances hazardous to health – COSHH).
Manufacturers'/suppliers' maintenance manuals	Provide information about how to set up, use, service and maintain machinery, plant and equipment.

INDUSTRY TIP

The Health and Safety Executive (HSE) has many powers, to enable it to enforce health and safety legislation in high-risk work environments. It can:

- enter any premises that the inspectors think it necessary to enter for the purposes of enforcing the Health and Safety at Work Act, and the relevant statutory provisions; inspectors may enter only at a 'reasonable time', unless they think there is a situation that may be dangerous; if they have 'reasonable cause to apprehend serious obstruction', inspectors may take a police officer with them on site
- order areas to be left undisturbed, take measurements, photographs and recordings, take samples, and take possession of, and carry out tests on, articles and substances that appear to have caused (or be likely to cause) danger
- require the production of, inspect and take copies of relevant documents
- require anyone it thinks might give it relevant information to answer questions and sign a declaration as to the truth of the answers
- require facilities and assistance to be provided
- seize and make harmless (by destruction if necessary) any article or substance that it has reasonable cause to believe is a cause of imminent danger of serious personal injury.

Source: hse.gov.uk

KEY TERM

KEY TERM

Approved Code of Practice (ACoP): a document, usually produced by the HSE, that provides practical advice on how to interpret regulations, for example COSHH, The Provision and Use of Work Equipment Regulations or Safe Use of Woodworking Machinery.

Use the internet to research the sources of information listed. You will find them very useful when preparing for the end-of-unit assessments or gathering work-based evidence for your apprenticeship. As you progress from your apprenticeship in the construction industry there will be many more opportunities to further your career. For example, you may choose to train as an assistant site manager, a site agent or a health and safety advisor.

INDUSTRY TIP

Approved Codes of Practice have a semi-legal status. This means that if you follow them you will be complying with the law; however, if you do not, you are not in breach of the law but would have to achieve or exceed the standards in them.

ACTIVITY

Use the sources of information listed in Table 1.1 to find out how you can control exposure to dust created by power tools in the workplace. Once you have completed your research, make a list of the control measures. How many of these measures do you use in your place of work?

Health and Safety Executive (HSE)

The **Health and Safety Executive (HSE)** is a body set up by the government to enforce the Health and Safety at Work Act 1974. HSE inspectors can be proactive in their role, meaning that they can randomly visit and inspect workplaces without an invitation. If an employer is in breach of health and safety law (it has failed in its duty to comply with it), then it may be

charged for the time that the HSE spends identifying the fault, making further visits, letter writing, etc. until the issues have been corrected.

KEY TERM

Health and Safety Executive (HSE): the main UK body responsible for the encouragement, regulation and enforcement of health, safety and welfare in the workplace. The HSE is also responsible for research into occupational risks in Great Britain.

If an HSE inspector visits a workplace, possible outcomes are:

- **Nothing** – everything that should be done is being done.
- **Investigation** – following an accident or dangerous event in the workplace.
- **Informal visit** – they may provide advice if small breaches of health and safety are discovered that are unlikely to cause immediate injury or ill health.
- **Improvement notice** – may be issued by an inspector if there is a more serious breach of health and safety law; employers will have at least 21 days for the corrective action to be taken or to appeal against the notice.
- **Prohibition notice** – if there is a breach of health and safety that is likely to cause **imminent danger** (a danger that will probably happen soon), then it is possible that this notice will be served; this means that the employer will have to stop the activity immediately to prevent any further risk.
- **Prosecution** – in either a magistrates' court, or the Crown Court for more serious breaches, possibly resulting in unlimited fines or imprisonment.

Companies that have had action taken against them by enforcement agencies, such as the HSE, have their details and the case information recorded on the HSE website. This information can have a negative impact on a company – for example, future clients may refer to it and make a judgement as to whether or not that employer is suitable for a contract with them. The impact of enforcement action against a company could result in closure of the business, leading to loss of employment for its staff.

Reporting of Injuries, Diseases and Dangerous Occurrences Regulations (RIDDOR)

All accidents at work should be reported and recorded. However, more serious events need to be reported to the enforcement agency the HSE under the **Reporting of Injuries, Diseases and Dangerous Occurrences Regulations (RIDDOR)**. The HSE reported that, in the year 2017/18, 144 people were fatally injured at work (a slight increase on the previous year), of which 38 were construction workers. Figure 1.1 shows that this is a large proportion of total fatalities at work.

KEY TERMS

Reporting of Injuries, Diseases and Dangerous Occurrences Regulations (RIDDOR): legislation that puts duties on employers, self-employed people and those in control of work premises to report certain serious accidents in the workplace, occupational diseases and specified dangerous occurrences (**near misses**).

Near miss: an incident that has happened, which had the potential to cause injury, illness or damage but did not (e.g. bricks falling from a scaffold platform and landing close to a worker).

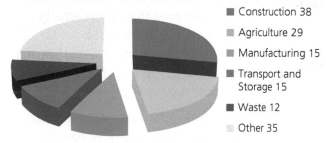

- ■ Construction 38
- ■ Agriculture 29
- ■ Manufacturing 15
- ■ Transport and Storage 15
- ■ Waste 12
- ▨ Other 35

▲ Figure 1.1 Fatal injuries to workers by main industry, 2017–18

Source: HSE

It is important to be aware of the main causes of fatal injuries at work so that employers can take the necessary precautions to protect their workers and also so that workers can be alert to the risks involved in particular activities. Figure 1.2 shows the main causes of fatal accidents to workers but the HSE reported that 100 members of the public were also killed as a result of work-related activities in 2017/18.

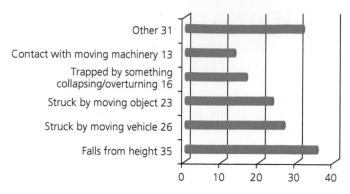

▲ Figure 1.2 Main causes of fatal accidents for workers (all industries), 2017–18

Source: HSE; reported fatalities via RIDDOR

ACTIVITY

Refer to Figure 1.2 and discuss with your peers possible causes of the other 31 fatal accidents.

An accident in the workplace can affect not only the injured person, but their family and friends, their employer and the person responsible.

For example, accidents may impact the **injured person** in the following ways:
- possible loss of life
- the time and expense of first-aid treatment and rehabilitation
- loss of wages
- potential disability
- depression or stress, loss of confidence
- possible increased alcohol consumption or drug use
- loss of sleep
- inability to continue in the same job role
- negative effect on family life.

Accidents may impact the **employer** in the following ways:
- the expense of providing employee sick pay
- lost production time
- damaged products, tools or equipment as a result of the accident, which may be costly to replace or may stop production until they are replaced
- time lost investigating the accident to prevent a reoccurrence
- negative consequences of enforcement action (e.g. legal action, prosecution or possible jail sentence)
- loss of staff morale, decrease in productivity and difficulty recruiting if workers don't feel safe
- missed completion dates for work
- lost reputation or reduction in public image.

Health and Safety at Work etc. Act (HASAWA, HSWA or HASWA)

The construction industry had a high rate of fatalities and injuries as a result of unregulated standards until the introduction of the Health and Safety at Work Act in 1974. The **Health and Safety at Work etc. Act (HASAWA, HSWA or HASWA)** is referred to as primary legislation (a main law) and states that the general duty of the employer is:

'To ensure, so far as is reasonably practicable, the health, safety and welfare at work of all his employees'

INDUSTRY TIP

The phrase 'so far as is reasonably practicable' means that the employer must assess the risk against the financial, time and convenience costs involved to control that risk. For example, if an employer has to replace a small section of damaged guttering on a low roof, the cost, time and effort involved in installing independent scaffolding would outweigh the risk. A logical alternative would be to use podium steps or a mobile tower scaffold.

The responsibilities of duty holders under HASAWA

Anyone with a legal duty under health and safety law is referred to as a **duty holder**. Everybody has a responsibility for health and safety in the construction industry, whether they are an employer, employee, self-employed person, a designer/manufacturer/importer/supplier or a controller of premises.

For example, employers must provide:
- safe plant and systems of work
- safe use, handling, storage and transportation of articles and substances
- information, instruction, training and supervision
- a safe workplace, and safe access to it and **egress** from it
- a safe working environment with adequate welfare facilities.

An employer also has a legal duty to prepare a written health and safety policy if it has five or more employees.

A **health and safety policy** is the foundation of a good management system. It is normally divided into three sections:

1 general statement of intent (the company's main objectives)
2 organisation (identifies the people responsible for health and safety within the company)
3 arrangements (how the policy is to be implemented and managed, e.g. arrangements for site inductions or risk assessments).

The policy should be signed and dated by the most senior person within the company to authorise it.

INDUSTRY TIP

The HASAWA states that an employer 'cannot charge an employee for things done to achieve legal compliance'. This means that items such as your safety footwear, hard hat and safety glasses should be provided free of charge, and replaced when they deteriorate or get damaged.

As duty holders under HASAWA, employees must:
- take reasonable care for the health and safety of themselves and of other persons who may be affected by their acts or omissions (something they did or did not do) at work
- co-operate with the employer to enable compliance with legal requirements.

KEY TERMS

Health and Safety at Work etc. Act 1974 (HASAWA, HSWA or HASWA): the primary piece of health and safety legislation that employers, the self-employed, employees, people in control of premises, manufacturers, designers, importers and suppliers of equipment have a duty to comply with.

Duty holder: anyone with a legal responsibility under health and safety legislation is referred to as a duty holder. For example, employers, employees and the self-employed all have legal responsibilities to comply with, therefore they are 'duty holders'.

Egress: a term used in construction law to describe a way out or exit. Employers have a legal responsibility to provide safe access and egress from your place of work.

Health and safety policy: a legal document written by employers with five or more employees outlining how health and safety will be managed on site.

As duty holders under HASAWA, controllers of premises (for example, a landlord or property management company) must ensure:

- the premises are safe
- the means of access and egress are safe
- any plant or substances provided by them for use in that premises are safe.

HEALTH AND SAFETY

You should listen to your employer and follow the site-specific rules. Every site is different and so are the hazards that a site presents; you should have been made aware of these hazards during your **site induction** (see pages 31–32). Here are some examples of how you can reduce risks.

- Don't cut corners, and follow your employer's risk assessments, which will identify areas of concern and how to manage them without risk to yourself or others.
- If you see an unsafe condition, report it to your supervisor.
- Don't use any equipment or machinery you haven't been trained on or authorised to use.
- Wear the PPE provided and don't be afraid to report damaged equipment.

KEY TERM

Site induction: a meeting or training session providing information to everyone entering a construction site on the risks that they are likely to face.

Site inductions help to fulfil an employer's legal duty to protect people at work from risks created by work activities. The induction may also tell workers about other key points of the site, such as permit to work systems. A carpenter or joiner entering a construction site may be informed about working at height, and the use of any leaning or stepladders they may be using to complete their activities. They will also be informed about sources of power on site – for example, 110 volts when using power tools.

As duty holders under HASAWA, designers, manufacturers, importers and suppliers' duties are to ensure that:

- any article or substance will be safe to use
- adequate testing takes place to ensure that it will be safe
- the end user is provided with information on its safe use
- the end user is provided with revisions of that information as necessary.

It's important to remember that the Health and Safety at Work etc. Act covers everyone at work and people who may be affected by work activities.

ACTIVITY

Using battery-powered tools on site will reduce the risk of electrocution and trips as a result of trailing leads. However, they don't have an unlimited power supply and will still need to be charged at some point. How do you charge your batteries on site? Are your battery chargers PAT tested? How do you dispose of your old batteries? Discuss with your peers where they charge their tools. Are you satisfied that these methods are the safest options? If you can think of a better or safer way, discuss your ideas with your supervisor or employer.

ACTIVITY

Take a look at Figure 1.3, which shows the positive impact the HASAWA has had since its introduction in 1974. At your place of work, consider all the things that you and your employer do as a result of current health and safety law. What might working conditions have been like before the introduction of the HASAWA? Use the internet to research films of working conditions on construction sites before 1974.

Watch a short video about a construction site in the 1960s: search YouTube for 'Life In London 1960s Construction Industry'. Now use what you have seen to critically analyse a typical construction site in the 1960s. Write a report for the site manager detailing all the hazards and breaches of current health, safety and welfare legislation shown. How have these been overcome on modern sites?

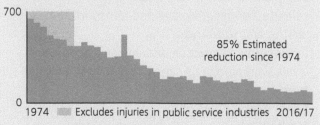

▲ Figure 1.3 Number of fatal injuries to employees, 1974 to 2016/17 (RIDDOR and earlier reporting legislation, Great Britain)

Source: HSE

Roles and responsibilities under health and safety regulations

Health and safety law (also referred to as 'legislation') can be confusing, especially for someone young, in training or new to the industry. The HASAWA is a very broad objective and open to interpretation – for example, 'An employer must provide adequate welfare facilities.' What does this mean? What your employer considers adequate and cost effective may not necessarily be suitable and sufficient. Therefore, secondary legislation is made under HASAWA to provide more specific duties to comply with. Table 1.2 shows some of the main regulations that you should be familiar with for your course, and while training and working in the construction industry.

▼ Table 1.2 The main regulations affecting the construction industry

Regulations	Abbreviation	Year of most recent revision of the regulations
Construction (Design and Management) Regulations	CDM	2015
Control of Asbestos Regulations		2012
Control of Noise at Work Regulations		2005
Control of Substances Hazardous to Health	COSHH	2002
Control of Vibration at Work Regulations		2005
Lifting Operations and Lifting Equipment Regulations	LOLER	1998
Manual Handling Operations Regulations	MHOR	1992
Personal Protective Equipment (PPE) at Work Regulations		1992
Provision and Use of Work Equipment Regulations	PUWER	1998
Reporting of Injuries, Diseases and Dangerous Occurrences Regulations	RIDDOR	2013
Working at Height Regulations	WAHR	2005

Key employee responsibilities

1　You must work safely and take care at all times.
2　You must make sure you do not put yourself or others at risk by your acts or omissions (something that hasn't been done).
3　You must co-operate with your employer in regard to health and safety. If you do not, you risk being removed from your place of work, injury (to yourself and others), prosecution and possible loss of employment.
4　You must use any equipment and safeguards provided by your employer. For example, you must wear, look after and report any damage to the personal protective equipment (PPE) that your employer provides.
5　You must not interfere or tamper with any safety equipment (e.g. removing guards from machinery or power tools).
6　You must not misuse or interfere with anything that is provided for employees' safety.

ACTIVITY

Study your key responsibilities as an employee, listed above. How many of them were you aware of before reading this list? Address each point in turn, and explain how you fulfil your responsibilities at your place of work.

Risk assessments and method statements

In the construction industry you will often hear the terms **risk assessment** and **method statement**. These are often referred to as RAMS (Risk Assessment and Method Statement). Employers have a legal duty to assess the risks to the health and safety of their

KEY TERMS

Risk assessment: a written document used to assess the level of risk in the workplace and the measures that should be taken to manage this risk. Risk assessments are a legal requirement under the Management of Health and Safety at Work Regulations.

Method statement: a document that provides a logical sequence of how to complete a task safely. Method statements are not a legal requirement, but are often used to manage hazards identified in risk assessments.

employees and those who may be affected by their work under the Management of Health and Safety at Work Regulations. Employers with five or more employees must ensure that 'suitable and sufficient' risk assessments are written. However, it is good practice and a requirement of many construction sites that risk assessments are provided for all work activities. The purpose of a risk assessment is to identify significant hazards and the measures that can be used to either eliminate them completely, or reduce the risks to an acceptable level.

IMPROVE YOUR ENGLISH

A term often used when referring to health and safety is 'mitigate', which means to reduce. For example, 'Identify the hazards and mitigate the risks to an acceptable level.'

INDUSTRY TIP

Suitable and sufficient

Many employers in the construction industry produce generic risk assessments and use them for every job they work on. Risk assessments should reflect the nature of the work being undertaken. Remember, all jobs are different and so are the hazards they present.

There is no legal requirement to complete method statements, although it is good practice. A method statement is a document that sets out a logical sequence to complete a job or process while managing the hazards that have been identified in the risk assessment. It should be clear, uncomplicated and easy to understand as it is for the benefit of those carrying out the work (and their immediate supervisors).

INDUSTRY TIP

You will often be shown relevant risk assessments and method statements (RAMS) before being allowed on to a construction site to undertake work activities, as part of your site induction. Failure to follow the control measures identified in the RAMS could result in you being removed from the site for breaches of health and safety or, worse, an accident.

Risk Assessment

Activity / Workplace assessed: Return to work after accident
Persons consulted / involved in risk assessment:
Date:
Reviewed on:

Location:
Risk assessment reference number:
Review date:
Review by:

Significant hazard	People at risk and what the risk is Describe the harm that is likely to result from the hazard and who could be harmed	Existing control measure What is currently in place to control the risk?	Risk rating Use matrix identified in guidance note Likelihood (L) Severity (S) Multiply (L) * (S) to produce risk rating (RR)				Further action required What is required to bring the risk down to an acceptable level? Use hierarchy of control described in guidance note when considering the controls needed	Actioned to: Who will complete the action?	Due date: When will the action be completed by?	Completion date: Initial and date once the action has been completed
			L	S	RR	L/M/H				
Uneven floors	Operatives Cut or broken leg/ankle/arm/wrist; bruising; head injuries	Verbal warning and supervision	2	1	2	M	None applicable	Site supervisor	Active now	Ongoing
Steps	Operatives Cut or broken leg/ankle/arm/wrist; bruising; head injuries; muscular skeletal injuries	Verbal warning	2	1	2	M	None applicable	Site supervisor	Active now	Ongoing
Staircases	Operatives Cut or broken leg/ankle/arm/wrist; bruising; head injuries; muscular skeletal injuries	Verbal warning	2	2	4	M	None applicable	Site supervisor	Active now	Ongoing

Likelihood				
		1 **Unlikely**	**2** **Possible**	**3** **Very likely**
Severity	**1** Slight/minor injuries/minor damage	1	2	3
	2 Medium injuries/significant damage	2	4	6
	3 Major injury/extensive damage	3	6	9

Likelihood
3 – Very likely
2 – Possible
1 – Unlikely

Severity
3 – Major injury/extensive damage
2 – Medium injury/significant damage
1 – Slight/minor damage

1   – Low risk, action should be taken to reduce the risk if reasonably practicable
2, 3, 4 – Medium risk, is a significant risk and would require an appropriate level of resource
6 & 9 – High risk, may require considerable resource to mitigate. Control should focus on elimination of risk,
            if not possible control should be obtained by following the hierarchy of control

▲ Figure 1.4 123-type risk assessment

There is no legal requirement to use a particular format for a risk assessment, however Figure 1.4 shows a useful example. In the diagram, the matrix below the assessment shows how the risk assessment is used. For example:

'Uneven floors' has been identified as a hazard, and given a score of between 1 and 3 to represent the likelihood of someone injuring themselves; in this case 2. The severity of a potential injury has also been quantified; in this case 1. These figures are multiplied together and plotted on the matrix:

(Likelihood) 2 × (Severity) 1 = 2

The table below the matrix indicates the action that should be taken based on the result of the risk assessment. The number 2 suggests a medium risk that requires an appropriate resource – for example, a recommendation that operatives wear safety footwear with a good grip and ankle support.

A risk assessment is a legally required tool used by employers to:
- identify work hazards
- assess the risk of harm arising from these hazards
- adequately control the risk.

There are five steps to a 'suitable and sufficient' risk assessment, as follows.

1 Identify the hazards: for example, working at height or the use of adhesives.

2 Decide who might be harmed and how: think about employees, visitors, members of the public or trespassers. Are there any additional needs, such as someone who is deaf or speaks English as a second language?

3 Evaluate the risk and control measures needed: how severe is the potential injury? How likely is it to happen? Even after measures have been put in place, sometimes there remains a 'residual' risk. You will have to decide whether this risk is high, medium or low.

4 Record your findings: all employers have a legal duty to produce written risk assessments if they have five or more employees; however, some contractors insist that all subcontractors have written risk assessments before being allowed on to their sites.

5 Review the assessment and update as necessary: risk assessments should be reviewed if there are any changes to the way the job is being completed or the tools or equipment used, or if any other significant changes are likely. From time to time they should also be revised and re-evaluated to ensure they reflect the way the work is to be carried out.

Under the Management of Health and Safety at Work Regulations 1999, where there is a risk, employers should apply the following 'General principles of prevention' to control the risk:

a) Avoid the risk where possible.

b) Evaluate risks (risk assess).

c) Combat risks at source (e.g. on-tool extraction is a better alternative than dust masks).

d) Adapt the work to suit the individual.

e) Adapt to technical progress (if new technology is developed to make the job safer it should be used).

f) Replace the dangerous with non-/less dangerous.

g) Develop a coherent (clear) overall prevention policy (covering technology, organisation of work, working conditions, etc.).

h) Give collective protective measures priority over individual protective measures (e.g. a collective measure such as a safety net will protect anyone who falls from height, while a harness will only protect the user).

i) Give appropriate instructions to employees.

Source: Management of Health & Safety at Work Regulations 1999

You may find the 'Principles of prevention' particularly useful when gathering your work-based evidence for your NVQ. Health and safety is often referenced throughout each unit in the qualifications.

INDUSTRY TIP

Once hazards have been identified through the risk assessment process, they should be controlled to an acceptable level by:

- eliminating the hazard completely if possible
- creating a safe place (e.g. using guards to prevent people coming into contact with moving parts)
- appointing a safe person (information, instruction training and supervision).

ACTIVITY

Reflect on a practical work activity that you have completed, and the hazards that you may have been faced with. Think about the control measures that were used and those that could have been used. Now write a risk assessment for that activity.

Control of Substances Hazardous to Health (COSHH) Regulations 2002

The Control of Substances Hazardous to Health (COSHH) Regulations 2002 control the use of dangerous substances in the workplace, e.g. preservatives, fuels (petrol and diesel), solvents, adhesives, cement and oil-based paint. These have to be transported, stored, used and disposed of without harming the environment or people.

Hazardous substances can be created in the workplace through work activities. For example, sanding timber creates airborne dust, disc-cutting concrete will create silica dust, and machinery or plant may create fumes. Some hazardous substances, such as asbestos or lead, may already be present on site, particularly if you are working on older buildings; these are not covered by COSHH because they have their own specific set of regulations.

INDUSTRY TIP

Asbestos is a naturally occurring mineral. Until 2000, it was extensively used in construction because it is flame retardant, doesn't rot and is extremely strong; in addition, it was often mixed with other materials such as cement for sheet roofing. However, breathing in airborne asbestos fibres can cause long-term ill health and even death. If you think your work is likely to disturb materials containing asbestos, STOP, warn others and speak to your supervisor immediately.

Safety data sheets

Employers have a duty to identify all hazardous substances that you are likely to come into contact with, how these can cause harm, and how the risks can be either avoided or controlled to an acceptable level. Manufacturers of hazardous materials and substances have a duty to produce product information and details of the risks associated with their products. This is normally done in the form of a **safety data sheet** (SDS). Employers can use the information contained in safety data sheets to produce COSHH assessments.

ACTIVITY

As a carpenter or joiner you will use many different types of hazardous materials that are regulated by COSHH, e.g. wood adhesives, expanding foam and gas fuel cells. Use the internet to find three examples of these materials, then locate their safety data sheets (SDSs) and download them. Analyse the SDSs and write a critical evaluation of them. Include in your evaluation measures that your employer takes to protect people at work from these hazardous substances. If you have difficulty finding the SDSs, use the following link to start you off: www.everbuild. co.uk/wp-content/uploads/2017/10/Everbuild%20 No%20Waste%20Expanding%20Foam%20V1.pdf.

The steps to be taken to complete a COSHH assessment are as follows:

1 Gather information about the substance, the work and working practices.
2 Evaluate the risks to health (either no significant risk, or potential exposure and significant risk).
3 Decide what actions need to be taken.
4 Keep a record of the assessment.
5 Review the assessment from time to time to keep it current.

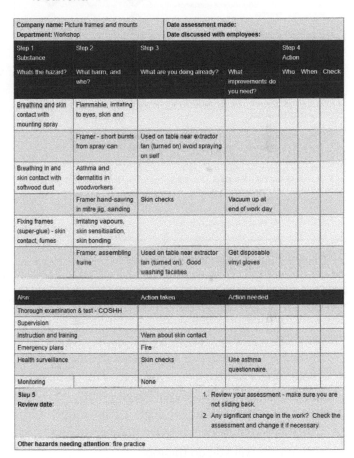

▲ Figure 1.6 Example COSHH assessment

When your employer assesses the risks of a potentially dangerous substance or material, it is important that they consider how operatives could be exposed to it. Listed below are the possible ways that hazardous substances could enter into the body, known as 'routes of entry':

- inhalation (e.g. breathing in dust)
- absorption (e.g. contact with the skin or via open cuts)
- ingestion (e.g. being swallowed)
- injection (e.g. via high pressure or sharp objects).

COCKBURN CEMENT

MATERIAL SAFETY DATA SHEET

Cement Kiln Dust

STATEMENT OF HAZARDOUS NATURE

This product is classified as hazardous according to criteria of United Kingdom Statutory Instrument

Cockburn Cement Limited
Russell Road
Munster
P.O. Box 38
Hamilton Hill
United Kingdom
Phone: (+44) 9411 1000

IDENTIFICATION

Product name:	Cement Kiln Dust
Other names:	CKD, HAD, Calfine
UN number:	None Allocated
Dangerous goods class and subsidiary risk:	None Allocated
Hazchem code:	None Allocated
Poisons schedule number:	None allocated
Use:	Cement Kiln Dust is used predominantly for acid neutralisation, pH control and as a raw material for use in fertiliser production where a solid residue is not a concern. Calfines can also be used in place of lime for soil stabilisation.

Physical Description/Properties

Appearance:	Off-white to grey coloured powder.
Boiling point / melting point:	Some components begin to melt above 1200°C
Vapour pressure:	Not applicable
Specific gravity:	2.4 to 2.8
Bulk density:	800 to 1800 kg/m³

Date of Issue: 01/04/19
Page: 1 of 5 pages

▲ Figure 1.5 Example of a COSHH data sheet

Safety data sheets

Products you use may be 'dangerous for supply'. If so, they will have a label that has one or more hazard symbols. Some examples are given here.

These products include common substances in everyday use such as paint, bleach, solvent or fillers. When a product is 'dangerous for supply', by law, the supplier must provide you with a safety data sheet. Note: medicines, pesticides and cosmetic products have different legislation and don't have a safety data sheet. Ask the supplier how the product can be used safely.

Safety data sheets can be hard to understand, with little information on measures for control. However, to find out about health risks and emergency situations, concentrate on:

- Sections 2 and 16 of the sheet, which tell you what the dangers are;
- Sections 4–8, which tell you about emergencies, storage and handling.

Since 2009, new international symbols have been gradually replacing the European symbols. Some of them are similar to the European symbols, but there is no single word describing the hazard. Read the hazard statement on the packaging and the safety data sheet from the supplier.

European symbols

Toxic | Very toxic | Harmful | Irritant
Highly flammable | Extremely flammable | Explosive | Dangerous to the environment
Oxidising | Corrosive

New International symbols

Hazard checklist

- ☐ Does any product you use have a danger label?
- ☐ Does your process produce gas, fume, dust, mist or vapour?
- ☐ Is the substance harmful to breathe in?
- ☐ Can the substance harm your skin?
- ☐ Is it likely that harm could arise because of the way you use or produce it?
- ☐ What are you going to do about it?
    - Use something else?
    - Use it in another, safer way?
    - Control it to stop harm being caused?

▲ Figure 1.7 Safety data sheet symbols

Control measures

The control measures below are listed in order of importance.

1 Eliminate the use of the harmful substance and use a safer one. For instance, swap high-**VOC** oil-based paint for a lower-VOC water-based paint.

2 Use a safer form of the product. Is the product available ready mixed? Is there a lower-strength option that will still do the job?

3 Change the work method to emit less of the substance. For instance, applying paint with a brush releases fewer VOCs into the air than spraying paint. Wet grinding produces less dust than dry grinding.

4 Enclose the work area so that the substance does not escape. This can mean setting up a tented area or closing doors.

5 Use an extraction system (**LEV**) in the work area.

6 Restrict access to the area to minimise the number of operatives exposed to the hazard.

7 Employers must provide appropriate PPE.

▲ Figure 1.8 Paint with high VOC content

New International symbols

Toxic | May explode when heated | Irritant

Causes fire | Explosive | Dangerous to the environment

Intensifies fire | Long term health hazard | Corrosive

▲ Figure 1.9 International COSHH symbols

HEALTH AND SAFETY

RPE is an abbreviation that you will come across during your training. It stands for respiratory protective equipment. This type of equipment is designed to protect the user from harmful substances such as dust, mist, gas or fumes. There are two main types of RPE: 'respirators' with filtering devices, and 'breathing apparatus'. A half/disposable dust mask is an example of a respirator. Breathing apparatus (BA) is available in a range of styles and needs a supply of breathing-quality air; this is usually supplied by an air cylinder or air compressor.

INDUSTRY TIP

Did you know that wood dust can cause skin disorders, asthma and nasal cancer? It is also flammable and capable of causing fire or explosions. Every year Fire and Rescue Authority personnel attend many fires started in extraction systems, so it is important that they are maintained and serviced regularly. You should also be aware that airborne dust can remain in the atmosphere for hours after you have finished sanding, so it is always better to control the hazard at source rather than wearing RPE that protects only the user.

KEY TERMS

VOC: the volatile organic compounds measure shows how much pollution a product will emit into the air when in use.

LEV: local exhaust ventilation system – a collective term referring to extraction systems, because not all airborne hazards are dust – they could also be mist, vapour, gas or fumes.

Reporting of Injuries, Diseases and Dangerous Occurrences Regulations (RIDDOR) 2013

Despite all the efforts put into health and safety, incidents still happen. The Reporting of Injuries, Diseases and Dangerous Occurrences Regulations (RIDDOR) 2013 state that employers must report to the HSE all accidents that result in an employee needing more than seven days off work. **Specified diseases** and dangerous occurrences must also be reported. A serious occurrence that has not caused an injury but had the potential to do so (a near miss) should still be reported. All accidents and near-miss accidents must be investigated so that lessons can be learned, to prevent a reoccurrence.

KEY TERM

Specified diseases: diseases that are reportable under the Reporting of Injuries, Diseases and Dangerous Occurrences Regulations (RIDDOR) 2013.

Below is a list of the types of reportable injury:

- Death of any person
- Specified injuries to workers:
  - fractures, other than to fingers and toes
  - amputations
  - any injury likely to lead to permanent loss of sight or reduction in sight
  - any crush injury to the head or torso causing damage to the brain or internal organs
  - serious burns
  - any scalping requiring hospital treatment
  - any loss of consciousness caused by head injury or asphyxia
  - any other injury arising from working in an enclosed space that leads to hypothermia or heat-induced illness, or requires resuscitation or admittance to hospital for more than 24 hours
- Seven-day-plus absence from work as a result of an injury
- Non-fatal accidents to non-workers (members of the public)
- Occupational diseases:
  - carpal tunnel syndrome
  - severe cramp of the hand or forearm
  - occupational dermatitis
  - hand–arm vibration syndrome
  - occupational asthma
  - tendonitis
  - any occupational cancer
  - any disease attributed to occupational exposure to a biological agent
- Injuries as a result of gas incidents.

Dangerous occurrences (near misses) are also reportable, even if nobody was injured.

INDUSTRY TIP

For more detailed information on RIDDOR visit the HSE webpage at www.hse.gov.uk/riddor.

HEALTH AND SAFETY

Leptospirosis (also known as Weil's disease) is a serious disease spread by rats and cattle. Initial symptoms resemble the flu: fever, headache and chills. One way to avoid contracting this disease is to use good methods of hygiene, such as washing your hands before smoking or eating.

HEALTH AND SAFETY

Not all construction sites permit smoking or the use of e-cigarettes. It is not a duty an employer has to fulfill, and is often discouraged because of the associated health, safety and welfare risks.

Although minor accidents and injuries are not reported to the HSE, records must be kept. Accidents must be recorded in the accident book. This provides a record of what happened and is useful for future reference. Trends may become apparent and the employer may take action to try to prevent a particular type of accident occurring again.

ACTIVITY

You are working on a construction site with another carpenter when they trip over a trailing lead and hit their head on the floor, causing a cut and bleeding. Explain to whom the accident should be reported and how it should be recorded.

Construction, Design and Management (CDM) Regulations 2015

The Construction, Design and Management (CDM) Regulations 2015 focus attention on the effective planning and management of construction projects, from the initial design concept through to maintenance and repair. The aim is for everyone to work together to make health and safety considerations an integral part of a project, rather than an inconvenient afterthought. The CDM Regulations reduce the risk of harm to those who work on or use the structure throughout its life, from construction through to demolition.

▲ Figure 1.10 The CDM Regulations play a role in health and safety during demolition

▲ Figure 1.11 The CDM Regulations protect workers from the construction to the demolition of projects of all sizes, from domestic to commercial

The CDM Regulations apply to all projects regardless of size, duration or nature of the work, from major house-building projects to small domestic jobs. Under the CDM Regulations, the following people or organisations have roles, and are referred to as 'duty holders':

- clients (including domestic)
- principal designers
- designers
- principal contractors
- contractors
- workers.

The purpose of the regulations is to use effective lines of communication about the risks and how they are managed to the people who need to know. The regulations also require competent people to be employed, to do the right job, at the right time. It is also a legal requirement that contractors or principal contractors engage and consult with the workers about the risks on site and how they are to be managed. Employers will need to provide information, instruction, training and supervision in the workplace, and provide further appropriate training if there are any shortfalls in skills and knowledge.

On construction projects where there is more than one designer, a principal designer is appointed by the client to oversee the pre-construction phase. A designer, such as an architect (if there is only one) or principal designer, has a duty under CDM Regulations to produce a health and safety file for the project. The health and safety file contains information about the structure of the building that could be used to safeguard anyone undertaking any future work on the building, such as maintaining, servicing, extending or demolition.

If there is more than one contractor, then a principal contractor is appointed. Either the contractor (if there is only one) or the principal contractor will manage the construction phase of a project to secure the health and safety of everyone involved. Under the CDM Regulations, contractors or principal contractors must prepare a construction phase plan. The plan should contain information such as the site rules, details of welfare facilities, etc.

Some projects will be referred to as 'notifiable' under CDM Regulations. This means that you must inform the Health and Safety Executive using an F10 form, available to download from its website. Projects are notifiable if:
- they last longer than 30 days and have more than 20 workers working simultaneously (at the same time), or
- they last longer than 500 person-days.

Duty holders' responsibilities

Duty holders' roles in accordance with CDM Regulations are listed in Table 1.3.

▼ Table 1.3 Duty holders' roles

CDM duty holders	Main duties
Commercial clients Organisations or individuals for whom a construction project is carried out that is done as part of a business	Make suitable arrangements for managing a project, including making sure that: • relevant information is prepared and provided to other duty holders • the principal designer and principal contractor carry out their duties • welfare facilities are provided.
Domestic clients People who have construction work carried out on their own home (or the home of a family member) that is not done as part of a business	Their client duties are normally transferred to: • the contractor for single-contractor projects • the principal contractor for projects with more than one contractor.
Designers Organisations or individuals who, as part of a business, prepare or modify designs for a building, product or system relating to construction work	When preparing or modifying designs, eliminate, reduce or control foreseeable risks that may arise during: • construction • the maintenance and use of a building once it is built.
Principal designers Designers appointed by the client in projects involving more than one contractor; they can be an organisation or an individual with sufficient knowledge, experience and ability to carry out the role	Plan, manage, monitor and co-ordinate health and safety in the pre-construction phase of a project.
Principal contractors Contractors appointed by the client to co-ordinate the construction phase of a project where it involves more than one contractor	Plan, manage, monitor and co-ordinate health and safety in the construction phase of a project. Make sure: • suitable site inductions are provided • reasonable steps are taken to prevent unauthorised access • workers are consulted and engaged in securing their health and safety • welfare facilities are provided.
Contractors Those who carry out the actual construction work; contractors can be an individual or a company	Plan, manage and monitor construction work under their control so it is carried out without risks to health and safety.

▼ Table 1.3 Duty holders' roles (continued)

CDM duty holders	Main duties
Workers Those working for or under the control of contractors on a construction site	Workers must: ● be consulted about matters that affect their health, safety and welfare ● take care of their own health and safety, and that of others who might be affected by their actions ● report anything they see that is likely to endanger either their own or others' health and safety ● co-operate with their employer, fellow workers, contractors and other duty holders.

Source: Health and Safety Executive website: CDM Regulations

▲ Figure 1.12 A client, a contractor and an operative consulting over building plans ahead of construction

Welfare facilities

Contractors and principal contractors will have to provide and maintain suitable welfare facilities under the CDM Regulations. The type and number of facilities will depend on several factors, such as the nature of the work being carried out (for example, workers who are outdoors, underground or on groundworks might need showers).

The workforce may be spread over a large site; therefore, several toilets may be appropriate to avoid walking long distances across a hazardous site.

Specialist work, such as asbestos removal, may involve segregation from other workers to prevent them from being contaminated.

The principal contractor will assess the exact welfare needs for the site as part of the CDM planning stage (called the construction phase plan). In general, the welfare facilities listed in Table 1.4 are usually provided on construction sites.

▼ Table 1.4 Welfare facilities usually provided on construction sites

Facility	Site requirements
Sanitary conveniences (toilets) 	● Portable chemical toilets are acceptable provided that it is not reasonably practicable to make other arrangements ● Suitable and sufficient toilets should be provided or made available ● Toilets should be adequately ventilated and lit, and should be clean ● Men and women can share the same toilet, if it is in a lockable room and partitioned off from any urinals ● Toilets must be well maintained in a clean condition
Washing facilities 	● A supply of clean hot and cold, or warm, water (which should be running water so far as is reasonably practicable) ● Soap or other suitable means of cleaning ● Towels or other suitable means of drying ● Sufficient ventilation and lighting ● Sinks large enough to wash face, hands and forearms

▼ Table 1.4 Welfare facilities usually provided on construction sites (continued)

Facility	Site requirements
Clean drinking water 	• Must be provided or made available • If water is stored, protect it from possible contamination • Should be clearly marked with an appropriate sign • Cups should be provided unless the supply of water is from an upward jet (e.g. a drinking fountain)
Changing rooms and lockers 	• Changing rooms must be provided or made available if operatives have to wear special clothing and they cannot be expected to change elsewhere • There must be separate rooms for, or separate use of rooms by, men and women where necessary • The rooms must have seating and include, where necessary, facilities to enable operatives to dry their special clothing, and their own clothing and personal effects • Lockers should also be provided
Rest rooms or rest areas 	• They should provide shelter from rain and wind • They should have enough tables and seating with backs for the number of operatives likely to use them at any one time • Arrangements must be made for heating water for drinks and warming food, e.g. microwave oven • Facilities should not be used to store materials, equipment or plant

ACTIVITY

Under CDM Regulations, employees must report anything they see that is likely to endanger either their own or others' health and safety. Analyse a construction activity that would fall into this category – for example, if some of the portable power tools you are using on site haven't been PAT tested for over a year and are starting to show signs of wear.

Provision and Use of Work Equipment Regulations (PUWER) 1998

The Provision and Use of Work Equipment Regulations (PUWER) 1998 place duties on:

• people and companies that own, operate or have control over work equipment
• employers whose employees use work equipment.

Work equipment can be defined as any machinery, appliance, apparatus, tool or installation for use at work. This includes equipment employees bring for their own use at work. The scope of work equipment is therefore extremely wide. The use of work equipment is also widely interpreted and, according to the HSE, means 'any activity involving work equipment and includes starting, stopping, programming, setting, transporting, repairing, modifying, maintaining, servicing and cleaning'. It includes equipment such as diggers, electric planers, stepladders, hammers or wheelbarrows.

Under PUWER, work equipment must be:

• suitable for the intended use
• safe to use
• well maintained
• inspected regularly.

HEALTH AND SAFETY

Abrasive wheels are used for grinding and cutting. Under PUWER, these wheels can be changed only by someone who has received training to do this. Wrongly fitted wheels can explode!

Regular inspection is important as a tool that was safe when it was new may no longer be safe after considerable use.

Additionally, work equipment may be used only by those who have received adequate instruction and training. Information regarding the use of the equipment must be given to the operator and equipment must be used only for what it was designed to do.

Protective devices (e.g. emergency stops and guards) must be used. Brakes must be fitted where appropriate to slow down moving parts to bring the equipment to a safe condition when turned off or stopped. Equipment must have adequate means of isolation. Warnings, either by signs or other means such as sounds or lights, must be used as appropriate. Access to dangerous parts of the machinery must be controlled. Some work equipment is subject to additional health and safety legislation, which must also be followed.

Manual Handling Operations Regulations 1992

Employers must try to avoid manual handling within reason if there is a possibility of injury. If manual handling cannot be avoided then they must reduce the risk of injury by using a risk assessment.

KEY TERM

Kinetic lifting: a controlled method of lifting that ensures that the risk of injury is reduced to an acceptable level.

The main reasons for manual handling injuries are normally related to:
- appropriate personal protective equipment (e.g. non-slip footwear) not being worn
- poor or incorrect grip on the load (e.g. incorrect hand positioning)
- trying to move loads that are too heavy (e.g. exceeding the recommended lifting weights identified in the risk assessment)
- incorrect lifting techniques used to lift or move the load (e.g. bent back while lifting from the ground).

Lifting and handling

Incorrect lifting and handling is a serious risk to your health. It is very easy to injure your back regardless of whether you are young and fit or an older experienced worker. An injured back can be very unpleasant, so it's best to look after it.

IMPROVE YOUR ENGLISH

A collective term used in the construction industry to describe pain and injuries caused to the body by manual handling is 'work-related musculoskeletal disorders' (MSDs).

How to lift and place an item correctly

If you cannot use mechanical means to move a load, it is important that you adopt the correct posture when lifting manually. The correct technique to do this is known as **kinetic lifting**. Always lift with your back straight, elbows in, knees bent and feet shoulder width apart.

▲ Figure 1.13 Safe kinetic lifting technique

Most workplace injuries are a result of manual handling. Remember, pushing or pulling an object still comes under the Manual Handling Operations Regulations.

When placing the item, again be sure to use your knees and legs to do the work, and beware of trapping your fingers. If stacking materials, be sure that they are on a sound, level base and on **bearers** if required.

INDUSTRY TIP

'Bearers' are usually short lengths of strong timber placed on the ground, or between packs of materials, to store materials on. This prevents the materials potentially being damaged on the surface of the ground and it also makes it easier to lift them if mechanical lifting aids are used, e.g. pallet truck, forklift truck or telehandler.

Heavy objects that cannot easily be lifted by mechanical methods can be lifted as a pair or team. It is important that one person in the team is the leader and that the lifting is done in a co-operative way. When lifting as a team, it is always best to use people of a similar build and height to distribute the weight and centre of gravity equally.

Control of Noise at Work Regulations 2005

Under the Control of Noise at Work Regulations 2005, duties are placed on employers and employees to reduce the risk of hearing damage to the lowest reasonable level practicable. Hearing loss caused by work is preventable. Hearing damage is permanent and hearing cannot be restored once lost.

Employers' duties

Employers must:

- carry out risk assessments and identify who is at risk
- eliminate or control their employees' exposure to noise at the workplace, and reduce the noise as far as practically possible
- provide suitable hearing protection
- provide health surveillance to those identified as at risk by the risk assessment
- provide information and training about the risks to their employees as identified by the risk assessment.

Employees' duties

Employees must:

- use personal hearing protectors provided to them by their employer
- report to their employer any defect in any personal hearing protectors or other control measures as soon as is practicable.

▲ Figure 1.14 Ear defenders

▲ Figure 1.15 Hard hat-mounted ear defenders, an example of compatible PPE

▲ Figure 1.16 Disposable ear plugs

There are many different types of hearing protection available, each offering different levels of comfort and ability to reduce noise levels. It is important that you understand how much the noise will be reduced by wearing your selected PPE, and whether it will still offer the same level of protection if you have long hair or are wearing glasses or safety goggles. Foam ear plugs are designed for single use and should be disposed of when they are removed from your ears. Sometimes these types of hearing protection perform better than ear defenders and they can be more comfortable, particularly in hot weather or when working in a confined space. The disadvantage of foam ear plugs is that they can cause infection if handled with dirty hands before they are inserted into the ears.

Safety signs and notices

Employers must provide safety signs in the workplace, if there are significant risks that can't be avoided or controlled in any other way. Safety signs are distinguished by their colour, shape and pictogram, and sometimes supported with supplementary information. There are four basic categories of safety sign under the Health and Safety (Safety Signs and Signals) Regulations 1996. These are shown in Table 1.5, along with supplementary information.

▼ Table 1.5 Safety signs

Type of sign	Description
Prohibition	Circular, white background with a black pictogram (or symbol) and red border.
	A sign prohibiting behaviour that could result in harm or injury.
Mandatory	Circular, blue background with a white pictogram (or symbol) and border.
	A sign instructing specific behaviour that you must do, e.g. eye protection must be worn.
Warning	Triangular, yellow background with a black pictogram (or symbol) and border.
	A sign providing a warning of danger or a hazard, e.g. slippery floor.

▼ Table 1.5 Safety signs (continued)

Type of sign	Description
Safe conditions 	Rectangular with a green background and white pictogram (or symbol). A sign providing information about safe conditions, e.g. emergency escape route, assembly point or first aid.
Supplementary information 	Rectangular signs, usually the same colour as the sign they are supporting. Some safety signs are generic and can be used in many different situations, therefore supplementary information is usually provided specific to the precaution.
Fire-fighting signs 	Rectangular, red background, white pictogram (or symbol) and border. These signs are used to identify the location of fire fighting equipment and fire alarm call points. They may also be used to locate other fire emergency devices, such as an emergency phone or ladder.
COSHH symbols 	Diamond, red border with a black pictogram (or symbol). These safety signs provide information relating to the use, storage, handling and transportation of hazardous materials and substances. Refer to COSHH Regulations, covered earlier in this chapter, for further examples of safety signs.

Safety signs must be displayed in prominent positions on construction sites and in workshops. It is important not to use too many signs, because they can cause confusion and may be disregarded.

ACTIVITY

The HSE publishes free guidance on the Health and Safety (Safety Signs and Signals) Regulations 1996. Use the following link to download a copy: www.hse.gov.uk/pubns/books/l64.htm.

Produce a suitable information board for your construction site, including all relevant safety signs under the Health and Safety (Safety Signs and Signals) Regulations 1996. Now take a picture with your mobile phone of an actual site information board and evaluate them both. Are they different? Why? Discuss your notice with your peers. How does yours compare with theirs?

Noise levels

Under the Regulations, specific actions are triggered at specific noise levels. These are referred to as 'action levels'. Noise is measured in decibels and shown as dB(A). The two main action levels are 80 dB(A), the lower action level, and 85 dB(A), the upper action level.

Requirements at the lower action level, 80 dB(A):
- Assess the risk to operatives' health and provide them with information and training.
- Provide suitable ear protection free of charge to those who request it.

Requirements at the upper action level, 85 dB(A):
- Reduce noise exposure as far as practicable by means other than ear protection.
- Set up an ear protection zone using suitable signage and segregation.
- Provide suitable ear protection free of charge to those affected, and ensure this is worn.

> **HEALTH AND SAFETY**
>
> A chop saw (mitre saw) can be as loud as 100 dB(A). Continued use of this type of power tool, even for short periods of time, can damage your hearing permanently.

Personal Protective Equipment (PPE) at Work Regulations 1992

Employees and subcontractors must work in a safe manner. Not only must they wear the PPE that their employers provide, but they must also look after it and report any damage to it. Importantly, employees must not be charged for anything given to them or done for them by the employer in relation to safety.

The main requirement of the Regulations is that PPE must be supplied and used at work wherever there are risks to health and safety that cannot be adequately controlled in other ways.

The Regulations also require that PPE is:
- included in method statements
- properly assessed before use to ensure it is suitable
- maintained and stored properly
- provided to employees with instructions on how they can use it safely
- used correctly by employees.

An employer cannot ask for money from an employee for PPE, whether it is returnable or not. This includes agency workers if they are legally regarded as employees. If employment has been terminated and the employee keeps the PPE without the employer's permission, then, as long as it has been made clear in the contract of employment, the employer may be able to deduct the cost of its replacement from any wages owed.

> **INDUSTRY TIP**
>
> Did you know that hard hats have an expiry date? Hard hats used after this date may not provide the same level of protection as they were designed to. If you have a hard hat that has expired speak to your employer, as they legally have to replace it free of charge.

Using PPE is a very important part of staying safe. For it to do its job properly it must be kept in good condition and worn correctly. If an article of PPE is damaged, it is important that this is reported and it is replaced.

▲ Figure 1.17 A site safety sign showing the PPE required to work there

You will be made aware of the mandatory PPE required for a site during your induction; most sites will also have safety signs to reinforce the site rules. It is common practice to wear three items of PPE (three points of contact):
1 safety footwear
2 hard hat
3 hi-vis vest.

Some construction sites will have five mandatory items of PPE (five points of contact), to include safety gloves and glasses. Risk assessments and method statements specific to a task may indicate that additional or specialist protective equipment is used. It is important to remember that there are many different types of PPE available, but in some cases certain types of PPE could lead to additional hazards – for example, if a carpenter wears gloves while fitting a kitchen worktop with a router, the gloves could become entangled in the moving parts of the router.

Table 1.6 shows the types of PPE used in the workplace and explains why it is important to store, maintain and use PPE correctly. It also shows why it is important to check and report any damage to PPE.

▼ Table 1.6 Correct use of workplace PPE

PPE	Correct use
Hard hat/safety helmet	Hard hats must be worn when there is danger of hitting your head or danger of falling objects. They can prevent a wide variety of head injuries. Most sites insist on hard hats being worn. They must be adjusted to fit your head correctly and must not be worn back to front! Check the date of manufacture as plastic can become brittle over time. Solvents, pens and paints can damage the plastic too.
Toe-cap boots or shoes	Toe-cap boots or shoes are worn on most sites as a matter of course and protect the feet from heavy falling objects. Some safety footwear has additional insole protection to help prevent nails going up through the foot. Toe caps can be made of steel or lighter plastic.
Ear defenders and plugs	Your ears can very easily be damaged by loud noise. Ear protection will help prevent hearing loss while using loud tools or if there is a lot of noise going on around you. When using ear plugs, always ensure your hands are clean before handling the plugs as this reduces the risk of infection. If your ear defenders fail to make a good seal around your ears or are damaged, have them replaced.
High-visibility (hi-vis) jacket	This makes it much easier for other people to see you, which is especially important when plant or vehicles are moving in the vicinity.
Goggles and safety glasses	These protect your eyes from dust and flying debris while you are working. It has been known for casualties to be taken to hospital after dust has blown up from a dry mud road. You only get one pair of eyes – look after them!

▼ Table 1.6 Correct use of workplace PPE (continued)

PPE	Correct use
Dust masks and respirators	Dust is produced during most construction work and can be hazardous to your lungs. It can cause all sorts of ailments, from asthma to cancer. Wear a dust mask to filter out this dust. You must ensure it is well fitted. Another hazard is dangerous gases such as solvents. A respirator will filter out hazardous gases but a dust mask will not. Respirators are rated P1, P2 and P3, with P3 giving the highest protection. Employers can measure the protection offered from half-face masks by undertaking 'face-fit testing'. Employees must attend this assessment when required by their employer.
Gloves	Gloves protect your hands. Hazards include cuts, abrasions, dermatitis, chemical burns or splinters. Latex and nitrile gloves are good for fine work, although some people are allergic to latex. Gauntlets provide protection from strong chemicals. Other types of gloves provide good grip and protect the fingers.
Sunscreen	Another risk, especially in the summer months, is sunburn. Although a good tan is sometimes considered desirable, over-exposure to the sun can cause skin cancers such as melanoma. When out in the sun, cover up and use sunscreen (i.e. sun cream) on exposed areas of your body to prevent burning.
Preventing HAVS	Hand–arm vibration syndrome (HAVS), also known as vibration white finger (VWF), is an industrial injury caused by using vibrating power tools (such as a hammer drill, vibrating poker or vibrating plate) for a long time. Such injury can be controlled by limiting the time for which such power tools are used.

INDUSTRY TIP

It's a common myth that you must be clean shaven before undertaking a face-fit test. Employers can request that you shave only to allow a better seal between the mask and your face, however you may have religious reasons for not wanting to shave.

The Work at Height Regulations (WAHR) 2005

If you are an employer and control work at height, then you will have duties under these regulations. Employees also have a duty to ensure they take reasonable care of themselves and other people who may be affected by their acts or omissions. They should also co-operate with their employer to enable it to fulfil its duties under the law.

Employers must ensure that work at height is risk assessed, properly planned, supervised and undertaken by competent workers. Although working at relatively low heights can reduce the risk, it is still possible to sustain an injury from a fall.

Employers' duties are to:
- avoid planning work at height if possible
- use equipment and other methods to prevent falls from height when work at height can't be avoided, e.g. use a harness and lanyard
- if the risk of falling cannot be completely eliminated, use equipment and measures to minimise the distance and consequences of a fall, e.g. use safety nets, a crash deck or air bags.

▲ Figure 1.18 Workers wearing safety harnesses and a lanyard on a mobile elevating work platform (MEWP)

Several points should be considered when working at height:
- How long is the job expected to take?
- What type of work will be undertaken? It could be anything from replacing a section of picture rail to constructing a roof.
- How is the access platform going to be reached? How many people will be using it?
- Will people be able to get on and off the access structure safely? Could there be overcrowding?
- Are there any risks to passers-by? Could debris or dust blow off and injure anyone on the road or path below?
- What are the conditions like? Extreme weather, unstable buildings and poor ground conditions need to be taken into account.

INDUSTRY TIP

Safety netting and air bags are referred to as 'soft landing systems'. Employers should have a recovery plan in place to rescue people from height in an emergency, and this plan should be practised.

INDUSTRY TIP

A crash deck is a temporary floor constructed to prevent falls from height and to provide a working platform. Alternatively, a birdcage scaffold could be used for the same purpose or greater heights if needed. Crash decks and birdcage scaffolds are commonly used by carpenters when constructing roofs.

▲ Figure 1.19 A crash deck being erected on site

▲ Figure 1.20 A birdcage scaffold

▲ Figure 1.21 A cherry picker can assist you when working at height

HEALTH AND SAFETY

Only trained, competent and authorised people should use any type of access equipment in the workplace.

Access equipment and safe working practice

The means of access should be chosen only after careful consideration of the risk assessment. Many accidents happen in the workplace every year as a result of falls from height. Employers should provide the safest means of access 'so far as reasonably practicable', not just the cheapest or what's available on site at the time.

Leaning ladders and stepladders

The HSE recommends that ladders should be used only for short-duration work (30 minutes or less) between movements. They are normally used for access onto a working platform, e.g. a scaffold. It is important to check the parts of a ladder before use:

- Check the stiles (sides) are in good condition, not split or bent.
- Check the rungs (steps) are not bent or missing.
- Check the feet don't have missing non-slip attachments.

INDUSTRY TIP

Wooden ladders are rarely used in the construction industry nowadays because they are heavy and easily damaged. By contrast, aluminium ladders are lightweight and extremely strong. Fibreglass ladders are preferred by electricians because they don't conduct electricity.

IMPROVE YOUR MATHS

The correct ratio for a leaning ladder is one unit out for every four units up. What angle is that?

On a roof ladder, the wheels are used to run the ladder up over the roof when positioning, and hook over the ridge).

▲ Figure 1.22 A roof ladder

INDUSTRY TIP

Ladders should not be rested on plastic guttering because the weight of the ladder and the person using it can cause the guttering to bend and break. A way to avoid this common cause of accidents is to use an aluminium 'ladder stand-off'. This will position the ladder away from the face of the wall to enable safe access for tasks such as maintaining guttering.

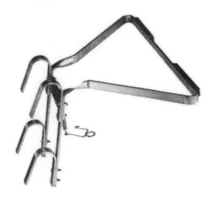

▲ Figure 1.23 Ladder stand-off

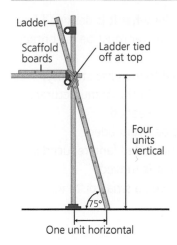

▲ Figure 1.24 Using a ladder correctly

Labels in figure: Ladder; Scaffold boards; Ladder tied off at top; Four units vertical; 75°; One unit horizontal

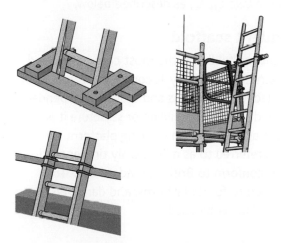

▲ Figure 1.25 Examples of how to secure a leaning ladder at the top and bottom.

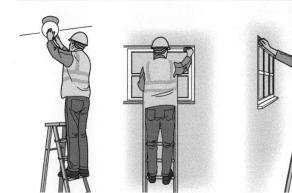

▲ Figure 1.26 Using a stepladder safely

▲ Figure 1.27 Using a stepladder correctly

Labels in figure: Working from the side can make stepladders unstable. Do not overreach; Prop; Don't stand on the top three steps; Stepladder is fully open; Locked open firm and level on the ground

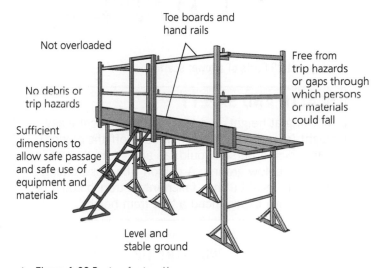

▲ Figure 1.28 Parts of a trestle

Labels in figure: Not overloaded; Toe boards and hand rails; Free from trip hazards or gaps through which persons or materials could fall; No debris or trip hazards; Sufficient dimensions to allow safe passage and safe use of equipment and materials; Level and stable ground

Tower scaffold

Tower scaffolds are usually used for light-duty work, such as painting and decorating, maintenance and plastering. They can be made from scaffold tubes, but most are made from galvanised steel or lightweight aluminium alloy sections. They are quick to erect by a competent person and can be mobile with the use of castors. PASMA is the lead trade association for mobile towers in the industry. Completion and achievement of a PASMA training qualification is widely recognised by employers, and evidenced with a competency card.

Outriggers can be added to a mobile tower if additional height is needed.

Tower scaffolds can be dangerous if used in high winds or poor weather conditions. You should always read the manufacturer's information for the recommended safe wind speeds and heights at which the tower can be used.

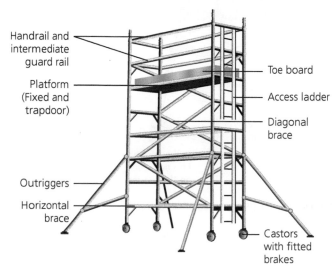

▲ Figure 1.29 Parts of a tower scaffold

Labels: Handrail and intermediate guard rail; Platform (Fixed and trapdoor); Outriggers; Horizontal brace; Toe board; Access ladder; Diagonal brace; Castors with fitted brakes

HEALTH AND SAFETY

'Working at height' is a term used to describe **any** height (even just a few centimetres off the ground) from which a person could fall and injure themselves, above or below ground. A person who was working at ground level but has fallen into a trench could be said to have experienced a 'fall from height'.

INDUSTRY TIP

You should only ever climb on the inside of a tower scaffold, to avoid altering the centre of gravity and making the structure unstable.

To use a tower scaffold safely:
- you must be trained and competent
- always read and follow the manufacturer's instruction manual
- use the equipment only for what it is designed for
- the wheels or feet of the tower must be in contact with a firm surface
- outriggers should be used to increase stability; the maximum height given in the manufacturer's instructions must not be exceeded
- the platform should not be overloaded
- the platform should be unloaded (and reduced in height if required) before it is moved
- never move a platform, even a small distance, if it is occupied.

Tubular scaffold

This comes in two types, as described below.

Independent scaffold

Independent scaffolding is the most commonly used in the construction industry. It consists of two sets of standards, ledgers, transoms and braces, 'independent' (or free-standing) of the building or structure it is used to gain access to. Each working platform within a scaffold is referred to as a 'lift'. Only timber scaffold boards that conform to British Standard **BS 2482** should be used to form platforms, and damaged boards should never be used.

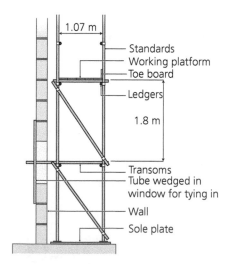

Labels: 1.07 m; Standards; Working platform; Toe board; Ledgers; 1.8 m; Transoms; Tube wedged in window for tying in; Wall; Sole plate

▲ Figure 1.30 Independent tubular scaffold

Putlog scaffold

Also known as a 'bricklayers' scaffold', this consists of a row of standards with a single ledger. A putlog is a term used to describe the transoms with one flattened end that are built into the mortar joints of the brickwork as the build is progressing. Putlog scaffolds allow the working platform to be as close to the building as possible without any standards getting in the way. As the scaffold is dismantled, the putlogs are removed and the holes in the mortar are filled in.

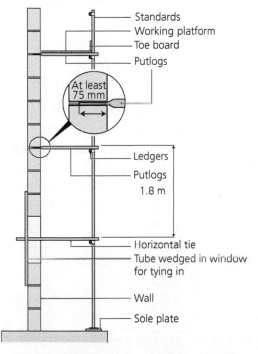

Standards
Working platform
Toe board
Putlogs
At least 75 mm
Ledgers
Putlogs 1.8 m
Horizontal tie
Tube wedged in window for tying in
Wall
Sole plate

▲ Figure 1.31 Putlog tubular scaffold

Tubular scaffolding (sometimes shortened to 'scaff') should be erected by specialist scaffolding companies and often requires structural calculations. Only trained and competent scaffold erectors should alter scaffolding. Scaffolding should be inspected after the initial erection and handover, at regular intervals and after poor weather conditions, such as a storm or high winds.

INDUSTRY TIP

If required, scaffolding should have loading bays designed and built into it to load materials safely onto the working platform. The bays should be suitably guarded when not in use. Any heavy loads (e.g. blocks and bricks) should be evenly distributed over the platform, adjacent to the standards.

Access to a scaffold is usually via a tied ladder with three rungs projecting at least 1 m above the step at platform level for a good handhold when exiting or starting off.

Many scaffolding systems now have to be sheeted to ensure that the work is contained in the area. Further information and guidance can be found on the National Access & Scaffolding Confederation (NASC) website: www.nasc.org.uk.

INDUSTRY TIP

You must be competent to erect, adjust and dismantle scaffolding. The Construction Industry Scaffolders Record Scheme (CISRS) is the preferred scaffolding qualification, and is widely recognised by many of the largest construction organisations in the country.

All scaffolding must have:
- no gaps in the handrails or toe boards
- a safe system for lifting any materials up to the working height
- a safe system for debris removal, such as a debris chute (Figure 1.32)
- brick guards fitted if materials are stored on the working platform above the height of the toe boards (150 mm).

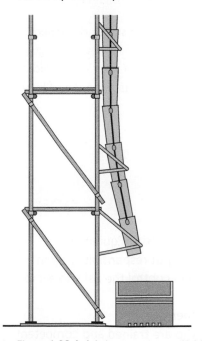

▲ Figure 1.32 A debris chute for scaffolding

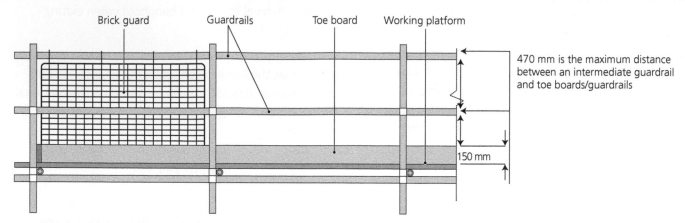

Minimum guardrail height 950 mm

Brick guard   Guardrails   Toe board   Working platform

470 mm is the maximum distance between an intermediate guardrail and toe boards/guardrails

150 mm

▲ Figure 1.33 A safe working platform on a tubular scaffold

Fall protection devices include:

- harness and lanyards (fall restraint – to minimise the distance of a fall)
- safety netting (fall arrest – to reduce the consequences of a fall)
- air bags (fall arrest – to reduce the consequences of a fall).

ACTIVITY

Imagine that, as a carpenter, you have been asked to do the tasks below while working at height. Suggest the safest and most practical type of access equipment in each case.

1 Fit new roof trusses on a two-storey detached property.
2 Replace a window on the first floor of a building.
3 Inspect the condition of the fascia and soffits on a bungalow.

Lifting Operations and Lifting Equipment Regulations (LOLER) 1998

The Lifting Operations and Lifting Equipment Regulations (LOLER) 1998 put responsibility upon employers to manage and control the risks to avoid damage or injury by:

- planning all lifts properly
- using only people who are trained and competent
- supervising lifts appropriately
- ensuring that lifts are carried out in a safe way.

LOLER also places a duty on responsible people to ensure that lifting equipment is fit for purpose, appropriate for the intended task and suitably marked with safe loading etc. Lifting equipment is also subject to thorough examinations and testing by a competent person, to ensure it doesn't deteriorate, break down prematurely or fail when in use.

▲ Figure 1.34 Using a scissor lift at height

Examples of lifting equipment include:

- mobile cranes
- truck-mounted loaders (the Swedish company HIAB is a widely recognised manufacturer)
- tower cranes
- mobile elevating work platforms (MEWP), e.g. scissor lifts, telescopic booms and vertical lifts
- passenger/goods hoists
- telehandlers.

2 SITE INDUCTIONS AND OTHER ON-SITE TRAINING

What do the professionals say?

'It can be frustrating going through inductions each time you start on a new site, but I realise they are necessary to keep me safe. If I have a serious injury at work, I won't be able to support my family.'

David Jones, self-employed site carpenter

Under the Health and Safety at Work Act 1974, employers have a duty to provide 'information and instruction'. This is usually done in the form of a site induction when you first arrive on site. Not all construction sites are the same and neither are the hazards on them, therefore everyone entering a site for the first time must attend induction training to be made aware of the risks. The length of time an induction takes will depend on a number of factors, including:

- size and complexity of the site
- nature of the work
- whether the person attending is young or inexperienced, a tradesperson or a **general operative**
- whether the person attending is a visitor (such as an architect or building inspector).

Temporary site visitors, such as external managers, building inspectors and architects, are likely to be experienced in the construction industry, and are less likely to undertake practical construction tasks while they are on site, so their site induction does not have to be as in depth as that aimed at a young person new to the industry, for example.

KEY TERM

General operative: a person working on a construction site without a specific set of trade skills. They are usually employed by a contractor to carry out labouring tasks, such as mixing plaster and mortar, keeping the site clean, and stacking and storing materials.

Induction topics

Site inductions are likely to cover:

- site-specific rules, e.g. mandatory PPE
- the process for accident reporting
- assembly points in case of emergency
- hazardous materials on site
- the location and use of welfare facilities, such as toilets, rest areas and showers.

Before allowing you onto a construction site, your employer will have worked with the health and safety team to plan and comple an assessment of any significant hazards. As part of the process they will have considered hazards relating to:

- the site layout (e.g. existing features such as protected trees, buildings or streams)
- services to the site (e.g. electricity and mains water)
- securing the site (e.g. lockable site gates and site notices)
- existing and temporary roads (e.g. site traffic management plan)
- storage areas (e.g. for flammable materials, bricks, timber)
- welfare facilities (e.g. toilets, canteen)
- hazardous materials already present on the site (e.g. asbestos)
- protection of the public from work activities (e.g. site fencing).

INDUSTRY TIP

Contractors have a legal responsibility to ensure that people on site are competent and safe. One way to measure competence is with qualifications, however health and safety knowledge needs to remain current. Most large construction companies in the UK require contractors to hold a competency card, such as the CSCS (Construction Skills Certification Scheme) card. To obtain an apprenticeship card you will have to confirm that you are registered for an apprenticeship. Generally, operatives will have to sit an online CITB health, safety and environment test before applying for a card, but as an apprentice you may be exempt if you have:

- a certificate for a one-day health and safety awareness course
- a health and safety unit included in your initial qualification or induction
- an email or letter from the managing agency of your apprenticeship confirming that you have met the managing agency's health and safety requirements.

Source: www.cscs.uk.com/card-type/apprentice

Apprentice CSCS cards are valid for four years and six months. There are a number of different cards available. More information can be found at www.cscs.uk.com/applying-for-cards/types-of-cards/.

Partner card schemes include CISRS (Construction Industry Scaffolders Record Scheme) and CPCS (Construction Plant Competence Scheme).

INDUSTRY TIP

The Health and Safety Executive (HSE) is a great source of information when dealing with matters relating to health and safety. An example of a simple site induction for smaller companies is available here: www.hse.gov.uk/construction/induction.pdf.

If you don't understand something you are told during a site induction, ask the person delivering the training. It is important for your safety and the safety of others that you are clear about the risks you will face on site and how to deal with them. Don't be surprised if your site manager or supervisor asks you some questions at the end of the induction, or even asks you to undertake a short test to measure the level of your understanding before you are allowed onto their site.

ACTIVITY

Write a brief induction for new students or employees at your place of work, college or training centre. Remember to include things like safety signs, assembly points, first aid and emergency exits.

Analyse the induction you have written. How does it compare with inductions you have been through? Would it be effective? How would you deliver it? Where would you deliver it? How long would it last? Could you engage an audience? If not, what could you do to make it more engaging? Use the internet to research site inductions to support your answers.

On-site training (toolbox talks)

A toolbox talk is a short training session, usually carried out on site, about a particular topic. For example, your employer may have noticed an increase in the number of reported near-miss incidents relating to stepladders within the last month, so they may invite employees who are likely to use stepladders on site to attend a training session in order to reduce the number of near misses.

ACTIVITY

As a Level 3 carpentry and joinery learner, you should be taking on more responsibility, and may even be managing people or small jobs in a supervisory role. Use this opportunity to write and present a toolbox talk to an individual or a small group of people on a trade-related topic of your choice, e.g. the safe use of portable power tools or hand tools on site.

Monitor the people to whom you have delivered the training over two weeks and reflect on the effectiveness of the session.

HEALTH AND SAFETY

If you are unsure about any health and safety issue always seek help and advice from your supervisor.

3 FIRST AID AND FIRST-AID KITS

If someone unexpectedly becomes ill or has an accident at your place of work, they will need to receive first aid to prevent their condition worsening. Employers have a duty to provide first aiders, facilities and appropriate equipment to ensure that employees receive immediate attention under the Health and Safety (First Aid) Regulations.

You should apply first aid to a casualty only if you have been trained in first aid and you are a **nominated first aider**. Even a minor cut or abrasion can become infected if it isn't treated properly. The purpose of first aid – known as the three Ps – is to:

1 preserve life (apply emergency first aid)
2 prevent deterioration (stabilise the patient's condition)
3 promote recovery (until the emergency services arrive).

Figure 1.35 shows the contents of a typical site first-aid kit.

Employers have a duty to assess the first-aid needs of the workforce, e.g. the number of employees, the geographic spread of the workforce, the nature of the work, lone working and adequate cover requirements for holidays/sickness. They must ensure all workers are aware of the provision and location of first-aid equipment. The contents of a first-aid box will reflect the type of work undertaken in the area in which it is located but it must never contain medicines or tablets – these should only be prescribed by a doctor.

INDUSTRY TIP

Lone working is when an individual is working remotely or away from their normal place of work, without any direct supervision, and is isolated from any other employees. There may not be anyone to help them if they fall ill, have an accident or are attacked, so employers should risk assess such activities to control the risk. Providing mobile phones and keeping in regular contact with the employer are some ways of managing the risk.

KEY TERM

Nominated first aider: a person who has been recognised by their employer as having the necessary knowledge and skills to perform the role of a first aider at their place of work.

INDUSTRY TIP

Are you aware of the location of the first-aid box at your place of work? Do you know how to raise the alarm in the event of an emergency?

Do you know who the first aider is, or what provision will be available if you are working in a remote location?

If the answer to any of these questions is no then you should check with your employer, who will be able to advise you.

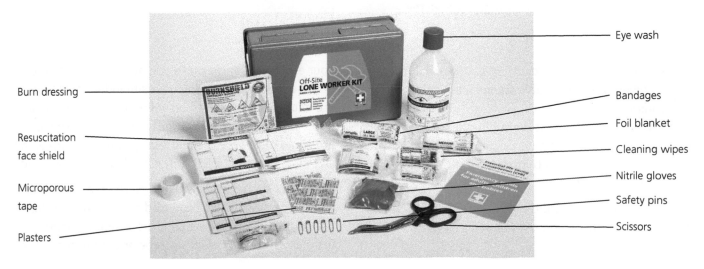

Eye wash

Burn dressing

Resuscitation face shield

Microporous tape

Plasters

Bandages

Foil blanket

Cleaning wipes

Nitrile gloves

Safety pins

Scissors

▲ Figure 1.35 Contents of a first-aid kit

4 FIRE

Arson (the criminal act of deliberately starting a fire) is the number-one cause of fires on construction sites. Other causes of fire include hot works such as welding, torch cutting or grinding. Hot works on construction sites should be controlled with a 'permit to work system'. The permit will control who is working, where the work is taking place and for how long. It will also ensure that only competent people undertake the work, with the correct equipment, supervision and methods to extinguish the fire in an emergency. The person undertaking the hot works must return to where they undertook the work after the work has been completed, to ensure there are no signs of potential fire.

▲ Figure 1.36 The fire triangle

INDUSTRY TIP

Permit to work systems are used to control dangerous and high-risk activities on construction sites, such as:
- hot works (welding and grinding)
- working at height
- breaking the ground (excavating – risk of coming into contact with electrical supplies)
- working with live electrical supplies
- working in confined spaces (with limited oxygen, potentially containing natural gases).

Fire needs three things to start; if just one of them is removed the fire will go out. This is the principle used with fire extinguishers – they either cool a fire to remove the heat or smother it to remove the oxygen.

The three things are:

1 oxygen – a naturally occurring gas in the air that combines with flammable substances under certain conditions
2 heat – a source of ignition, such as a hot spark from a grinder or naked flame
3 fuel – things that will burn, such as acetone, timber, cardboard or paper.

If you have heat, fuel and oxygen you will have a fire. Remove any of these and the fire will go out.

Preventing the spread of fire

Good housekeeping (being tidy) will help prevent fires starting and spreading. For instance:
- Wood offcuts should either be stored neatly to reuse or disposed of in a designated bin.
- Always replace the cap on unused fuel containers when you put them away, otherwise they are a potential source of danger.
- Dispose of empty fuel cells (used to power nail guns) responsibly in a designated waste bin.
- Flammable liquids (not limited to fuel-flammable liquids), such as oil-based paint, thinners and oil, must be store in a locked metal cupboard or shed (in accordance with COSHH).
- Smoking around flammable substances should be avoided.
- Dust can be explosive, so when doing work that produces wood dust it is important to use some form of extraction (local exhaust ventilation system – LEVS) and to have good ventilation.

Fire extinguishers and their uses

Your employer should make you aware of the location of fire extinguishers during your site induction. Use a fire extinguisher on a fire only if you have been trained to do so; using the wrong type of fire extinguisher on a fire could make it worse (e.g. using water or foam on an electrical fire could lead to the user being electrocuted).

Employers should have assessed all the fire hazards at your place of work and produced a fire plan. The plan will contain information about fire assembly points, fire alarms, smoke detectors, type and locations of fire extinguishers, fire marshals, escape routes and storage of flammable materials. Besides fires extinguishers, fire blankets, hose reels, sprinkler systems and fire buckets can be used to fight fires.

Table 1.7 shows the different classifications of fire and which extinguisher to use in each case.

▼ Table 1.7 Types of fire extinguisher and their uses

Class of fire	Materials	Type of extinguisher
A	Wood, paper, hair, textiles	Water, foam, dry powder, wet chemical
B	Flammable liquids	Foam, dry powder, CO_2
C	Flammable gases	Dry powder, CO_2
D	Flammable metals	Specially formulated dry powder
E	Electrical fires	Dry powder, CO_2
F	Cooking oils	Wet chemical, fire blanket

▲ Figure 1.38 CO_2 extinguisher

▲ Figure 1.39 Dry powder extinguisher

INDUSTRY TIP

All fire extinguishers are red. Their contents are identified by the coloured labelling and writing on the container. It's a good idea to become familiar with the different types of extinguisher and the classifications of fires; it could save your life one day!

▲ Figure 1.40 Water extinguisher

▲ Figure 1.41 Foam extinguisher

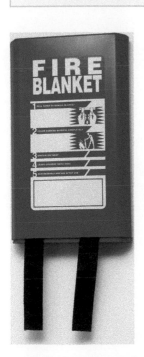

▲ Figure 1.37 A fire blanket

Emergency procedures

In an emergency, people tend to panic. Would you know what to do in an emergency, for example, if you discovered a fire, a bomb or some other security problem? It is vital to be prepared and it is your responsibility to know the emergency procedures on your work site.

If you discover a fire, or other emergency, you will need to raise the alarm:

- Tell a nominated person. Who is this?
- If you are first on the scene, phone the emergency services on 999.
- Be aware of the alarm signal. Is it a bell, a voice or a siren?
- Where is the assembly point? You will have to proceed to this point in an orderly way. Leave all your belongings behind as they may slow you or others down.
- At the assembly point, someone will ensure everyone is out safely by taking a count or register. Do you know who this person is? If during a fire you are not accounted for, a fire-fighter may risk their life to go into the building to look for you.
- How do you know it's safe to re-enter the building? You will be told by the fire marshal. It's very important that you do not re-enter the building until you are told to do so.

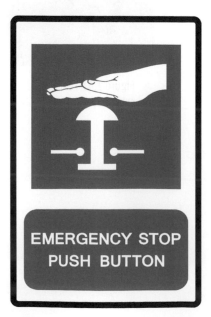

▲ Figure 1.42 Emergency procedure sign

INDUSTRY TIP

You will not be expected to tackle a fire yourself unless you have been trained to use an extinguisher, it is relatively small or it is blocking your escape route. Did you know that more people die from smoke inhalation than from a fire itself?

▲ Figure 1.43 Fire assembly point sign

ACTIVITY

Locate all the first aid boxes and provision at your place of work, college or training centre. Do you consider there to be enough first-aid kits? How often are they checked? Do you know who the first aider is? Do you know how to contact them? What impact could a depleted first aid box have?

Test your knowledge

1 When should you attend a site induction?

A Every day

B Once a week

C When you start at a new site

D Only if you don't have a CSCS card

2 Which one of the following does not legally have to be provided by an employer for your welfare?

A A smoking shelter

B Drinking water

C A canteen

D Facilities to dry your hands

 3 You are about to extend the roof on an old property but you suspect that the soffits may contain asbestos. What action should you take?

A Carefully remove the old soffits, taking care not to break them and release the hazardous fibres.

B Stop work, warn others and report your concerns to your supervisor.

C None. If there is any asbestos present, it will be low risk and unlikely to harm you.

D None. Asbestos is not used any more, therefore there is no hazard present.

4 What type of safety sign is this?

A Warning

B Prohibition

C Mandatory

D Safe conditions

5 When are you considered to be working at height?

A At any distance from which a fall is likely to cause harm

B When working on access equipment

C 500 mm above the ground

D 2 m above the ground

6 While at work, you have had a visit from the Health and Safety Executive (HSE). Your employer has been issued with a Prohibition Notice. What does this mean?

A Something is unsafe but you can carry on using it as long as it is put right within 21 days.

B It's a certificate issued to all workplaces that achieve high standards of health and safety.

C Your employer has been issued with a fine for failing to meet the minimum standards for health and safety in the workplace.

D Something is unsafe and likely to cause an injury, so its use must stop immediately until it is put right.

7 Why is it considered dangerous to work in an excavation?

A You could be forgotten about when the site is secured at the end of the day.

B The sides of the excavation could collapse.

C It could be very cold in the excavation during the winter.

D Your high-visibility clothing could get excessively dirty, making it harder for you to be identified on site.

8 Why is it dangerous to use machinery, equipment and tools that produce high levels of vibration for long periods of time?

A They could cause hand–arm vibration syndrome (HAVS).

B They could cause Weil's disease.

C The vibration could cause long-term damage to your eyes.

D They could cause leg–foot vibration syndrome (LFVS).

9 How often does the HSE recommend that power tools used on construction sites are PAT tested?

A Annually

B Every three months

C Every time they are serviced

D Only if they are to be sold

10 A fire needs three elements to burn. Which one is missing from the 'fire triangle' illustrated below?

11 You have been asked to remove a guardrail from a scaffold to allow materials to be lifted onto the working platform. Explain what action you should take.

12 Your employer has asked you to attend a face-fit test. Explain the purpose of the test and why you should attend.

13 Between 2012 and 2017 there were 196 fatalities in the construction industry (source: RIDDOR). Of these deaths, 49 per cent were caused by falls from height. To the nearest person, how many people is that?

14 When must an employer provide hearing protection upon request, and when is it mandatory to wear it?

Practical activity

Part 1

Produce a plan of your place of work and indicate on it the control measures that have been taken to protect those in it, such as:

- safety signs and information boards
- fire exits and emergency escape routes
- fire extinguishers
- first-aid facilities
- security (boundary, temporary fencing, hoarding)
- pedestrian walkways (segregation of plant, vehicles and people)
- hazardous material storage (COSHH)
- welfare facilities (safe zone)
- site office
- environmental control, such as waste materials (segregation of waste into designated skips or bins)
- location of PPE stores
- working at height and access equipment.

Also consider any points that have not been listed above, and areas that you think could be improved. You might want to talk with your tutor or employer about any concerns you have – you could be saving a life.

Part 2

In this chapter we have looked at the CDM regulations and duty holders' responsibilities. As a duty holder, your employer will have implemented some additional health and safety control measures at your place of work to comply with these regulations. Find out and list the site-specific measures taken by your employer that relate to CDM regulations.

CHAPTER 2

PRINCIPLES OF ORGANISING, PLANNING AND PRICING CONSTRUCTION WORK

INTRODUCTION

As you progress in your career and gain responsibility within the industry you will be involved with interpreting a range of information. Understanding and communicating this information are crucial to the success of a building project. Mistakes made in interpreting information always prove costly in terms of labour, building materials and time. Whether you are reading a drawing, cross-checking information on a schedule, planning building work or calculating a price, each process requires a methodical approach and should be easy for those using the documentation to interpret. Much of this information is provided to you by the **contract documents**.

LEARNING OBJECTIVES

By reading this chapter you will learn about:

1 how the construction industry is regulated
2 energy efficiency and sustainable materials for construction
3 estimating quantities and pricing work for construction
4 planning work activities for construction
5 written and oral communication on site
6 different types of drawn information and associated software.

1 HOW THE CONSTRUCTION INDUSTRY IS REGULATED

Health and safety regulations

You may consider health and safety to be something that you think about only when you are about to do something – for example, working at height or using a power tool. However, effective health and safety control is considered and planned from the initial building design stages through to completion. Even the maintenance, servicing and demolition of the structure is planned to keep people safe from harm. As a carpenter or joiner, it may not be your duty under the various types of law to prepare risk management plans; however, it is important for you to understand your role in helping your employer to fulfil their duties.

In Chapter 1, we looked at the main health and safety regulations you should be aware of. In this chapter, we will think about how these regulations affect the way risk is managed throughout a project, rather than the practical ways of ensuring health and safety on site. Refer back to Chapter 1 to see how much information you remember.

KEY TERM

Contract documents: these comprise the working drawings, schedules, specifications, bill of quantities and contracts.

Planning permission and building control

New buildings and substantial changes to existing properties are likely to require approval from the local authority before any work can go ahead. This is referred to as 'planning permission', and aims to prevent developers or individuals from building whatever they what, where they want, without regard for the environment or how the work may affect other people. Permission may be requested by completing and submitting a planning application form, along with a fee and all details for the proposed building project, e.g. design drawings. This information is then used by the local authority's planning department to decide whether the project should go ahead.

Planning permission may be either for outline or detailed planning consent.

The outcome of an application could be:
- permission granted in full (consent)
- permission granted with planning conditions
- permission refused.

If the application is granted (called planning consent), the building work can commence without delay. If permission is 'granted with conditions', it is approved in principle as

long as minor changes advised by the local authority are made in the building phase. For example, a window in a property may have to be removed or the glass obscured to prevent people seeing into a bathroom.

Local authorities are keen to protect areas of natural beauty, historic buildings and land of architectural interest. When the planning department considers an application it may also look at how the proposed development will fit into its surroundings. For example, new homes with a modern design (two storey, flat roof, thin-coat masonry walls, etc.) wouldn't be in keeping in a **conservation area** containing single-storey homes with thatched roofs.

▲ Figure 2.1 A conservation area

If a planning application is refused, changes need to be made to the proposal before it is resubmitted. If it is refused again, then the person or organisation making the application can appeal in front of the planning committee against the decision.

It is not advisable to complete any work that requires planning permission without the approval of the local authority, because you could be legally made to undo or dismantle it.

Minor building work is sometimes permitted without making an application for planning permission, under what are known as **permitted development** rights.

An application may be made for the alteration, change of use or demolition of a listed building (usually an old building that is of historical or architectural interest). This is called an application for **listed building** consent. Before carrying out any building work on an old property, it's important to check whether it is listed; this information can be obtained from your local authority. Work on a listed building may not be carried out unless listed building consent has been granted by the local authority.

ACTIVITY

Use the following link to find all of the Grade 1 listed buildings within a 20-mile radius of where you live: https://historicengland.org.uk/listing/the-list/.

If possible, visit some of the buildings, or research them online if they're not in a public place, and explain why you think they are Grade 1 listed.

KEY TERMS

Permitted development: the right to construct without obtaining planning permission. This is sometimes used to convert a loft or garage, or to add a single-storey extension or porch to an existing building.

Listed building: a building that is protected by law and has restrictions on what you can and can't do to develop it. Listed buildings are categorised as Grade I (the most significant), Grade II* or Grade II. Before carrying out any building work on an old property, it is important to establish whether it is 'listed', by contacting your local authority.

INDUSTRY TIP

You can access further information on the planning process and permitted development online at www.planningportal.co.uk.

ACTIVITY

Use the internet or local newspaper to find three recent planning applications. Follow the applications through to the decision process to see what the outcomes were for each. If any of the applications were refused, analyse the decisions. Was there anything that could have been designed differently for the application to have been successful?

Building Regulations

All building work must be carried out to a professional standard so that it can resist the loads imposed upon it, and provides a comfortable and safe environment for those expected to use it. The minimum standard for building work in England is contained in the Building Regulations. These are supported with a number of Approved Documents that provide practical advice on building standards. Table 2.1 shows the subjects covered by each Approved Document.

▼ Table 2.1 What is covered by Building Regulations Approved Documents?

Building Regulations Approved Document	Content and relevance to carpentry work
A	Structure: walls, floors and roofs
B	Fire safety in residential homes, flats, schools, colleges and offices, e.g. fire doors and frames
C	Site preparations and resistance to contaminants and moisture
D	Toxic substances
E	Resistance to sound: acoustic walls, floors and doors
F	Ventilation in windows and roof spaces
G	Sanitation, hot water safety and water efficiency
H	Drainage and waste disposal
J	Combustion appliances and fuel storage systems
K	Protection from falling, collision and impact: staircases, ramps and guarding
L	Conservation of fuel and power
M	Access to and use of buildings
P	Electrical safety
Q	Security in dwellings
R	High-speed electronic communications networks
7	Material and workmanship

2 ENERGY EFFICIENCY AND SUSTAINABLE MATERIALS FOR CONSTRUCTION

Sustainable development

There is a finite (limited) supply of some of the most commonly used building materials, and it is our responsibility to make the most of what we have. By using sustainable products, we can control the rate of consumption of resources and conserve our natural assets. Constructing buildings using these methods will pay off. Many organisations can help us achieve this. For example, the Energy Saving Trust (EST) is a social enterprise with a charitable foundation. It offers impartial advice to communities and households on how to reduce carbon emissions, use water more sustainably and save money on energy bills.

- EST endorsed products: setting standards for best-in-class for products such as boilers and glazing.
- EST listed: products are listed in a directory and checked against quality and safety standards. Aimed at housing associations and other house builders.
- Verified by EST: test reports are verified by EST to provide assurance on products' energy-saving claims.

ACTIVITY
Plastic is made from oil. List all of the items you can think of that are used in the building of a house that use this finite resource.

The code for sustainable homes

In recent years, people have become more aware of energy efficiency in their own homes, and how poorly insulated properties and wasted energy can and do cost money.

ACTIVITY
Go to www.energysavingtrust.org.uk to see how an existing cavity wall can be insulated.

Simple changes in the way we live can make a massive difference, not only financially but to our environment. When purchasing a new house, you will look not only at cosmetic elements of the building, but also at how it performs and how much it will cost to run. Of course, Building Regulations will dictate the minimum building standards, but what sets one building apart from another? House builders are now constructing to voluntary standards to attract new buyers.

The UK government used to have a 'code for sustainable homes', which was a national standard for the sustainable design and construction of new homes. Local Authorities previously had powers to require Code as a planning requirement and it was also used for funding requirements (e.g. Registered Social Landlords under the Affordable Funding Programme 2015-2018). The Code for Sustainable Homes was withdrawn in 2015, amalgamating some parts into the Building Regulations.

The Building Research Establishment (BRE) offers **BREEAM** certification, which is an international sustainability assessment method for master planning projects, infrastructure and buildings. Impartial trained and licensed BREEAM assessors check each stage of the build, from the built environment, design and construction through to the final stages of operation and refurbishment. BREEAM rating benchmarks are awarded based on the percentage score achieved in the assessment. These are rated from acceptable to pass, good, very good, and excellent to outstanding.

KEY TERM
BREEAM: Building Research Establishment Environmental Assessment Method, a certification of the assessment of a building's environmental, social and economic sustainability performance.

▲ Figure 2.2 The BREEAM logo

Passive houses

Houses that have been designed and built to rigorous energy efficiency standards, that require little energy to heat them in the winter months and keep them cool in the summer, may be assessed against voluntary standards and classified as 'passive houses'. Passive houses are designed with features such as triple glazing, highly insulated doors and **heat recovery units**, and are completely airtight. As part of the assessment to classify a building as a passive house, it must undertake an air test to ensure that it is as airtight as it can possibly be. Areas that generally cannot be sealed from the outside of the building include extraction/heat recovery vents and chimneys. There are very few passive houses in the UK, although for many designers and builders this is the benchmark for building standards.

▲ Figure 2.3 A passive house

KEY TERM

Heat recovery units: units installed in buildings to extract the stale air from the inside and replace it with fresh air from the outside. The heat from the stale air is trapped and used to heat the colder fresh air being introduced into the building.

Building Regulations

Approved Document Part L, 'The Conservation of Fuel and Power in Buildings', is the standard that applies to construction projects that are new, extended, renovated, refurbished or involve a change of use.

To achieve compliance with Part L, the standard approach is to follow the guidance set out in the government's Approved Documents, of which there are four:
1 ADL1A New Dwellings
2 ADL1B Existing Dwellings (extensions, renovations, change of use or energy status)
3 ADL2A New Non-domestic Buildings
4 ADL2B Existing Non-domestic Buildings (extensions, renovations, change of use or energy status).

▲ Figure 2.4 Building Regulations Approved Documents relating to Part L

The route to compliance for new buildings and extensions is with the national calculation methodology (NCM) software, which calculates a dwelling or building's carbon dioxide emission rate and compares it with the target emission rate, also calculated by the software for the same building. The relevant calculations are:
● SAP (Standard Assessment Procedure) for ADL1
● SBEM (Simplified Building Energy Model) for ADL2.

Specialist companies can carry out these calculations for you or you can use guidance documents to show you how to meet the requirements of the Approved Documents. Architects will produce their construction drawings and specifications to meet these requirements, and these will be checked by the local authority's Building Control department when a planning application is submitted or Building Regulations approval is sought.

Thermal insulation and heat loss

Heat can be lost through buildings, as shown in Figure 2.5. The improved insulation of these areas will increase the energy efficiency of buildings.

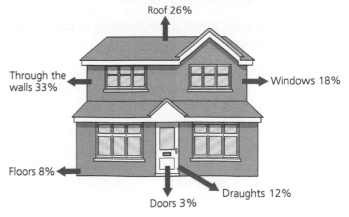

▲ Figure 2.5 Sources of heat loss from a house

43

In addition, heat flows out of a building, as shown in Figure 2.6, so it makes sense to insulate the structure to minimise this.

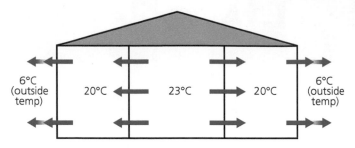

▲ Figure 2.6 Heat flowing from a building

Highly insulated structures will help to:
- prevent heat loss
- reduce the size of heat-providing appliances
- reduce costs to the user
- help the environment
- reduce the country's energy demands.

INDUSTRY TIP

Invest in the best insulation affordable. It will save you money on energy in the long term.

Thermal transmittance (U-values)

Thermal transmittance, also known as U-value, is the rate of transfer of heat through a structure or, specifically, through one square metre of a structure divided by the difference in temperature across the structure. It is expressed in watts per metre squared Kelvin, or W/m²K. Well-insulated parts of a building have a low thermal transmittance, whereas poorly insulated parts of a building have a high thermal transmittance. The lower the U-value, the greater the insulation properties of the structure.

Table 2.2 shows the thermal conductivity of commonly used building materials.

The Approved Documents inform what U-value a structure must meet in order to comply with the Building Regulations.

ACTIVITY

Go to www.planningportal.co.uk and look up the minimum U-value required for a roof.

▼ Table 2.2 Example U-values for various structures

Structure	U-value in W/m²K
Single-glazed windows, allowing for frames	4.5
Double-glazed windows, allowing for frames	3.3
Double-glazed windows with advanced coatings and frames	1.2
Triple-glazed windows, allowing for frames	1.8
Triple-glazed windows with advanced coatings and frames	0.8
Well-insulated roofs	0.15
Poorly insulated roofs	1.0
Well-insulated walls	0.25
Poorly insulated walls	1.5
Well-insulated floors	0.2
Poorly insulated floors	1.0

▲ Figure 2.7 Infrared image of heat escaping from a house

Examples of U-values required in a modern building

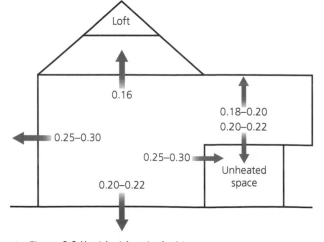

▲ Figure 2.8 Heat lost in a typical house

Table 2.3 shows typical values of specific structures.

▼ Table 2.3 Typical values of specific structures

Element	Construction type	U-value (W/m²K)
Solid wall 	Brickwork 215 mm, plaster 15 mm	2.3
Cavity wall 	Brickwork 103 mm, clear cavity 50 mm, lightweight concrete block 100 mm	1.6
Cavity wall 	Brickwork 103 mm, insulation 50 mm, lightweight concrete block 100 mm, lightweight plaster	0.48
Cavity wall 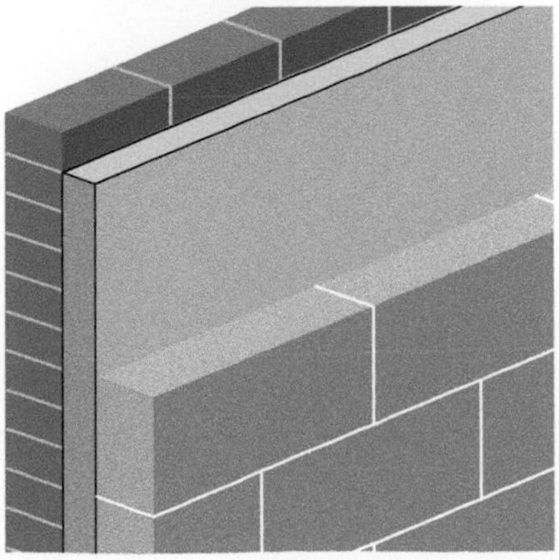	Brickwork 103 mm, insulation 100 mm full cavity fill, lightweight concrete block 100 mm, lightweight plaster	0.28

▼ Table 2.3 Typical values of specific structures (continued)

Element	Construction type	U-value (W/m²K)
Cavity wall – timber frame 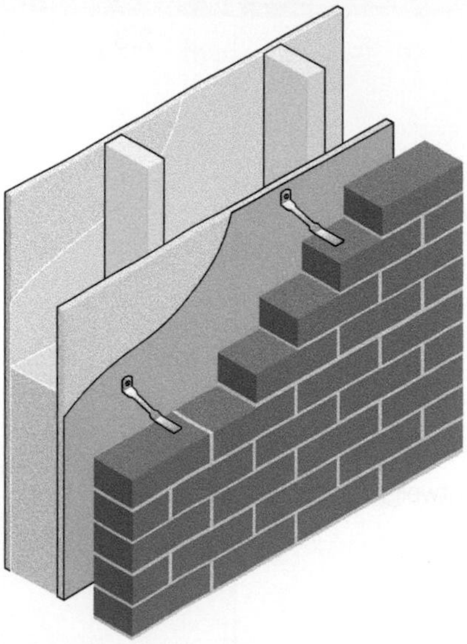	Brickwork 103 mm, clear cavity, 140 mm studwork filled with PIR (polyisocyanurate, see Table 2.4) insulation	0.28
Timber frame and clad 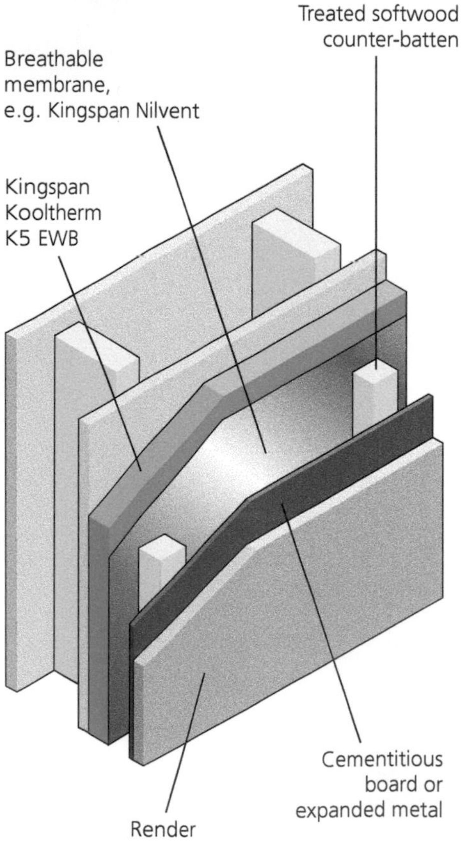 Breathable membrane, e.g. Kingspan Nilvent Kingspan Kooltherm K5 EWB Treated softwood counter-batten Render Cementitious board or expanded metal	Studwork, sheathing, battens finished with either tiles, render or cladding.	0.28

▼ Table 2.3 Typical values of specific structures (continued)

Element	Construction type	U-value (W/m²K)
Pitched roof 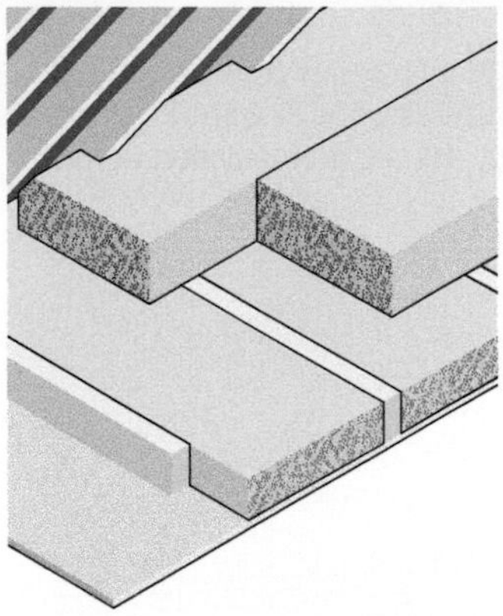 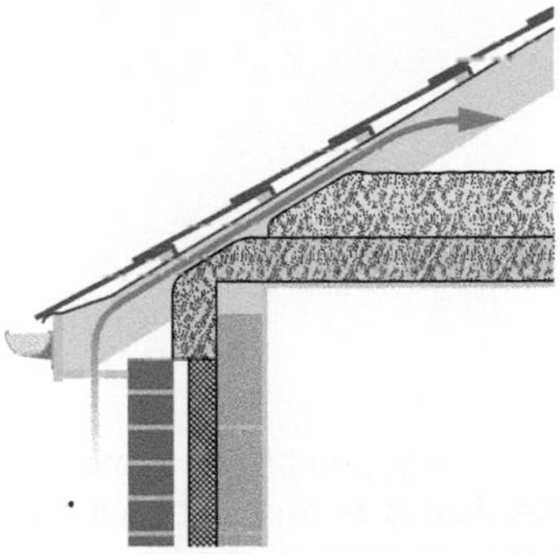	Tiles on battens, felt, ventilated loft airspace, 100 mm mineral wool between joists, 170 mm mineral wool over joists, plasterboard 13 mm	0.16
Warm deck flat roof 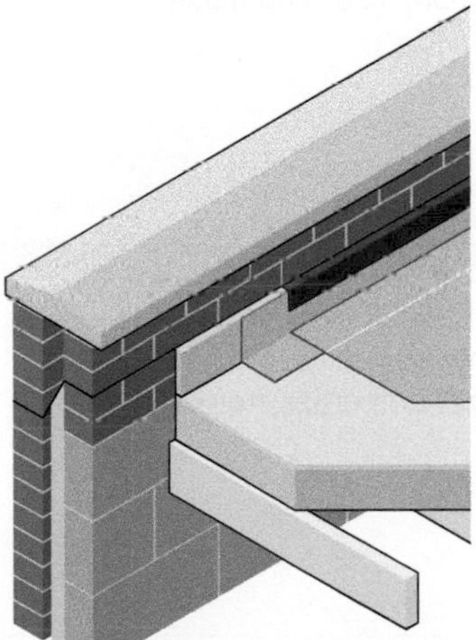	150 PIR over joists, 13 mm plasterboard	0.18

ACTIVITY

Research two additional types of wall construction that will meet the current Building Regulations for a residential property.

Insulation materials

The types of materials listed in Table 2.4 are available to designers to reduce heat loss in order to achieve the Approved Document requirements.

▼ Table 2.4 Insulation materials

Description	Cost	Type of insulation
Blue jean and lambswool  	£	Lambswool is a natural insulator. Blue jean insulation comes from recycled denim.
Fibreglass/mineral wool 	£	This is made from glass, often from recycled bottles or mineral wool. It holds a lot of air and therefore is an excellent insulator. It is also cheap to produce. It does however take up a lot of space as it needs to be thick to comply with Building Regulations. Similar products include plastic fibre insulation made from plastic bottles, and lambswool.
PIR (polyisocyanurate) 	££	This is a solid insulation with foil layers on the faces. It is lightweight, rigid and easy to cut and fit. It has excellent insulation properties. Polystyrene is similar to PIR. Although polystyrene is cheaper, its thermal properties are not as good.
Phenolic insulation board 	££	Phenolic insulation boards are among the most effective rigid boards available. The boards are manufactured with a plastic foam core, sandwiched between two flexible facing layers. The boards are laminated with either aluminium foil faces, plasterboard or glass tissue. The rigid boards can withstand consistent temperatures up to 120°C.

▼ Table 2.4 Insulation materials (continued)

Description	Cost	Type of insulation
Multi-foil 	£££	A modern type of insulation made up of many layers of foil and thin insulation layers. These work by reflecting heat back into the building. Usually used in conjunction with other types of insulation.
Double glazing and draught-proofing measures 	Double glazing ££ Draught-proofing measures £	The elimination of draughts and air flows reduces heat loss and improves efficiency.
Loose-fill materials (polystyrene granules) 	£	Expanded polystyrene beads (EPS beads) are also used as cavity wall insulation. These are pumped into the wall cavity after being mixed with an adhesive that bonds the beads together to prevent them spilling out of the wall. This type of insulation can be used in narrower cavities than mineral wool insulation and can also be used in some stone-built properties.
Expanded polystyrene (EP) fibreglass 	£	A graphite-impregnated expanded polystyrene bead board, designed to provide enhanced thermal performance. Lightweight, and easy to handle, store and cut on site. The expanded polystyrene foam is about 95% air and only 5% plastic.
Autoclaved aerated concrete blocks 	£	Autoclaved aerated concrete blocks are excellent thermal insulators and are typically used to form the inner leaf of a cavity wall. They are also used in the outer leaf, where they are usually rendered.

▼ Table 2.4 Insulation materials (continued)

Description	Cost	Type of insulation
Materials formed on site (expanded foam)	££	Spray foam insulation is an alternative to traditional building insulation such as fibreglass. A two-component mixture composed of isocyanurate and polyol resin comes together at the tip of a gun and forms an expanding foam that is sprayed onto the underside of roof tiles, onto concrete slabs, into wall cavities or through holes drilled into the cavity of a finished wall.
Double and triple glazing	Double glazing ££ Triple glazing £££	Single glazing has a U-value of 5 and older double-glazing about 3. New modern double glazing has a U-value of 1.6, which is mainly due to the fact that the cavity is gas filled, which improves the efficiency of the units. An added advantage of triple glazing over double is the improved reduction of external noise.

INDUSTRY TIP

Planning restrictions, for example to a listed building, may mean it is not possible to remove existing single-glazed windows. However, secondary glazing can be fitted between the reveals behind it (subject to approval). This glazing is fitted in slimline frames, usually with sliding sashes between them, to allow access to the window on the external side of the building.

▲ Figure 2.9 Secondary glazing

Energy-saving measures
Renewable energy sources

The majority of homes across the country are supplied with utilities such as oil or gas for heating and cooking, and electricity for heating, lighting and powering many appliances that we use, e.g. computers, televisions and games consoles. The Earth has a limited supply of non-renewable gas and oil, and eventually these resources that we heavily rely upon will run out. It is difficult to say exactly how much gas and oil is left in the world, but we do know that most of the sources on land are now exhausted; some experts predict that, at the current rate of consumption, only 53 years' worth are left.

INDUSTRY TIP

Oil, coal and gas are natural resources that we use for fuel. These 'fossil fuels' are formed from the remains of plants and animals from millions of years ago that come from the Earth's crust. They are a source of non-renewable energy, meaning that once they have been used they will take millions of years to recover. This is one reason why you should source your building materials locally, to preserve the Earth's limited natural resources.

Given that energy prices are extremely high and there are global shortages of finite resources, people are considering other forms of renewable energy. Examples include:

- solar panels (thermal) – using the sun's energy (heat) to create electricity
- photovoltaic (PV) – using the sun's energy (light) to generate a flow of electricity, usually in specially designed panels and stored in batteries
- ground source heat pumps – long pipes are buried either horizontally or vertically in the ground, and a mixture of water and antifreeze is circulated through them; this harnesses the natural heat in the ground and increases the temperature; the heat is used for hot water and underfloor heating
- air source heating – air is absorbed into a fluid; this is then passed through a compressor and the temperature increased to provide heating and hot water
- wind turbines – these generate electricity as wind moves propeller-like blades connected to a generator
- tidal – there are many methods of creating usable energy from the movement of water, from creating electricity from water passing through a dam, to tidal stream generators placed below the surface of the water; tidal stream generators are very similar to wind turbines, although they use the movement of the change in tides (kinetic energy) rather than wind.

Examples of sustainable design features

We all have to embrace reductions in energy consumption, for example, by buying only energy-efficient light bulbs or 'A'-rated electrical appliances to reduce fuel bills.

A lot of energy is used to filter, clean and purify water we use regularly for flushing the toilet and watering the garden, etc. Water for these purposes can be recycled using a rainwater harvesting system. Rainwater can be collected from the roof in the guttering at the eaves, or from permeable pavement and garden lawns, and stored in holding tanks either underground, in the loft or above ground. The saved water is filtered and pumped to where it is required (note: this is not to be used for drinking water).

Local and sustainable materials

A building that is sustainable must be constructed using local sustainable materials. These are materials that can be used without having an adverse effect on the environment, and that do not have to travel far to be used. It is essential that these materials are renewable, non-toxic and therefore safe for the environment. Ideally, they will be recycled, as well as recyclable.

Consideration should also be given to the extent to which a building material will contribute to the maintenance of the environment in years to come. Alloys and metals will be more damaging to the environment over a period of years as they are not biodegradable, and are not easily recyclable, unlike wood, for example. It is also important to consider the extent to which the material can be replenished. If the material is locally sourced and is likely to be found locally for the foreseeable future, transporting it will be kept to a minimum, reducing harmful fuel emissions.

Building materials can be sustainable if they are chosen carefully. For example, the manufacture of concrete uses considerable fuel and produces a lot of carbon dioxide (a gas that contributes to damaging the climate and our environment). On the other hand, trees absorb carbon dioxide through the process of photosynthesis (converting light energy to chemical energy) as they grow and can be grown sustainably. Trees not only look attractive, but can also be used to make a range of products (e.g. furniture, building products, paper products, medicines, cosmetics and rubber). Some timber, however, is harvested from rainforests without thought for the surrounding environment, the life it supports or the fact that some species are close to extinction.

Brick manufacturers are doing a lot to ensure that the manufacture of bricks is sustainable. They achieve this by continuous improvement of their extraction and manufacturing processes, and by providing products that contribute to sustainable construction. The clay is sourced from areas local to manufacturing plants, so minimising the transport required, meaning that less carbon is burned.

51

Managed timber sources

Managed forests, where new trees are planted after others are harvested, provide a sustainable source of timber. The Forest Stewardship Council (FSC®) is an international non-profit organisation, dedicated to promoting responsible forestry. The FSC® certifies forests all over the world to ensure they meet the highest environmental and social standards.

The system has two key components, as follows.

1 Forest Management and Chain of Custody certification: this system allows consumers to identify, purchase and use timber and forest products produced from well-managed forests.

2 The FSC®'s 'tick tree' logo is used on product labels to indicate that the products are certified under the FSC® system. When you see the FSC® logo on a label you can buy timber and other wood products, such as paper, with the confidence that you are not contributing to the destruction of the world's forests.

▲ Figure 2.10 Converted timber stamped to show certification

▲ Figure 2.11 FSC® logo

ACTIVITY

Research three finite and three non-finite building materials. Write a conclusion outlining the impact that using the finite resources you have found will have on the planet.

ACTIVITY

Conduct a survey of the waste materials created at your place of work, and indicate what percentage are recycled, reused or sent to landfill. Evaluate the results and suggest ways to reduce the amount of waste sent to landfill. Would your suggestions be financially viable for the business?

Examples of recycled building materials

- Crushed concrete or bricks for hardcore
- Reuse of tiles or slates
- Bricks cleaned up and reused
- Steel sections shot blasted and refabricated
- Crushed glass recycled as sand or cement replacement or for the manufacture of kitchen worktops
- Reuse of doors
- Panel products with chipped recycled timber
- Reused timber sections or floorboards
- Reuse of period architectural features
- Reuse of period fixtures and fittings (ironmongery, sash weights, etc.)

INDUSTRY TIP

Using recycled materials is not only good for the environment, but can look better than new materials. For example, a roof extension with weathered (recycled) roof tiles will look seamless, compared with placing new roof tiles against the existing ones.

▲ Figure 2.12 Recycled bricks

Energy performance certificates (EPCs)

Energy performance certificates (EPCs) are needed whenever a property is:

- newly built
- placed on the market for sale
- placed on the market as a rental property.

EPCs were introduced as a result of a European Union Directive relating to the energy performance of buildings and have been a legal requirement since 2008 for any property, whether commercial or domestic, that is to be sold or rented. Since April 2012, legislation has made it illegal to market a property without a valid EPC.

EPCs give potential buyers an upfront look at how energy efficient a property is, how it can be improved and how much money this could save.

The document is valid for ten years and shows how good – or bad – the energy efficiency of a property is. It grades the property's energy efficiency from A to G, with A being the highest rating.

A brand new home is likely to have a high rating. An older home is likely to be around D or E. EPCs for a property can be obtained through specialist companies. A surveyor will be sent out to take details of the structure, the heating system and even the light bulbs. A software program or tables can be used to obtain a holistic (overall) value to grade the property.

▲ Figure 2.14 An energy-efficient light bulb

Case study: Clare

▲ Figure 2.15 A site manager

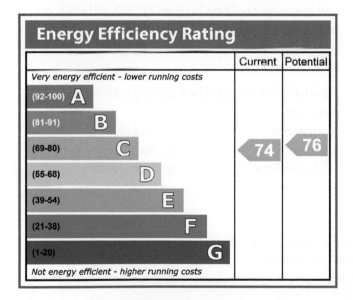

Energy Efficiency Rating

		Current	Potential
Very energy efficient - lower running costs			
(92-100) A			
(81-91) B			
(69-80) C		74	76
(55-68) D			
(39-54) E			
(21-38) F			
(1-20) G			
Not energy efficient - higher running costs			

▲ Figure 2.13 EPC rating chart

The chart in Figure 2.13 shows an example of a current rating and the potential improvement once changes have been made. The energy performance certificate also lists ways to improve the rating – such as installing double glazing, or loft, floor or wall insulation.

Clare has been appointed site manager for the building of four two-bedroom bungalows. The company she works for has a history of constructing a range of standard dwellings. This is Clare's first role as a project manager and she is keen to show that she is capable of the task. To ensure everything runs smoothly and on time, she gathers information from previous projects and plans the programme using this information. There are differences, however, as this site is quite remote and the long-range weather forecast for the winter of the build is poor.

Clare researches all the local suppliers and asks them to quote for the staged delivery of the materials required for the contract. She has the option of using company staff or local labour. Company staff will work out more expensive as accommodation costs will have to be paid. Local labour will be cheaper but workers will not have the experience of this specific work and the quality of their work will be unknown until they start. On balance, Clare chooses to use known company labour. An added advantage is that if workers are staying away from home they are more likely to put in extra hours if required.

Finally, Clare has to decide whether some of the smaller plant required should be purchased or hired. She gets quotes and decides to use one of the material suppliers she has already contacted as she gets a preferential deal because they are also supplying the building materials. All Clare's planning will pay off and should provide enough information to allow her to put together an accurate programme for the building work.

3 ESTIMATING QUANTITIES AND PRICING WORK FOR CONSTRUCTION

Before you can estimate quantities of construction materials, you need to know how to prepare a materials list using a schedule.

Schedules

Drawings alone will not contain all the information required to carry out operations efficiently. Drawings will show the positions of doors, windows, lintels and reinforcing, for example, but will not give specific detail.

Some of this information can be extracted and shown within a schedule, which is often read in conjunction with the range drawing. It provides an easy-to-handle and readable table. Schedules may be produced by a **quantity surveyor** as a part of their role.

KEY TERM

Quantity surveyor: a job role that involves producing the bill of quantities, and working with a client to manage costs and contracts.

A schedule records repeated design information that applies to a range of components or fittings, such as:
- windows
- doors
- reinforcement.

A schedule is used mainly on larger developments and contracts where there are several designs of house and each type has different components and fittings. This avoids a house being given the wrong components or fittings. On a typical plan, the doors and windows could be labelled D1, D2, W1, W2, etc., and these components would be included in the schedule, which would provide additional information about them.

Schedules will provide the following information about products:
- quantity
- colour
- dimensions
- location
- installation details
- manufacturer.

ACTIVITY

Copy out the following table and produce a door schedule for at least five differently sized/types of doors within your college/training centre. An example has been included to start you off:

Master internal door schedule							
Ref:	Door size:	Door type:	Self-closing:	Finish:	Glazed:	Ironmongery:	Location:
Example D1	762×1981×44	FD30	Yes	Natural Beech	No	Satin aluminium	Entrance

Calculating costs and pricing work

The calculation of costs for building work is carried out by an estimator. Estimators build their tender figures using the information contained in the contract documents (drawings, bill of quantities and specifications).

Bill of quantities

The quantity surveyor will have produced the bill of quantities. This describes each item on **dimension paper**, and the quantity of that item as **taken off** the drawing. For many years, the descriptions of the items listed followed the SMM7 (Standard Method of Measurement 7) rule book; however, this was superseded in July 2013 by the NRM2 (New Rules of Measurement 2). This is the standard for most architect-designed buildings. Standardising the description and quantities of common features, such as brickwork, means that contractors can be certain of correctly interpreting requirements when quoting for this work. This makes the process fair, as every contractor is pricing for the same items.

The Royal Institute of Chartered Surveyors (RICS) also publishes another document, NRM3. This provides a standard set of measurement rules for procurement, cost planning and estimating construction projects and maintenance work.

NRM provides an industry recognised standard for benchmarking and consistency, and helps to prevent disputes.

ACTIVITY

Use the following link to read about and compare SMM7 and NRM2: www.designingbuildings.co.uk/wiki/Comparison_of_SMM7_with_NRM2. How are they different?

KEY TERMS

Dimension paper: paper with vertically ruled columns, on which building work is described, measured and costed.

Taken off: materials measured from the contract drawings.

Bill of Quantities		
	Ground floor	
12.95 6.23	13.500 Less external walls 2/0.275 0.550 12.95 6.7780 Less external walls 2/0.275 0.550 6.23 Reinforced in situ concrete (1:3:6) horizontal work in structures n.e. 300 mm thick. × 0.10 = m³ & Mesh reinforcement Ref A2252 weighing 3.95 kg/m² with 150 mm minimum side and end laps. & Imported hardcore bed over 50 mm, but not exceeding 500 mm deep compacted in layers. 150 mm maximum thickness. × 0.15 = m³ & 1200 gauge polythene horizontal damp proof membrane exceeding 500 mm wide. & 50 mm thick horizontal rigid sheet insulation laid on concrete.	Note: The damp proof membrane and damp proof course are lapped on the inner skin of the external wall.

▲ Figure 2.16 Extract from a bill of quantities

In addition to the standard items of work listed, the bill of quantities will contain information on the following.

- Preliminaries: time-related costs such as management and supervision costs, **setting-up costs**, etc., rather than the costs of the actual building work.
- Preambles: an explanation of a document usually found in bills of quantities and specifications. They usually refer to a tender letter that describes the materials requirements for construction, and gives the quantity, type and characteristics of the materials required.
- Provisional sum: undefined bills of work without an accurate estimate. It must include the type of work, how the work is to be done and the limitations likely to be encountered (such as the work hours being limited to night only).
- Prime cost (PC): this covers work undertaken by nominated subcontractors and suppliers, and **statutory undertakings**; costs are based on quotations for items of work that can be populated into the quote.

Specification

This provides essential information to the contractor about the materials, finish and workmanship required for the building construction (for example, the type of brick or timber required). If this information was noted on the drawing, it would clutter it and make it difficult to read as there would be more text than drawing.

On small projects, the specification is written in the notes column above the title block on the drawing. On larger projects, the specification is produced in a separate document.

Often, the description in the specification will be linked to the relevant British Standard.

The tender process

The purpose of the tendering process is to provide the client with several estimates for the proposed work. The client will ask known contractors to submit a **tender** (called a closed tender) or, for competitive tendering work, the client will advertise the work and any contractor can submit quotes (this is known as open tender). Using the contract documents, the estimator will begin to build a tender cost. Each line of the bill of

Specification		
102 External cavity walling		
• Walling below ground:		
- Type:	Cavity wall, concrete filled.	▼
- Masonry units:	Common bricks.	▼
- Mortar:	Class M6 mortar.	▼
• DPC at ground floor:	Flexible cavity trays.	▼
• Walling above ground:		
- External leaf above ground:		
Masonry units:	Facing bricks.	▼
Bond or coursing:	Flemish bond.	▼
- Internal leaf above ground:		
Masonry units:	Aerated concrete blocks.	▼
- Mortar:		
Type:	Class M4 mortar.	▼
Joint profile to		
external faces:	Bucket handle.	▼
- Wall ties:	Insulation retaining wall ties.	▼
- Cavity insulation:	Full fill cavity insulation.	▼
- Ventilation components:	Air bricks and sub-floor ventilation ducts.	▼
• Openings:		
- Lintels:		
Type:	Manufactured stone lintels.	▼
Cavity tray cover:	Flexible cavity trays.	▼
- Cavity closers:	Flexible insulated dpcs.	▼
- Sills:		
Type:		▼
DPC below:	Manufactured stone sills.	
• Abutments:	Natural stone sills.	
Cavity trays and dpcs:	Precast concrete sills.	▼
Flashings built into masonry:	As drawings.	▼

▲ Figure 2.17 An example of a specification

quantities will be costed. The figure in the rate column will be an 'all-in rate' including the following.

- Labour rates: these will be calculated not at the operatives' hourly rate, but including costs such as annual holiday pay, employers' National Insurance contributions, pension contributions, sickness pay, bonuses, travelling time, etc.

- Overheads: the costs of things needed to run the construction site, for example electricity, gas, phone, insurance, plant and equipment, depreciation on equipment and buildings, factory supplies and personnel (other than direct labour).
- Contingencies: a small percentage added to cover unforeseen costs that may arise during the contract.
- Profit: the amount of money gained after direct costs have been paid. The rate applied will depend on a number of factors, including current workload, competition and the complexity of the project.
- VAT on materials.

KEY TERMS

Setting-up costs: costs involved in preparing the site for construction work, such as the costs of hoarding, temporary services and temporary site accommodation.

Statutory undertakings: the services that are brought to the site, such as water and electricity.

KEY TERMS

Tender: a process of formally estimating the costs of potential building work, presented by a number of contractors to the client.

Unit rate: the labour rate + the material rate, resulting in a figure that can be used in quotations and tender documents.

Calculating a unit rate

IMPROVE YOUR MATHS

Calculate a **unit rate** for brickwork using the following information:

- facing brickwork (in half-brick walling) laid in 1:5 cement mortar
- gang of two bricklayers, one labourer
- hourly rate – bricklayer rate £15.00, labourer £8.00
- production rate – 50 bricks per hr (per bricklayer)
- 60 bricks per m²
- cost of bricks per 1000 = £400 (includes 5 per cent waste)
- pre-mixed mortar on site £30/330-litre tub (60 litres of mortar per m² of brickwork).

Example

To calculate the unit rate for the example in the Improve your maths feature above:

Labour rate

Labour rate $= (2 \times 15) + 8 = £38$ total hourly rate

$$\text{Production rate} = \frac{\text{bricks laid/hr}}{\text{bricks/m}^2} = \frac{100}{60}$$
$$= 1.7 \text{ m}^2/\text{hr}$$

$$\text{Labour rate} = \frac{\text{total hourly rate}}{\text{production rate (m}^2/\text{hr)}}$$
$$= \frac{38}{1.7} = £22.35/\text{m}^2$$

Materials rate

Cost of bricks/m^2: $\frac{400}{1000} = 0.4$ (each brick costs 40p)

0.4×60 (bricks/m^2) = cost of bricks/m^2 = £24

Cost of mortar/m^2: 1 litre/brick $\times 60 = 60$ litres/m^2

$\frac{30}{330} \times 60 = £5.45$

Cost of materials = 24 + 5.45 = £29.45

Unit rate

Total unit rate = labour rate + materials rate

$$= 22.35 + 29.45 = \textbf{£51.80m}^2$$

The total unit rate would now be used in the rate column and multiplied by the quantity column to produce the cost for this item. See the example in Table 2.5.

▼ Table 2.5 Extract from a bill of quantities

Number	Item description	Unit	Quantity	Rate	Amount £	pp
E.47	Half-brick walling	m²	75	51.80	3885	00

Each line of the bill of quantities should be completed and the last columns totalled to provide the final estimated cost. Once this has been completed, it should be reviewed by the management team before it is submitted to the client for consideration. It is common for a client to ask for at least three

estimates if it's a big project; an estimator's job can involve a lot of work, with very few contracts obtained. One in ten jobs won would be a good success rate.

ACTIVITY

Produce a costing for the practical assignment you are currently working on. Your working out should include:

- labour cost(s)
- materials, including waste
- consumables (e.g. glue, saw blades, sandpaper)
- VAT
- an additional 6 per cent for overheads (e.g. van, insurances).

INDUSTRY TIP

Setting up a database of standard all-in rates can save a lot of time when estimating.

Other ways of estimating costs

Traditionally, many companies use a building price book, which is a complete guide for estimating, checking and forecasting building work. The figures in these books are established all-in rates, which are updated every year. To calculate the cost of proposed work, you simply find the description of work requiring costing and use the figures given.

There are now many software packages that allow fixed unit rates to be used. You simply enter the quantity in the correct row and the software will total up the work as you go along.

What is the difference between a quotation and an estimate?

- A **quotation** (quote) is a document submitted as a formal response to a request for the cost of specific work, based on the contract documents supplied. It is a promise to do work at an agreed price. It should set out exactly what work will be done for that agreed price. Acceptance of this quotation by the client or their representative creates a legally binding agreement between the two parties. Any additional work requested by the client will not form part of this quote and should be costed separately (this is called a variation order).

- An **estimate** is a best guess as to how much specific work will cost or how long it will take to complete. Unlike a quote, it is not an offer to carry out this work for a fixed cost. This means that the job could cost more or less than the estimated cost. Any charges expected to be above the estimate should be flagged as early as possible to avoid disputes later.

INDUSTRY TIP

Always obtain or provide estimates or quotes in writing, regardless of the size of the job. Disputes between the client and contractor after work has been completed are more difficult to settle if there is no documentation to refer to.

▲ Figure 2.18 Quotes

A & E BUILDERS LTD.

JOB ESTIMATE   DATE:

MATERIALS

ITEM	QUANTITY	UNIT COST	SHIPPING	AMOUNT

DIRECT MATERIALS COST

LABOUR

NOTES

DIRECT LABOUR COST

DIRECT MATERIAL COST	
DIRECT LABOUR COST	
COST OF GOODS	

▲ Figure 2.19 An estimate

Selecting suppliers

Suppliers are often sourced and selected during the pre-contract planning phase. Local suppliers are asked to provide quotes and outline their ability to supply materials to a given schedule. It will not always be possible to source specific or specialist items locally; therefore designers may nominate suppliers in the specification. It pays to set up an early working relationship with suppliers to ensure a good service.

Preferred suppliers

Suppliers can become 'preferred' in a number of ways. For example, your organisation may have used them before, they may have approached you or your technical colleagues with details of their proposition, they may have made a previous successful tender, or they may have been recommended by a similar organisation. The term preferred supplier does not in itself guarantee a level of business, but should be thought of as a guide when considering a sourcing strategy. If there are several suppliers from which you can source the same materials, the preferred supplier may be the one that consistently gives the best price, or the one that is most reliable and always delivers on time.

Estimating quantities of construction materials

At Level 2 you will have carried out calculations for quantifying materials (see *The City & Guilds Textbook: Level 2 Diploma in Site Carpentry and Bench Joinery*). These workings are generally based on linear, squared or volume calculations.

Units of measurement

The construction industry uses metric units as standard; however, you will occasionally come across imperial units. Material sizes in particular are often still referred to in imperial units even though they are now sold in metric units – for example, 8 ft × 4 ft sheets of ply, where the correct size is 2440 mm × 1220 mm, or '3 by 2' timber (referencing inches) where the correct size is 63 mm × 38 mm CLS (Canadian lumber stock).

ACTIVITY

Research the following on the internet:

- How many millimetres are there in a foot?
- How many inches are there in a metre?
- How many square metres are there in a standard sheet of chipboard flooring?

INDUSTRY TIP

You are expected to have an excellent understanding of metric units of measurement, but it is useful if you can read or reference imperial measurements as well.

▼ Table 2.6 Metric and imperial measurement units

Units for measuring	Metric units	Imperial units
Length	millimetre (mm) metre (m) kilometre (km)	inch (in) or ″, e.g. 6″ (6 in) foot (ft) or ', e.g. 8' (8 ft)
Liquid	millilitre (ml) litre (l)	pint (pt)
Weight	gram (g) kilogram (kg) tonne (t)	pound (lb)

▼ Table 2.7 Metric measurement quantities and examples

Units for measuring	Quantities	Example
Length	There are 1000 mm in 1 m There are 1000 m in 1 km	1 mm × 1000 = 1 m 1 m × 1000 = 1 km 6250 mm can be written 6.250 m 6250 m can be written 6.250 km
Liquid	There are 1000 ml in 1 l	1 ml × 1000 = 1 l
Weight	There are 1000 g in 1 kg There are 1000 kg in 1 t	1 g × 1000 = 1 kg 1 kg × 1000 = 1 t

INDUSTRY TIP

Remember that buying a large quantity of a product can often result in a cheaper rate per unit.

Linear calculations

Linear means how long several items would measure from end to end if laid in a straight line. Examples of things that are calculated in linear measurements are:

- skirting board
- timber
- foundations
- wallpaper.

Figure 2.20 gives a reminder of how to work out linear calculations.

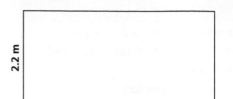

▲ Figure 2.20 Calculating skirting quantities

Figure 2.21 gives a reminder of how to calculate the area of the most common shapes.

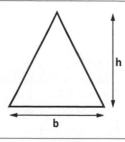	**Triangle** Area = ½ × b × h b = base h = vertical height	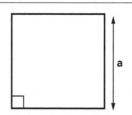	**Square** Area = a² a = length of side
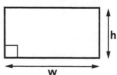	**Rectangle** Area = w × h w = width h = height	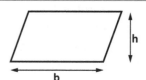	**Parallelogram** Area = b × h b = base h = vertical height
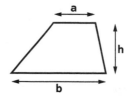	**Trapezium** Area = ½(a +b) × h h = vertical height	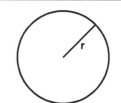	**Circle** Area = π × r² Circumference = 2 × π × r r = radius
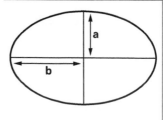	**Ellipse** Area = πab	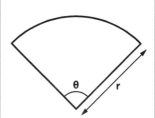	**Sector** Area = $\dfrac{\theta}{360} \times \pi r^2$ θ = angle in degrees r = radius

Note: a = length of side
b = base
h is at *right angles* to b:

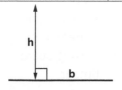

▲ Figure 2.21 Calculating area from basic shapes

IMPROVE YOUR MATHS

What is the area of this rectangle?

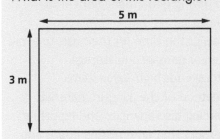

What is the area of this circle? Remember the value of Π (pi) is 3.142.

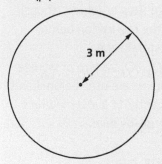

Calculating volume

Volume is measured in cubes (or cubic units) and is written as m³ in metric. For example, 36 cubic metres would be written as 36 m³.

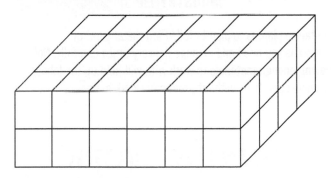

▲ Figure 2.22 A cuboid

We can count how many cubes are in the rectangular prism shown in Figure 2.22, but it is quicker to multiply its length, width and height. The volume of a rectangular prism is length × width × height. To calculate the volume of a rectangular prism we need to do two multiplications. We calculate the area of

one face (or side) and then multiply that by its height. The rectangular prism in Figure 2.22 has a volume of 48 cubic units. The example in Figure 2.23 shows the method for calculating this volume.

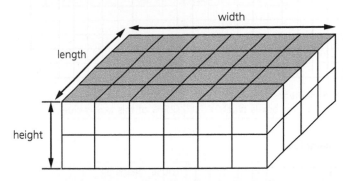

▲ Figure 2.23 Example of calculation

IMPROVE YOUR MATHS

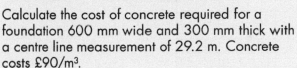

Calculate the cost of concrete required for a foundation 600 mm wide and 300 mm thick with a centre line measurement of 29.2 m. Concrete costs £90/m³.

IMPROVE YOUR MATHS

Calculate the cost of timber required for 240 m of 225 mm × 50 mm softwood floor joisting. Softwood costs £450/m³.

Wastage

No matter how carefully we work, there will be some waste involved in the construction process, for example offcuts of wood left over, extra mortar that has already been mixed or part-full tins of paint. The estimate will include an allowance for this waste. However, avoidable wastage can have a considerable effect on the overall profit a company makes (or loses).

How much we allow for natural waste will depend on the product. Typically, we would allow an additional 5 per cent for bricks, blocks and timber for construction purposes. This does not work for all materials – for example, if six rolls of wallpaper are required for a room, 5 per cent would be 0.3 of a roll. As we can't order part of a roll, we would have to order a complete roll.

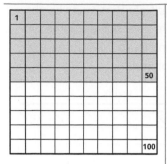

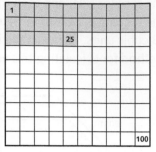

So **50%** means 50 per 100 (50% of this box is blue)

And **25%** means 25 per 100 (25% of this box is blue)

Examples

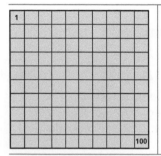

100% means all.

Example:

100% of 80 is $\frac{100}{100} \times 80 = $ **80**

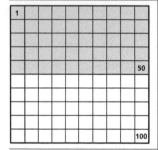

50% means half.

Example:

50% of 80 is $\frac{50}{100} \times 80 = $ **40**

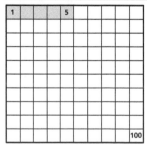

5% means 5 per 100.

Example:

5% of 80 is $\frac{5}{100} \times 80 = $ **4**

▲ Figure 2.24 Calculating waste percentages

IMPROVE YOUR MATHS

Calculate the following:

a 40 per cent of 240

b 30 per cent of 300

c 57 per cent of 1140.

Waste does not just include materials; it also includes idle labour. Waste can largely be minimised by using common sense and having good management of resources. This includes:

- ordering materials just in time for their use, to avoid them being stolen or damaged in storage
- having secure storage for high-value items
- using a robust method of checking-in materials on delivery, recording discrepancies and reporting errors; this will ensure that any missing goods are delivered in time for their planned use
- using an internal requisition system for materials required (common at large sites or joiners' shops), where a supervisor signs the requisition before goods are given out by a storekeeper
- ensuring a first-in first-out (FIFO) system of storage is used, where the oldest items are used before items that have arrived more recently, to avoid materials being stored beyond their use-by date.

INDUSTRY TIP

Disposal of waste from a site can be very expensive, and can pollute the environment.

INDUSTRY TIP

'To fail to plan, is to plan to fail.' Plan the amount of materials you need and arrange delivery times carefully. Not having enough materials to work with will hold up the labour, or leave operatives without productive work to do.

4 PLANNING WORK ACTIVITIES

Planning construction works

To ensure that work can be completed on schedule within the given budget, the building process needs to be planned. Good management is essential for a successful build; good or bad management will determine whether a profit is made and whether the company can remain in business.

Planning should take place at various stages during the contract, as follows.

- **Pre-tender planning** includes deciding whether the work can be undertaken and whether there is the capability to achieve it within the time frame. An estimate is then prepared for the tender.
- **Pre-contract planning** occurs after the project has been won and this information contributes to the contract with the client. The contractor may have up to six weeks to plan for the commencement of work on site, during which time they will organise the:
  - placing of orders for subcontractors
  - planning of the site layout in terms of temporary site buildings, storage of resources, traffic routes, position of crane, etc.
  - laying on of temporary services
  - preparation of the work programme
  - production of method statements
  - sourcing of suppliers and labour.
- **Contract planning** is required to order the work in a logical sequence, determine labour, resource and plant requirements, maintain control and ensure work progresses as planned to meet the handover date.

Generally, it is only the contract planning that concerns us at this level, not in terms of producing a contract plan but in terms of knowing what one looks like and interpreting the information contained in it.

Project drawing:
1 Planning drawings: to a small scale.
2 Construction drawings: to a larger scale.

Schedules:
Information in a table form, taken from drawings and specifications, e.g. a door or decoration schedule.

Building contract:
Gives details of start/end dates, conditions of work, methods of payment, etc.

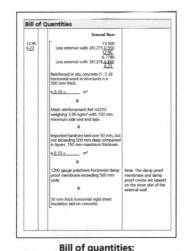

Bill of quantities:
Written document prepared in accordance with the new Rules of Measurement Standard Information, used to provide pricing for a quote.

Specification:
Written document providing specific details of the materials and workmanship required.

▲ Figure 2.25 Flowchart of contract documents

METHOD STATEMENT

Revision Date:	Revision Description:		Approved By:
Work Method Description	Risk Assessment	Risk Levels	Recommended Actions* (Clause No.)
1.			
2.			
3.			
4.			

RISK LEVELS: Class 1 (high) Class 2 (medium) Class 3 (low) Class 4 (very low risk)

Engineering Details/Certificates/Work Cover Approvals:		Codes of Practice, Legislation:	
Plant/Equipment:		Maintenance Checks:	

Sign-off

Print Name:	Print Name:	Print Name:	Print Name:
Signature:	Signature:	Signature:	Signature:
Print Name:	Print Name:	Print Name:	Print Name:
Signature:	Signature:	Signature:	Signature:

▲ Figure 2.26 Method statement

INDUSTRY TIP

Every job and every site is different so method statements should always be site specific. They should be compiled by a competent person who is familiar with health and safety guidelines, the process of works and the site characteristics.

Site layout and organisation

As a carpenter/joiner you may be involved in the planning or setting up of the site layout. On large sites this may form part of the pre-contract planning, while on smaller sites this would be planned by the site manager.

The purpose of site planning is to ensure that the layout, position and routes between temporary site buildings, stationary plant, storage areas, cranes and welfare facilities are as strategic and convenient as possible, with the overall aim of providing the best conditions for maximum economy, continuity, safety and tidiness. Site routes should be kept clear and should not impede construction of the building. Planning should minimise movement of materials to avoid double handling. Considerable additional costs can be incurred if portable buildings or materials have to be relocated. A site plan should be produced to show where all these elements are to be positioned.

Fencing or hoarding may be required and, if it has to be positioned on or over a public highway, a hoarding licence must be obtained from the local authority and displayed on the hoarding. Each authority has its own rules about how the hoarding is to be lit, decorated and formed. The hoarding must display warning signs at the site entrance, providing information to visitors and workers alike.

▲ Figure 2.27 Safety signs and notices

ITEMS included on SITE LAYOUT PLAN

Site security fencing.

Entrance gates.

A Welfare facilities.

B Site offices.

C Stores - lock up.

D Storage racking for finishing materials.

E Brick storage area on hardstanding.

F Formwork/reinforcement fabrication areas.

G General hardstanding area - formed up on commencement of contract.

H Area for subcontractor's accommodation and storage.

I Bagged aggregates and cement storage

● Mortar mixing area.

⊠ Position of tower crane

▯ Car parking spaces.

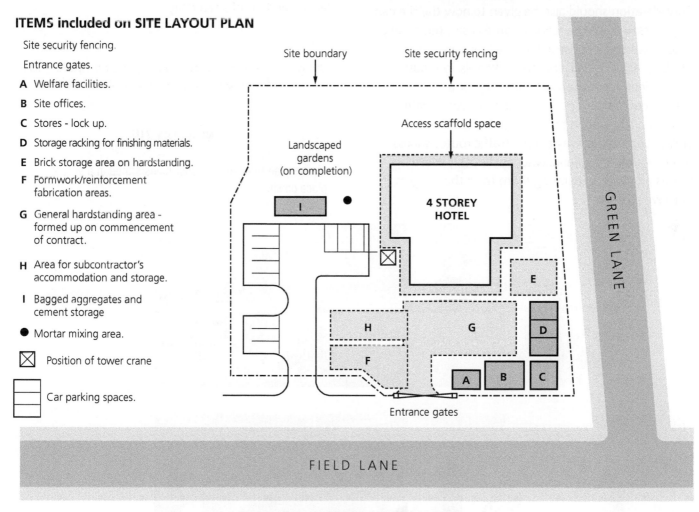

SITE LAYOUT PLAN: 4 STOREY HOTEL

▲ Figure 2.28 Typical site layout

▲ Figure 2.29 Multi-storey site offices/welfare facilities

Access and traffic routes

Consideration should also be given to how the site can be accessed from the road. In some cases, temporary roads have to be constructed to allow access to the site by plant and delivery lorries. Permission must be obtained from the local authority for access over or encroachment on a public footpath. Traffic routes on site can be two-way, but often there is only room for one-way traffic. Traffic routes should be clearly identified and, where possible, pedestrian routes provided, separating them from the dangers of moving plant.

Temporary site buildings

These are required for the management team, welfare facilities (including canteen and toilets) and storage. Where space on site is limited, temporary buildings can be stacked, with suitable stair access. Larger sites have gatekeepers who monitor and record all comings and goings of staff, materials and visitors to the site. The buildings should contain the facilities to fulfil their function effectively. Offices should be well lit and contain all required services, such as plumbing and Wi-Fi. The canteen should be suitably furnished, adequately heated and with facilities for drying clothes. Larger sites may have a separate facility for storing and drying clothes. The toilets should be cleaned and disinfected daily.

ACTIVITY

Using the internet, research the types of temporary site accommodation available and the costs to hire them.

INDUSTRY TIP

Always plan to have high-cost items stored in a secure place on site.

▲ Figure 2.30 Typical temporary site buildings: (a) site manager's office; (b) rest facilities; (c) clothes storage and drying facilities; (d) meeting room facility

Materials storage and handling

Planning will help minimise wastage and losses arising from careless handling, poor storage, theft and double handling. Storage containers are required for high-value and fragile items, and for those that deteriorate when exposed to the weather. Open storage areas are required for bulk items such as timber, bricks, drainage pipes and roof trusses. Some items can be stored in the building as it is completed. It is essential that all materials are stored as specified by the manufacturers to ensure they remain fit for purpose, and do not deteriorate or become damaged.

Stationary plant

Careful planning and positioning of the site crane will allow materials to be offloaded, stored and taken to their final position efficiently, without double handling. The crane is often centrally positioned, to encompass the whole site in its radius. Mortar/concrete mixing or plant should be placed where the aggregate is stored.

▲ Figure 2.31 Construction crane

Planning work activities

Programmes of work

There are several methods for programming building work. The most common is to use a type of bar chart (often called a Gantt chart). Each company is likely to use its own form of this chart. Critical path analysis is another method but is used less as it is more difficult to interpret the information.

Gantt chart programming

The programme of work is the key to a successful and efficiently run contract, and will help the site manager and supervisors to follow a set plan of action. The programme will show:

- the start date
- the sequence in which the building operations are to be carried out
- an estimated time for each operation
- the labour required
- the plant required
- when materials need to be delivered
- the contract end date
- any public holidays.

Preparing the programme

The programme is prepared based on past experience, method statements and the measured rates from the estimate. In order to prepare a basic programme for all trades, times are often based on a previous similar project.

Table 2.8 gives an example of a basic programme for a small building contractor constructing a four-bedroom, brick-built house in 15 weeks.

▼ Table 2.8 Example programme

Total time to build	100%	15 weeks (75 days)
Start to DPC	15%	11 days
DPC to watertight	45%	34 days
Internal work and finishing	40%	30 days

The programme shown in Table 2.8 is based on the breakdown of tasks illustrated in Figure 2.32.

Operation number	Description	Trade	Comment
1	Site preparation and setting out	Labourer, carpenter	Start to DPC
2	Excavation and concrete to foundations and drains	Labourer	
3	Brickwork to DPC	Bricklayer, labourer	
4	Back fill and ram	Labourer	
5	Hardcore and ground floor slab	Labourer, bricklayer	
6	Brickwork to first lift	Bricklayer, labourer	DPC to watertight
7	Scaffolding	Subcontractor, labourer	
8	Brickwork to first floor	Bricklayer, labourer	
9	First floor joisting	Carpenter, labourer	
10	Brickwork to eaves	Bricklayer, labourer	
11	Roof structure	Carpenter, labourer	
12	Roof tile	Subcontractor, labourer	
13	Windows fitted	Carpenter, labourer	
14	Carpentry first fix	Carpenter, labourer	Internal work and finishing
15	Plumbing first fix	Subcontractor, labourer	
16	Electrical first fix	Subcontractor, labourer	
17	Services	Subcontractor, labourer	
18	Plastering	Subcontractor, labourer	
19	Second-fix carpentry	Carpenter, labourer	
20	Second-fix plumbing	Subcontractor, labourer	
21	Second-fix electrical	Subcontractor, labourer	
22	Decoration	Subcontractor, labourer	
23	External finishing/snagging	Subcontractor, labourer	

▲ Figure 2.32 A chart showing the sequence of building operations

These items of work can now be used on the bar chart (Figure 2.33).

As you can see, the bar chart shows the sequence of operations and the labour and plant requirements very clearly. The second row of each activity line is used to measure progress against the planned activities. This will be shaded in another colour, usually green, as the build progresses. Careful monitoring of this progress line and early intervention to deal with any slippage will ensure the contract remains on time. Generally, unless insufficient

time has been allowed, inclement (bad) weather, staff sickness or lack of materials will cause delays. In order to catch up, it is likely that extra labour will be brought in, a more reliable supplier will be found, or both.

ACTIVITY

Using the programme chart shown in Figure 2.33, for how many weeks in total are the following trades required on site?

a Bricklayers

b Carpenters

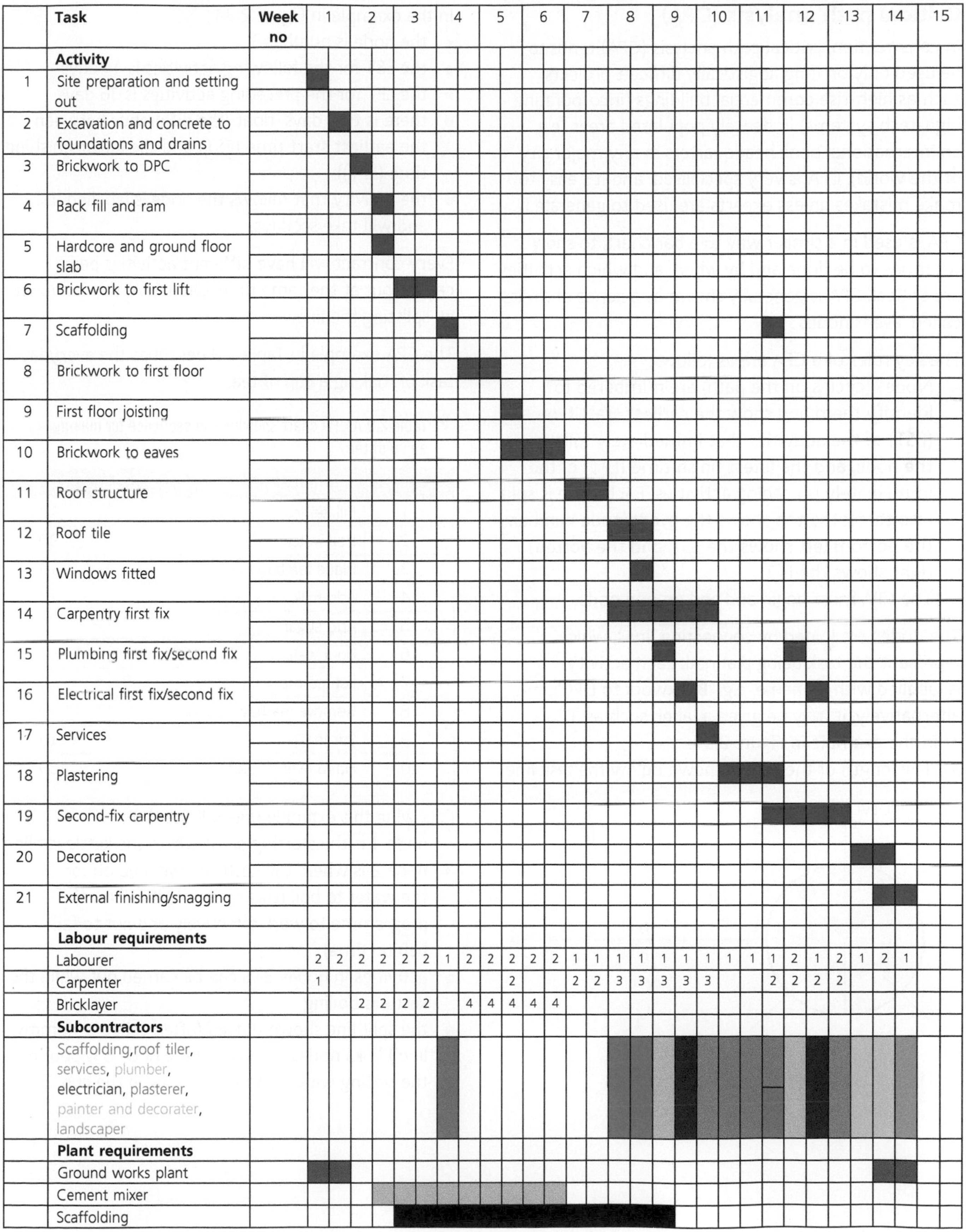

	Task	Week no	1	2	3	4	5	6	7	8	9	10	11	12	13	14	15
	Activity																
1	Site preparation and setting out		█														
2	Excavation and concrete to foundations and drains			█													
3	Brickwork to DPC			█													
4	Back fill and ram				█												
5	Hardcore and ground floor slab				█												
6	Brickwork to first lift				█												
7	Scaffolding					█							█				
8	Brickwork to first floor					█											
9	First floor joisting						█										
10	Brickwork to eaves						█	█									
11	Roof structure								█								
12	Roof tile									█							
13	Windows fitted									█							
14	Carpentry first fix									█	█						
15	Plumbing first fix/second fix										█			█			
16	Electrical first fix/second fix										█			█			
17	Services											█	█				
18	Plastering												█	█			
19	Second-fix carpentry													█	█		
20	Decoration															█	
21	External finishing/snagging															█	
	Labour requirements																
	Labourer		2 2	2 2	2 2	1 2	2 2	2 2	1 1	1 1	1 1	1 1	1 1	2 1	2 1	2 1	
	Carpenter		1			2			2 2	3 3	3 3	3		2 2	2 2		
	Bricklayer			2 2	2 2	4	4 4	4 4									
	Subcontractors																
	Scaffolding, roof tiler, services, plumber, electrician, plasterer, painter and decorater, landscaper					▓				▓	▓	▓	▓	█	▓	▓	
	Plant requirements																
	Ground works plant		█ █													█	
	Cement mixer			▓	▓	▓	▓	▓									
	Scaffolding				█	█	█	█	█	█							

▲ Figure 2.33 Planned activities, labour and plant shown on a programme/progress chart

Critical path analysis (CPA)

In construction, critical path analysis (CPA) tends to be used only on long, logistically difficult projects such as high-rise commercial buildings (incorporating high-tech systems) in densely populated areas (e.g. an international bank headquarters in any major city of the world). CPA is very specialised, and it's easy to make mistakes unless experts are used to generate it.

CPA is used in a similar way to a bar chart, to show what has to be done and by when, and which activities are critical. CPA is generally shown as a series of circles called 'event nodes'.

The key rules of a CPA are as follows:
- Nodes (circles on the path) are numbered to identify them and show the earliest start time (EST) of the activities that immediately follow the node, and the latest finish time (LFT) of the immediately preceding activities. Each node is split into three: the top shows the event/node number, the bottom left shows the EST, and the bottom right shows the LFT.
- The CPA must begin and end on one node.

The nodes are joined by connecting lines, which represent the task being planned. Each activity is labelled with its name, e.g. 'Brickwork to DPC'; alternatively, it may be given a label, such as 'D', as in the example in Figure 2.34.
- The length of the task is shown below the task line.

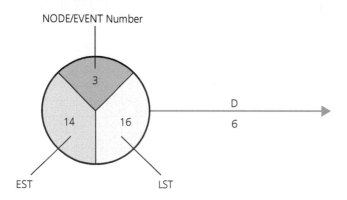

▲ Figure 2.34 Example of a CPA node

In the example in Figure 2.34:
- the node is number 3
- the EST for the following activities is 14 days
- the LFT for the preceding activities is 16 days
- there is two days' float (the difference between the earliest start time (EST) and the latest finishing time (LFT))
- the activity that follows the node is labelled 'D' and will take six days.

Every contract will have different activities being carried out at the same time. On a CPA, this is shown by splitting the line.

The CPA example in Table 2.9 describes the everyday task of making a cup of tea.

▼ Table 2.9 A CPA chart showing the sequence for making a cup of tea

	Method statement	Secs
1	Fill kettle	10
2	Boil water	90
3	Place tea bag in cup	10
4	Add milk	10
5	Add sugar	10
6	Pour water	10
7	Let tea brew	30
8	Remove tea bag from cup	10
9	Stir tea	10
10	Hand over to client	160

Following this example you will see that:
- node 1 is the starting point where the kettle is filled
- node 2 is where the kettle is switched on for the water to boil (this node is 'critical' for the programme to work effectively and not to fall behind time)
- activities in nodes 3–5 can be carried out while the kettle is boiling
- the split line shown dotted (often called a 'dummy line') links nodes 2 and 6, where node 6 is pouring the boiling water onto the tea bag in the cup.

ACTIVITY

Study this CPA diagram for the process of making a cup of tea, as detailed in Table 2.9.

Make a cup of tea

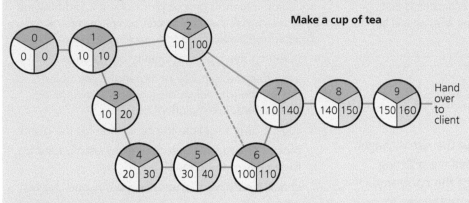

▲ Figure 2.35 CPA diagram

Following this example, produce a CPA for the start-to-DPC section of the bar chart we looked at earlier (Figure 2.33).

Plant

During the pre-contract planning stage, a decision will be made as to what items of plant will be purchased or hired. Some plant may already be owned by the company (cement mixers, power tools, etc.), while others may not have been purchased due to their storage requirements and/or whether their initial cost can be recovered over time.

▲ Figure 2.36 Hired plant

Hire or purchase?

- A crane, for example, would usually be hired as it would be cheaper to hire than purchase and does not have any storage requirements when not in use.
- Hired electrical equipment will always be supplied tested (**PAT**) and ready to use.
- The hiring can be terminated when the items are not in use, saving costs.
- There are no ongoing maintenance costs for hired equipment.
- Repeated use of hired equipment may be more expensive than the initial cost of outright purchase.
- You could purchase the equipment and sell it once the contract has finished.

KEY TERM

PAT (portable appliance testing): there is a legal requirement for all portable electrical appliances to be tested on a regular basis (dependent on use). PAT testing itself isn't a legal requirement, but is one method of managing the process. Most companies test appliances annually.

Buying materials and stock systems

The programme of work should allow the site manager or company buyer to plan for the advance ordering of materials (depending on the size of the company). It is good practice during pre-contract planning to source suppliers that can provide all the materials required for the contract. In some cases, these may be nominated suppliers. Most sites have limited space for storage so materials will be delivered on a regular basis. Some materials may have a long **lead time**. It is essential to know what lead times are required for all the materials needed for the project. Often discounts can be obtained if one builder's merchant is used to supply all materials required for a contract; knowledge of the reliability of this supplier is an important factor to avoid delays. As mentioned earlier, 'just-in-time' delivery methods work well to avoid wastage, damage and the need for storage space.

KEY TERM

Lead time: the delay between the initiation and execution of a process. For example, the lead time between placing an order for a staircase from a joinery manufacturer and its delivery may be anywhere from two to eight weeks.

ACTIVITY

Speak to your employer about the process used to requisition materials at your place of work and how this is recorded. Ask them if you can complete a small order of materials needed for a task, making sure that you:

- research the type and quality of materials
- calculate the quantity of materials; you can do this on the job, taken off working drawings or from schedules/specifications
- source and negotiate the best price for the goods
- place the order yourself, either by email, face to face or over the phone
- check the delivery against the order and delivery note
- check for signs of damage, missing parts, etc.
- sign the delivery note.

5 WRITTEN AND ORAL COMMUNICATION

In Chapter 2 of *Level 2 Diploma in Site Carpentry and Architectural Joinery*, we looked at common methods of communicating information on site and the use of standard documentation to help to make communication clear, simple and accurate. As you will see, there are a lot of documents in every workplace, and these need to be managed carefully.

Site documentation

Table 2.10 shows most of the standard forms of documentation found on site or in a joiner's shop.

▼ Table 2.10 Site documentation

Type of documentation	Description
Timesheet **Timesheet** Employer: CPF Building Co. Employee Name: Louise Miranda Week starting: 1/6/19 Date: 21/6/13 	Day	Job/Job Number	Start Time	Finish Time	Total Hours	Overtime
---	---	---	---	---	---
Monday	Penburthy, Falmouth 0897	9am	6pm	8	
Tuesday	Penburthy, Falmouth 0897	9am	6pm	8	
Wednesday	Penburthy, Falmouth 0897	8.30am	5.30pm	8	
Thursday	Trelawney, Truro 0901	11am	8pm	8	2
Friday	Trelawney, Truro 0901	11am	7pm	7	1
Saturday	Trelawney, Truro 0901	9am	1pm	4	
Totals				43	3 | Employee's signature:_____ Supervisor's signature: _____	Used to record the hours completed each day; usually the basis on which pay is calculated. Timesheets can also be used to work out how much the job has cost in working hours, and used for future estimating work when working up a tender.
Job sheet **CPF Building Co** **Job sheet** Customer name: Henry Collins Date: 9/12/19 Address: 57 Green St Kirkham London Work to be carried out Finishing joint work to outer walls Instructions Use weather struck and half round	Gives details of a job to be carried out, sometimes with material requirements and hours given to complete the task.
Variation order, confirmation notice, architect's instruction **CPF Building Co** **Variation order** Project Name: Penburthy House, Falmouth, Cornwall Reference Number: 80475 Date: 13/11/19 From: _____ To: _____ | Reason for change:	Tick
Customer requirements	☑
Engineer requirements	☐
Revised design	☐	 Instruction: Entrance door to be made from Utile hardwood with brushed chrome finished ironmongery (changed from previous detail, softwood with brass ironmongery). Signature _____	Sometimes, alterations are made to the contract, changing the work that needs to be completed, e.g. a client may wish to move a door position or they might request a different brick finish. This usually involves a variation to the cost. This work should not be carried out until a variation order and a confirmation notice have been issued. Architect's instructions are instructions given by an architect, first verbally and then in writing to a site agent, as work progresses and questions arise over details and specifications.

▼ Table 2.10 Site documentation (continued)

Type of documentation	Description
Requisition order	Filled out to order materials from a supplier or central store. These usually have to be authorised by a supervisor before they can be used.

CPF Building Co
Requisition order

Supplier Information: Construction Supplies Ltd   **Date:** 9/12/19

Contract Address/Delivery Address: Penburthy House, Falmouth, Cornwall

Tel number: 0207294333

Order Number: 26213263CPF

Item number	Description	Quantity	Unit/Unit Price	Total
X22433	75 mm 4 mm gauge countersunk brass screws slotted	100	30p	£30
YK7334	Brass cups to suit	100	5p	£5
V23879	Sadikkens water based clear varnish	1 litre	£20	£20
Total:				£55.00

Authorised by: Denzil Penburthy

Delivery note	Accompanies a delivery. Goods must be checked for quantity and quality against the delivery note and order before the note is signed. Any discrepancies will be recorded on the delivery note. Goods that are not suitable (because they are not as ordered or because they are of poor quality) can be refused and returned to the supplier. It's not uncommon for deliveries to be made, or attempted, to the correct company but at the wrong site. This costly mistake can be avoided if the delivery address is also checked on the delivery note.

Construction Supplies Ltd
Delivery note

Customer name and address:	Delivery Date: 16/12/19
CPF Building Co	Delivery time: 9am
Penburthy House	
Falmouth	
Cornwall	Order number: 26213263CPF

Item number	Quantity	Description	Unit Price	Total
X22433	100	75 mm 4 mm gauge countersunk brass screws slotted	30p	£30
YK7334	100	Brass cups to suit	5p	£5
V23879	1 litre	Sadikkens water based clear varnish	£20	£20
			Subtotal	£55.00
			VAT	20%
			Total	£66.00

Discrepancies: ....

Customer Signature: ....

Print name: ....

Date: ....

Invoice	Sent by suppliers to companies that have credit accounts, or issued as a receipt of payment if payment is made at the time of collection or ordering. It lists the services or materials supplied, along with the price the contractor is requested to pay.

Davids & Co
Invoice

Invoice number: 75856   **Date:** 2nd April 2019
PO number: 4700095685

Company name and address:	Customer name and address:
Davids & Co	CPF Building Co
228 West Retail Park	Penburthy House
Ivybridge	Falmouth
Plymouth	Cornwall

VAT registration number: 663694542

For:

Item number	Quantity	Description		Unit Price
BS3647	2	1 tonne bag of building sand	£30	
CM4324	12	25kg bags of cement		£224
			Subtotal	£2748.00
			VAT	20%
			Total	£3297.60

Please make cheques payable to Davids & Co
Payment due in 30 days

▼ Table 2.10 Site documentation (continued)

Type of documentation	Description
Statement **Davids & Co** Statement Customer name and address: / Customer order date: 28th May 2019 CPF Building Co Penburthy House Falmouth Cornwall <table><tr><th>Item number</th><th>Quantity</th><th>Description</th><th>Unit Price</th><th>Date Delivered</th></tr><tr><td>BS3647</td><td>2</td><td>1 tonne bag of building sand</td><td>£30</td><td>3/6/19</td></tr><tr><td>CM4324</td><td>12</td><td>25kg bags of cement</td><td>£224</td><td>17/6/19</td></tr></table> Customer Signature: .... Print name: .... Date:	Every month a supplier will issue a statement that lists all the materials purchased or hired for that month. The statement usually lists the invoice numbers against the items in the order they have been delivered or collected from the supplier. Statements are issued only to companies that have a credit account. There will be a time limit within which to pay – usually the end of the following month. Sometimes there will be a discount for quick payment or penalties for late payment.
Site diary 	This will be filled out daily. It records anything of significance that happens on site, such as deliveries, absences or specific events, e.g. a delay due to the weather.
Method statement **METHOD STATEMENT** CONTRACT: Drysdale Avenue SHEET: 2 of 14 CLIENT: Mrs Sherbert DATE: 07/12/2019 **MATERIALS** <table><tr><th>OP No.</th><th>OPERATION</th><th>METHOD & SEQUENCE</th><th>PLANT</th><th>LABOUR</th><th>REMARKS</th></tr><tr><td>1</td><td>Prelims.</td><td></td><td></td><td></td><td></td></tr><tr><td></td><td>1.1</td><td>See site layout plan for access position, remove all rubbish.</td><td>JCB</td><td>3 labourers</td><td>Segregate all rubbish into appropriate skips</td></tr><tr><td></td><td></td><td>Erect site hoarding and decorate.</td><td>–</td><td>3 carpenters 2 painters 3 labourers</td><td>Check with L/O for hoarding requirements</td></tr><tr><td></td><td>1.2</td><td>Level ground and create hardstandings for temporary site buildings.</td><td>JCB</td><td>3 labourers</td><td>Scan for services</td></tr><tr><td></td><td>1.3</td><td>Position temporary site buildings</td><td>Crane</td><td>Subcontractor</td><td>–</td></tr><tr><td></td><td>1.4</td><td>Connect temporary services.</td><td>–</td><td>Subcontractor</td><td>Liaise with service providers</td></tr><tr><td></td><td>1.5</td><td>Setting out for drains and roads.</td><td>–</td><td>Site agent 3 labourers</td><td>Liaise with service providers</td></tr><tr><td>2</td><td>Drains and roads</td><td></td><td></td><td></td><td></td></tr></table>	There are two types of method statement in common use. The first includes a risk assessment. This would be written during the pre-contract stage of the work programming. The second is more traditionally used to estimate the time that each operation in the building process will take, and is related to the bill of quantities. It is used as a guide to completing the work programme.

→

▼ Table 2.10 Site documentation (continued)

Type of documentation	Description
Risk assessment **Risk Assessment** Assessment carried out by: _____ Date assessment was carried out: _____ Date of next review: _____ 	Area/ activity	Hazard	Risk to (list persons)	Current pre-cautions	Action	Action required by	Date for required action	Complete
---	---	---	---	---	---	---	---
									An assessment of the hazards and risks associated with an activity, and the ways in which they should be monitored and reduced. Applying risk assessments and maintaining high standards of health and safety is the responsibility of everybody working in the construction industry.
Permit to work **PERMIT TO WORK** 1. Area 2. Date 3. Work to be Done 4. Valid From 5. Valid To 6. Company 7. Person in Charge 8. No of People 9. Safety Precautions 10. Safety Planning Certificate (cancelled if alarm sounds) I have inspected the above job which has been safely prepared according to requirements of a safety planning certificate Signed .... 11. Approval of Permit to Work I am satisfied that this permit is properly authorised, that safe access is provided, and that all persons affected by this job have been informed Signed .... 12. Electrical Equipment All power has been isolated/locked/tagged/tried* Circuits are live for troubleshooting only Signed .... 13. Acceptance of Permit to Work I/we * have read and understood the above precautions and will observe them. All equipment complies with relevant standards. I understand the site emergency plan. Signed .... 14. Completion of Permit to Work I/we* certify that this job is complete/incomplete*, all guards have been replaced and secured and all equipment has been removed. The job site has been left clean and tidy. Signed .... 15. Renewal of Permit to Work (same day only) Approved until .... Signed .... Approved until .... Signed .... If the alarm sounds: 1. Stop Work 2. Make equipment safe 3. Leave the building by the nearest exit 4. Make your way to the main car park If you discover a fire: 1. Break fire point 2. Leave the building 3. Ring 999 and give name, position, description etc 4. Report to incident controller in Main car park Do not re-enter any building until you are told it is safe	A permit to work is a document that gives authorisation for certain people to carry out specific hazardous work within a set period of time. It sets out the precautions required to complete the work safely, based on the risk assessment.

The nature of communication is changing rapidly from paper-based to electronic, therefore it is important that contractors keep up to date with such developments to enable them to work effectively. An example of new technology used in construction to communicate can be seen when 'augmented reality' (AR) is used in connection with 'building information modelling' (BIM). Search for 'GAMMA AR Construction with augmented reality' on YouTube to watch a video of this.

Manufacturers' instructions

It is important to follow manufacturers' instructions for your own safety and to protect your product or purchase. Some instructions are in warning form while others explain how to use a product. Warning instructions such as 'do not place in contact with fire' mean that the product is flammable; not following this instruction could cause an explosion that may injure people.

Other examples of manufacturers' instructions are information:

- on the back of a paint tin showing drying times, types of solvents required for cleaning brushes, etc.
- supplied as a user manual for powered equipment such as a mixer or chop saw
- on how to set up and use access equipment
- on how to use a brick cleaner safely or how to use a two-part epoxy adhesive.

Not following these instructions can lead to the invalidation of a **warranty**, misuse of the product or the product not performing as expected.

KEY TERM

Warranty: an assurance of performance and reliability given to the purchaser of a product or material. If the item fails within the warranty period, the purchaser is entitled to a replacement or to have it repaired, depending on the terms of the warranty.

Other documents used on site

Agenda

Before any meeting, an agenda should be drawn up, detailing the matters to be discussed.

ACTIVITY

Research what information needs to be included in an agenda and how it is laid out. Imagine you are going to have a meeting at your college or workplace to discuss the training needs of your class or colleagues. Draw up an agenda, making sure you include all the necessary information so that people attending know where and when the meeting is, and what it will cover.

Distribute the agenda (by email) in advance, so that members of the group have a chance to prepare for the meeting. After the meeting, follow up any action points.

Minutes

Minutes (key points) of formal meetings provide a record of what happened at the meeting, what actions are required, who is responsible for carrying them out

and by what date. A set of minutes normally includes the following information:

- the time, date and place of the meeting
- a list of the people attending
- a list of absent members of the group
- approval of the previous meeting's minutes, and any matters arising from those minutes
- for each item in the agenda, a record of the principal points discussed and decisions taken, and who is responsible for actioning any items discussed
- any other business (AOB)
- the time, date and place of the next meeting
- the name of the person taking the minutes.

INDUSTRY TIP

Any action or target for someone to complete should be SMART:

- **S**pecific – contain the detailed requirements of the task
- **M**easurable – ensure the completion of the task can be measured in terms of performance
- **A**chievable – able to be completed with the resources and time available
- **R**ealistic – within the capabilities of the person
- **T**ime bound – have a completion date for the task.

Memo

A memo, short for memorandum, is a brief document or note informing or reminding the receiver of something that is happening or something that has to be done. Memos have now mostly been replaced by emails.

Memo

Hi Clare, Just to let you know that the client will be joining us at the review meeting on site next Monday.

Kind regards, Martin.

▲ Figure 2.37 Memo

Organisational chart

An organisational chart is a diagram that shows the structure of an organisation, and the relationships between the roles, positions and jobs within the organisation (Figure 2.38).

Site induction

Before starting work on site (or even when visiting) everyone should receive a site induction. This should give information on the following:

- a general introduction to the project and staff
- details of welfare facilities, site access and parking arrangements
- designated first aiders, what to do in the case of an accident, etc.
- procedures for reporting accidents and near misses
- fire procedure and muster points
- notification of daily hazards such as working at height or moving vehicles
- notification about method statements and risk assessments
- site rules including those concerning drug or alcohol abuse, bullying, etc.
- any other site-specific requirements.

There should be a record kept of who has been inducted and when. It is in everyone's interest to ensure that working on site is as safe as possible. We have looked at site inductions in detail earlier in this book, so revisit Chapter 1 to remind yourself of the key points.

Toolbox talks

As also mentioned in Chapter 1, a 'toolbox talk' is a short presentation given to a workforce on a single aspect of health and safety. A toolbox talk might be prepared by a company's safety officer, a manufacturer's representative (for example, about a product being used) or a site supervisor. It should provide timely safety reminders and outline any new hazards that the workforce may come into contact with. This may be new materials to be used, changes to or new risk assessments arising from an accident or near miss, or advice for working in inclement weather.

The HSE has prepared several toolbox talk presentations and PDFs; these will save companies the time and effort required to write them from scratch.

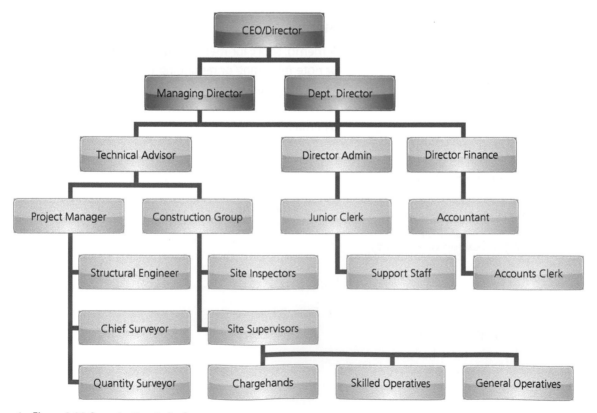

▲ Figure 2.38 Organisational chart

ACTIVITY

Prepare a toolbox talk in the form a three- or four-slide presentation on a topic of your choice. Perhaps you could base it on a near miss accident that has recently happened at your place of work. Show your presentation to your employer for some feedback; maybe it could be used to train staff in the future.

Defects survey

Prior to a final certificate being written to sign off a contract, a defects survey should be carried out. Larger contracts should have a defects liability period (sometimes called a rectification period). This is a set time after a construction project has been completed during which a contractor is responsible for remedying any defects. A typical defects liability period lasts for 12 months. A sum of money (typically 5 per cent) is normally retained ('retention') until after this period has come to an end. This will ensure that any necessary work is carried out.

The defects survey could be done by a building surveyor, clerk of the works or maintenance manager. It will identify any:

- poor standards of work
- poor quality of materials used
- damaged materials (damaged during construction, not afterwards)
- human error.

A comprehensive list or table is collated on a schedule of outstanding works (commonly called a snagging list). The work is systematically worked through, often by a snagging supervisor.

Schedule of outstanding works		Harold Court, Clacton-on-Sea
Location	**Defect**	**Remedial action**
Front entrance	Entry phone not working properly	Contact Door Services Ltd to send out an engineer to solve problem and recommission
First flight	Loose vinyl tiles on treads 4 and 6	Remove and bond new tiles to both steps
Front door to flat 3	Door twisted and not closing correctly	Contact Speedy Joinery to request replacement

▲ Figure 2.39 Example snagging list

Types of communication

The importance of good communication cannot be overemphasised. Using effective communication skills is crucial to relationships and to success at work. We use many different forms of verbal communication, non-verbal communication (e.g. body language and facial expressions) and written communication. Drawings are probably the most common means of giving information in our industry. The information shown on a drawing will communicate its contents to a speaker of any language. It is important for any communication to convey information without ambiguity (uncertainty).

Writing letters

Sometimes it is necessary to put in writing a permanent record of a complaint or concern, for example when a situation may result in a court case. Such a letter should then be sent using a method that requires the recipient to sign to say they have received it. Letters are almost exclusively written on computers now. They should be set out with the following layout and information.

▲ Figure 2.40 Operative typing a letter

- Your address: also known as the 'return address', this comes first (leave this off if you're using letter-headed paper).
- The date: directly below your address, include the date on which the letter was written.
- The greeting: after the recipient's address, leave a line's space, then put 'Dear Mr Jones', 'Dear Bob' or 'Dear Sir/Madam', as appropriate.
- The subject: you may want to include a subject for your letter – this is often helpful to the recipient to identify the order number or job concerned. This should be centred on the page.
- The text of your letter: this should be …
  - Concise: letters that are clear and brief can be understood quickly.
  - Authoritative: letters that are well written and professionally presented have more credibility and will be taken more seriously.
  - Factual: accurate and informative letters enable the reader to see immediately the relevant details, dates and requirements, and to justify action to resolve the complaint.
  - Constructive: letters with positive statements, suggesting concrete actions, encourage action and quicker decisions.
  - Friendly: letters with a considerate, co-operative and complimentary tone are prioritised because the reader responds positively to the writer and wants to help.
- The closing phrase and your name and signature: after the main body of the text, your letter should end with an appropriate closing phrase, such as 'Yours sincerely'. Leave several blank lines after the closing phrase so that you can sign the letter after printing it, then type your name. You can put your job title and company name on the line below this.

> **INDUSTRY TIP**
>
> If your letter is addressed to someone whose name you know (e.g. 'Dear Mrs Smith'), end it with 'Yours sincerely'. If your letter is addressed to someone whose name you don't know (e.g. 'Dear Sir/Madam'), end it with 'Yours faithfully'.

A similar format to the letter in Figure 2.41 can be used to communicate with a supplier or contractor about defective or damaged materials or delivery problems. It is important to state clearly what action you expect and in what time frame.

> **ACTIVITY**
>
> Using the information on writing letters, reply to the letter in Figure 2.41, outlining what you will do as the joinery contractor to solve the issue.

Business emails

Emails are a common way to communicate information quickly and accurately. A professional email address will look similar to this: 'yourname@companyname.com'. Anything other than this format can give a poor impression of a company. The tone of a business email will be more professional and structured than that used in a personal email. A subject line is usually added to the top of the email to make referencing and prioritising straightforward for the recipient.

> **ACTIVITY**
>
> Make a list of the differences between a business email and a personal email. Think about the type of language and vocabulary you would use, the tone of voice, and how you might start and end the email. Then write an email, as if you were E.G. Martin, responding to the letter you wrote in the previous activity and explaining that the problem is now solved.

MB Construction
1 High Street
London
EC12 D34

01.12.2019

Mr Dawson
Scales Joinery Ltd
14 Riverside Way
Dalston Lane
Hackney E1

RE: 3 Harold Court (order 1629)

Dear Mr Dawson

Following our recent phone conversation I am writing to confirm that on further inspection the front door to 3 Harold Court is still not closing correctly. Our snagging supervisor has checked to see that the frame has been fixed plumb and informs me that it has. He has tried to ease the rebates to allow it to close more easily but the owner is still having difficulty.

Can I request that when you are delivering the paladin bin store doors you take the opportunity to have a look? If you can reliably confirm that your company will be able to rectify this problem without it re-occurring, please go ahead. Otherwise I would request that it is replaced by 14.01.2020.

Yours sincerely,

E G Martin

E G Martin
Managing Director

▲ Figure 2.41 An example letter

Verbal communication

Talking face to face or on the phone is still the most common form of communication. Unfortunately, there is rarely a record of these conversations, so you need to remember the following:

- Think before you speak, so you get across exactly what you want to say.
- Be clear and concise.
- Ask for confirmation that what you have said has been understood.

Body language

Body language refers to different forms of non-verbal communication. This often reveals an unspoken attitude or intention.

Examples include:

- rolling your eyes (to an onlooker, meaning 'here we go again')
- yawning (indicating boredom or tiredness)
- hands in pockets (indicating lack of interest)
- crossed arms (indicating disagreement with what is being said)
- smiling (indicating happiness)
- frowning (indicating unhappiness).

Everyone has the right to be treated with respect and this will create a productive environment. Aggression breeds aggression, leads to poor working relations and levels of production, and gives a poor image to customers.

▲ Figure 2.42 Types of body language

⑥ UNDERSTANDING AND USING DRAWINGS AND ASSOCIATED SOFTWARE

This section will discuss the different types of drawn information used and found in construction. In *The City & Guilds Textbook: Level 2 Diploma in Site Carpentry and Bench Joinery* we looked at some of this information. Now we will recap this and look in a little more detail at how drawings are produced.

Drawings are required at every stage of building work. After the building has been designed and final designs agreed with the client, drawings are required to apply for planning consent and Building Regulations approval. These drawings will show the size, position and general arrangement of the proposed construction, and allow the local authority planning committee to decide whether approval should be given and whether the proposed building meets the current Building Regulations.

Building information modelling (BIM)

Building information modelling (BIM) is a collaborative digital tool, represented in the form of a 3D model of a building or structure. The digital representation is used by the whole project team to handle and share information before building. Accurate models provide detailed project information that can be used to manage risks at the initial design stage. They are also used to create specifications, schedules, programmes of work and costings for individual elements of the project or as a whole. BIM can provide virtual and mixed reality tours of a project at the design stage. It can even be used to add digital images of people, furniture and decor to provide perspective to the model. Earlier in this chapter we looked at a YouTube video illustrating how BIM can be used in conjunction with augmented reality (AR) for building simulation and project management. Since the innovation of BIM, the government has been a driving force in promoting its use in the construction industry so that the whole sector can become world leaders in this cutting-edge technology.

▲ Figure 2.43 A building designed using BIM technology

Types of drawing required by the local authority

Location plan

This shows the proposed development in relation to surrounding properties. It must be based on an up-to-date map and an identified standard metric scale (typically 1:1250 or 1:2500). The site of the proposed development needs to be outlined in red, and any other land owned by the applicant that is close to or adjoining the site needs to be outlined in blue. **Scale** is usually measured and drawn using a 'scale rule'.

> **KEY TERM**
>
> **Scale:** a reduction in size by a given ratio, e.g. 1:5 on a location plan means that the measurement on the scale is five times smaller than the actual structure and area shown. In this case, if the full-size object is 500 mm long, the scale-size object shown on the drawing will be 100 mm long.

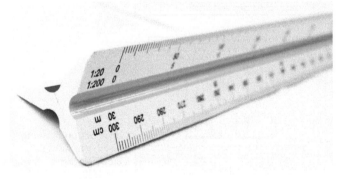

▲ Figure 2.44 Scale rule

> **ACTIVITY**
>
> Go to www.planningportal.co.uk and research what mandatory documents are required in a planning application. List them and give a short summary of each one.

Site plan

This shows the proposed development in relation to the property boundary (sometimes called a 'block plan'). Site plans are typically submitted at a scale of either 1:200 or 1:500, and should include the following information:

- the size and position of the existing building (and any extensions proposed) in relation to the property boundary
- the position and use of any other buildings within the property boundary
- the position and width of any adjacent streets and adjoining roads
- the position of existing significant or protected trees
- contours of the land, and floor levels within the property.

General arrangement drawings

These include elevations, floor plans and sections of the proposed building.

Elevations

These show the external appearance of each face of the building, including features such as slope of the land, doors, windows and the roof arrangement at a scale of 1:50 or 1:100, depending on the size of the project and the paper it is printed on.

Floor plans

These are used to identify the layout of the internal walls, doors and stairs, and the arrangement of the bathrooms and kitchen. Again, these will be drawn to a scale of 1:50 or 1:100.

Sections

These are used to show vertical views through the building, including room heights, and floor and roof constructions. These are generally drawn to a scale of 1:50.

Additional drawings required to construct the building

Construction drawings

These provide the detailed information required by individual trades to carry out their construction work. They are on a larger scale and may include any or all of the following.

Assembly drawings

These are used to show how various components fit together. They are generally drawn at scales of 1:20, 1:10 and 1:5.

Component/range drawings

These supply the information required by manufacturers producing components for the finished building, e.g. purpose-made doors or kitchen units. A range drawing could also show a manufacturer's standard range of products, e.g. doors and windows. These are often shown at scales of 1:10 and 1:20. A range of doors or windows might be indicated on the floor plans with a simple code such as D1, D2, W1, W2, and on a door or window schedule which will provide information about ironmongery and any glass requirements.

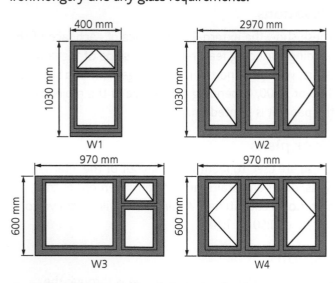

▲ Figure 2.45 Example of a window range drawing

INDUSTRY TIP

Sketches are quick and can often be used to communicate information more effectively than a description in words.

▲ Figure 2.46 Sketch of a cross-halving joint

Detail drawings

Detail drawings show very accurate large-scale details of the construction of a particular item such as a window or stair construction. Examples are joinery detailing and complex brickwork features. Typical scales are 1:1 (full size), 1:2 and 1:5.

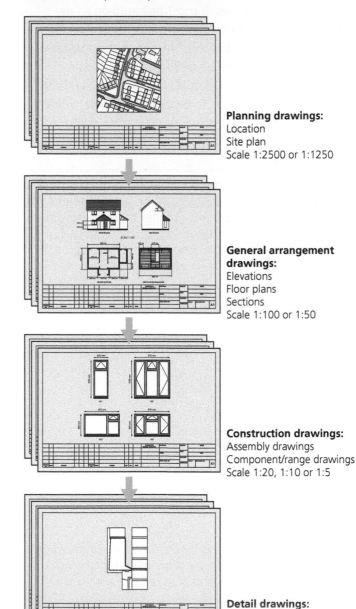

Planning drawings:
Location
Site plan
Scale 1:2500 or 1:1250

General arrangement drawings:
Elevations
Floor plans
Sections
Scale 1:100 or 1:50

Construction drawings:
Assembly drawings
Component/range drawings
Scale 1:20, 1:10 or 1:5

Detail drawings:
Large-scale sections
Scale 1:5 or 1:2

▲ Figure 2.47 Flowchart showing progression of planning drawings

Figure 2.47 highlights the importance of the use of various project drawings at each stage of a development; in essence, the further into a project you get, the more detail is required. For example, at the planning stage, the local authority will need a location plan so it has a holistic view of the project to base its decision on.

Production of drawings

Traditionally, all drawings would have been produced by a draughtsperson by hand on large parallel-motion drawing boards. Some drawings are still produced in this way but most are now produced electronically using CAD (computer-aided design).

▲ Figure 2.48 Parallel-motion drawing board

Computer-aided design

This is a method of producing drawings and designs using a design and drawing software package. These packages have been around for about 35 years now and have significantly improved during this period. They are used by millions of people worldwide. Anything can be drawn using these software packages in either two or three dimensions.

All architects or **architectural technicians** will produce drawings using CAD, as it has a number of advantages over hand-produced drawings.

> **KEY TERM**
>
> **Architectural technician:** a technician who is employed by an architectural practice to produce construction drawings.

- High-quality drawings can be produced.
- Drawings can easily be magnified, manipulated and amended.
- Standard details can be saved and reproduced as any new contract requires.
- Drawing layers can be produced, enabling particular details to be extracted.
- Objects and structures can be viewed from any angle.
- Drawings can be electronically archived without taking up valuable office space.
- Drawings can be uploaded and shared online.
- Three-dimensional virtual models and **walkthroughs** can be created.
- CAD can be used with compatible software products to prepare schedules and specifications, and with CAM (computer-aided manufacturing) to produce products by machine, for example staircases.

> **KEY TERM**
>
> **Walkthrough illustration:** an animated virtual-reality sequence as seen at the eye level of a person walking through a proposed structure.

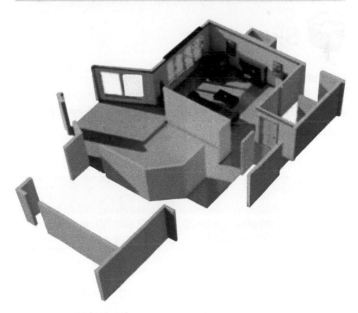

▲ Figure 2.49 A typical 3D model of the first floor of a house.

▲ Figure 2.50 Typical CAD workstation

While the advantages of CAD are significant there are also a few disadvantages:

- the initial outlay for the equipment
- the training of staff to use the software
- the cost of updating the software.

There are many CAD packages that can be purchased or downloaded for free. Autodesk produces a range of programs that are used universally. AutoCAD is very popular (but costs upwards of £1000 for a subscription). SketchUp, formerly owned by Google, is a free-to-download program.

▲ Figure 2.51 SketchUp software logo

Table 2.11 lists a range of CAD software packages that you might come across.

▼ Table 2.11 Examples of CAD software packages

Software package	Operating system	Used for:
AutoCAD	Windows, macOS, iOS, Android	2D and 3D architecture, drafting
Inventor by Autodesk	Windows	3D drafting
SolidWorks	Microsoft Windows	2D/3D hybrid
TurboCAD	Windows, macOS	2D and 3D design and drafting
ArchiCAD	Windows, macOS X	Building information visualisation and modelling functions, 2D and 3D drafting

All drawings are produced on standard-sized sheets of paper (see Table 2.12).

▼ Table 2.12 Standard paper sizes

Drawing paper		Other-sized paper	
Name	Size (mm)	Name	Size (mm)
A0	1189 × 841	A5	210 × 148
A1	841 × 594	A6	148 × 105
A2	594 × 420	A7	105 × 74
A3	420 × 297	A8	74 × 52
A4	297 × 210		

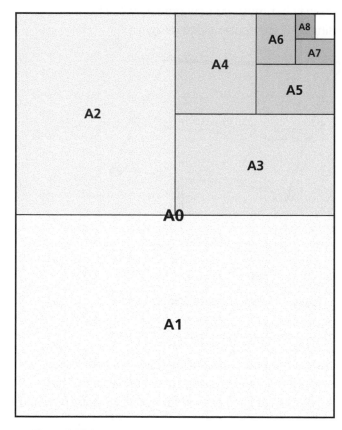

▲ Figure 2.52 Standard drawing sheet sizes

Drawing projection methods

Orthographic projection

Most drawings are produced in two dimensions (length and width). These are called orthographic drawings. First angle orthographic projection is used in the UK. This shows how the views are projected onto a flat surface surrounding the object. The surface is then folded flat to show the views in first angle projection on a drawing.

Most of the drawings we use in construction are produced in orthographic projection. These show a front elevation, end elevation and plan view of the proposed building. Additional view can show more detail as required. In order to produce these drawings, the following information is required:

- overall sizes of the proposed building
- room dimensions
- position and sizes of doors and windows
- position of internal walls
- brickwork, carpentry and joinery specifications
- building regulations requirements.

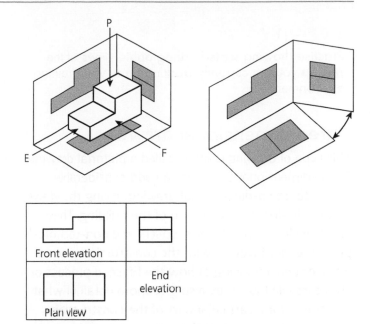

▲ Figure 2.53 First angle orthographic projection

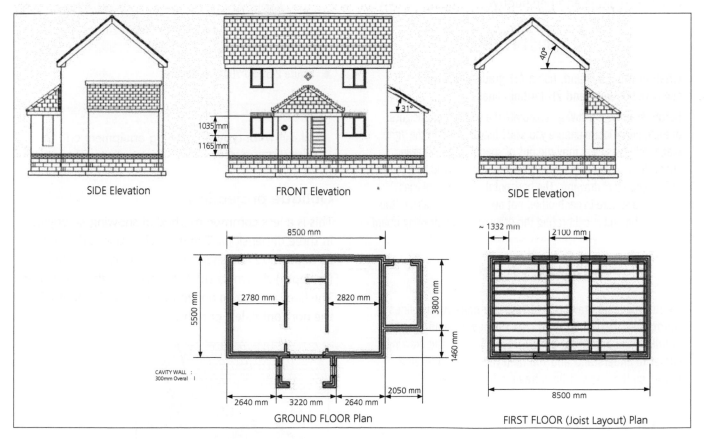

▲ Figure 2.54 Orthographic projection of a house

Three-dimensional drawing

This type of drawing is often called a pictorial drawing. Three-dimensional drawings can add considerable clarity to a two-dimensional drawing, giving the viewer a more life-like representation of the building. They give the client, operatives, suppliers and non-technical personnel not involved with the construction process a better understanding of how the finished product or structure will look; they also give more detail of what is required for a particular part of the construction. There are many types of three-dimensional drawing, but the most common types used in construction are isometric, oblique and perspective.

▲ Figure 2.55 An isometric drawing of a house

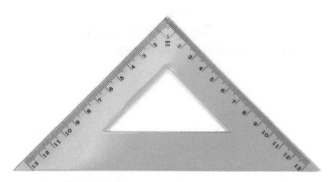

▲ Figure 2.56 30°/60° set square

Oblique projection

This is a less common method of showing an object in three dimensions. The front elevation is drawn to its actual size and shape, with the third dimension (or depth) shown to either the left or right as required. The lines are drawn horizontally, vertically or at 45° to the horizontal, left or right.

Isometric projection

This is the most common method and produces a life-like representation of the subject. It is often used to show an overall view of the object and forms the basis of most sketches. All lines are drawn vertical or at 30° to the horizontal, left or right (using a standard 30°/60° set square).

▲ Figure 2.57 An oblique projection of a house

Perspective drawings

These give the most realistic view of a building. Lines are drawn vertically or drawn back to vanishing points (VP). The vanishing points can be positioned at any height but most commonly at the viewer's eye level.

The same information is required to produce three-dimensional drawings as for two-dimensional drawings:

- dimensions
- scale required
- axis (30°, 45° or vanishing)
- position and required view of object.

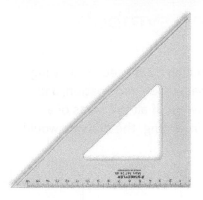

▲ Figure 2.58 45° set square

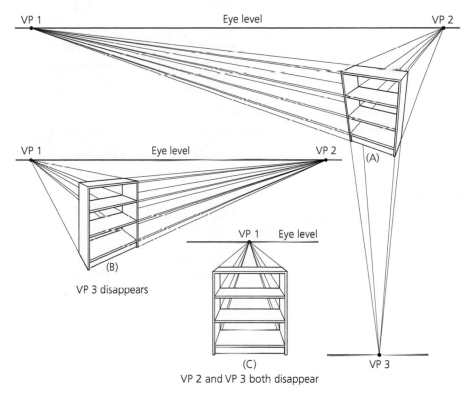

▲ Figure 2.59 A perspective drawing

▲ Figure 2.60 An example of a perspective drawing

INDUSTRY TIP

Before you start a job you have not done before, always try to visualise the finished outcome, as this may prevent mistakes. Producing sketches can help you to do this.

Basic drawing symbols (hatchings)

Standard symbols (hatchings) are used on drawings as a simple means of passing on information, and to avoid cluttering drawings. If all the parts of a building were labelled in writing, the drawing would become very crowded and mistakes could be made when reading it. Additionally, it is important to use standard symbols that mean the same thing to everyone. Professionals normally use British Standard symbols, which everyone in the construction industry should understand. Figure 2.61 shows some of these standard symbols.

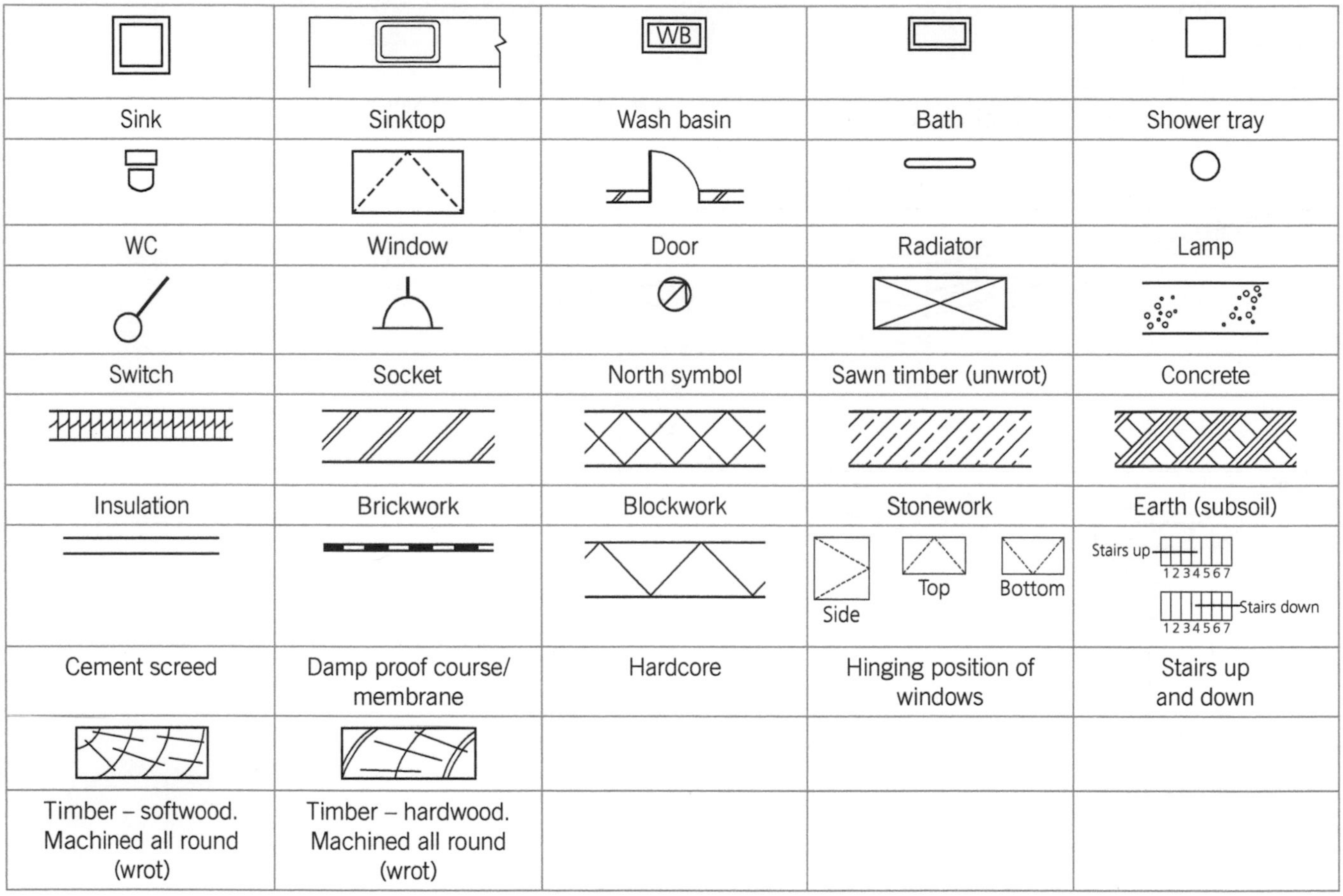
WB
Sink
Sinktop
Wash basin
Bath
Shower tray
WC
Window
Door
Radiator
Lamp
Switch
Socket
North symbol
Sawn timber (unwrot)
Concrete
Insulation
Brickwork
Blockwork
Stonework
Earth (subsoil)
Top
Bottom
Side
Stairs up
1 2 3 4 5 6 7
Stairs down
1 2 3 4 5 6 7
Cement screed
Damp proof course/ membrane
Hardcore
Hinging position of windows
Stairs up and down
Timber – softwood. Machined all round (wrot)
Timber – hardwood. Machined all round (wrot)

▲ Figure 2.61 British Standard drawing symbols

Test your knowledge

1 Which one of the following is an advantage of using CAD?

A Software is not required.

B Objects can be reproduced quickly.

C Lines can be drawn quickly with set squares.

D It can't be used with augmented reality.

2 Which one of the following scales is used to produce a location plan?

A 1:5

B 1:100

C 1:1250

D 1:4500

3 Which one of the following three-dimensional drawings has the lines drawn back at 30°?

A Oblique

B Isometric

C Perspective

D Orthographic

4 Which one of the following is a document that contains a promise to carry out work for a specific sum?

A Estimate

B Quotation

C Specification

D Bill of quantities

5 Who usually produces the bill of quantities?

A Builder

B Architect

C Clerk of works

D Quantity surveyor

6 A detailed description of how a task will be carried out is called a:

A programme

B specification

C risk assessment

D method statement.

7 The time it takes for a product to be manufactured is called the:

A quote

B estimate

C lead time

D time delay.

8 Which one of the following is a disadvantage of hiring plant and equipment?

A Costs are reduced in the long term.

B Repeated use of hired equipment may be more expensive than the initial cost of outright purchase.

C It is returned at the end of the contract.

D Replacement parts have to be purchased.

9 Which one of the following items will be required to document a client's request for a change to the work required?

A Memorandum

B Specification

C Variation order

D Method statement

10 Which Approved Document covers 'Conservation of fuel and power'?

A K

B L

C M

D N

11 Explain the 'tendering' process.

12 Calculate the cost of a replacement door and ironmongery in the room you are in. Use whatever resources you need to find the price of the materials. Your hourly rate will be the year of your birth. For example, if you were born in 2002, your hourly rate will be £20.02. You should include £18.74 including VAT for the hire of a cordless plane. Add 6 per cent for overheads, 21 per cent profit and the current rate of VAT.

13 List as many forms of communication as you can think of, and write an advantage and disadvantage for each method.

14 Research a health and safety topic and prepare a toolbox talk that could be used at your place of work.

15 List four sources of renewable energy.

CHAPTER 3

SETTING UP AND USING FIXED AND TRANSPORTABLE MACHINERY

INTRODUCTION

This chapter will provide you with the skills and knowledge to safely inspect, perform basic maintenance on, set up and use both fixed and transportable woodworking machinery. The skills developed will enable you to safely use appropriate safety aids when producing joinery components using the following woodworking machines:

- circular saws (rip and crosscut)
- narrow bandsaws
- surface planers
- thicknessers
- combination planers
- chisel morticers
- router tables and spindle moulders.

LEARNING OBJECTIVES

By reading this chapter you will learn how to:

1 understand and use legislation and documentation relating to the safe use of woodworking machinery
2 carry out the inspection and maintenance of fixed and transportable machinery
3 use sawing machines
4 use planing machines
5 use a morticing machine.

1 LEGISLATION AND DOCUMENTATION RELATING TO THE SAFE USE OF WOODWORKING MACHINERY

Legislation

To enable the safe use of fixed and transportable woodworking machinery, you must comply with multiple pieces of legislation. The list below outlines the principal pieces of legislation that apply to woodworking machinery:

- Provision and Use of Work Equipment Regulations 1998 (PUWER)
- Approved Code of Practice for Safe Use of Woodworking Machinery (ACoP)
- Control of Noise at Work Regulations 2005
- Control of Substances Hazardous to Health 2002 (COSHH)
- Health and Safety at Work etc. Act 1974.

As well as legal guidance and regulations, other sources of information include:

- Health and Safety Executive (HSE) woodworking information sheets
- British Woodworking Federation (BWF) safety cards.

Provision and Use of Work Equipment Regulations 1998 (PUWER)

The primary piece of legislation that governs the use of work equipment, which includes woodworking machines, is the Provision and Use of Work Equipment Regulations 1998, commonly known as PUWER. These regulations require the prevention or control of risks to people's health and safety posed by work equipment that they use while at work. PUWER deals with all types of work equipment and industries in general, rather than dealing with woodworking machines specifically. This means that the interpretation of PUWER in relation to the safe use of woodworking machines can be difficult and confusing. As a result, the HSE produced an Approved Code of Practice (ACoP) for safe use of woodworking machinery. This guidance gives specific advice on the precautions to take in order to ensure safe use of woodworking machines.

Although the ACoP for safe use of woodworking machines is only guidance and not a regulation or legally binding, it is accepted that, if you follow the ACoP for the safe use of woodworking machines, you will be complying with PUWER and as a result will be complying with legally binding regulations and safe working practices.

ACTIVITY

Go to the Health and Safety Executive (HSE) website at www.hse.gov.uk and download the Approved Code of Practice for woodworking machines. Identify the five main topic areas that PUWER covers. You can use this as an additional source of information and guidance in relation to the machines covered in this chapter.

Approved Code of Practice for Safe Use of Woodworking Machinery (ACoP)

The Approved Code of Practice and guidance is provided to help put into practice PUWER regulation as it applies to woodworking machines. It applies to most woodworking machinery but not hand-held tools, and includes tasks involving wood, corkboard, fibreboard and composite materials. It gives practical advice on the safe use of woodworking machinery, and covers the provision of information and training, as well as aspects of guarding. Further detailed interpretation of the ACoP relating to specific woodworking machinery will be given later in this chapter within the individual machine descriptions and safe working practices.

INDUSTRY TIP

To help support the ACoP for Safe Use of Woodworking Machinery, the HSE published a series of wood information sheets. These sheets contain practical advice and guidance on the safe use of woodworking machinery.

ACTIVITY

Display copies of the HSE Approved Code of Practice woodworking information sheets at each relevant machine to reinforce safe working practices.

Using the Approved Code of Practice, outline the requirements for the suitability of woodworking machinery to carry out particular jobs safely, and how this affects your choice of machine for machining operations.

Control of Noise at Work Regulations 2005

The aim of the Control of Noise at Work Regulations is to ensure that workers' hearing is protected from excessive noise at their place of work, which could cause them to lose their hearing and/or to suffer from **tinnitus**. Woodworking machinery produces high levels of noise and specific machine controls need to be put in place; as a last resort ear protection should be used to help control exposure to high noise levels. Noise control and protection are dealt with later in the chapter.

KEY TERM

Tinnitus: a permanent ringing or hissing in the ears, often caused by exposure to loud noise such as that from woodworking machinery.

Control of Substances Hazardous to Health 2002 (COSHH)

The COSHH regulations require employers to control substances that are hazardous to health. In the case of woodworking machinery, this is typically wood dust or dust produced by timber-based products. Employers of people engaged in the use of woodworking machinery can prevent or reduce workers' exposure to hazardous substances by:

- finding out what the health hazards are
- deciding, through risk assessment, how to prevent harm to health
- providing control measures to reduce harm to health, and making sure they are used
- keeping all control measures in good working order
- providing information, instruction and training for employees and others
- providing monitoring and health surveillance in appropriate cases
- planning for emergencies, e.g. a worker could get wood preserver splashed into their eyes.

You should always try to prevent exposure to hazardous substances at source. For example:

- Can you avoid using the hazardous substance or can you use a safer process, e.g. using a ducted extraction system as opposed to dust bags, or using a vacuum cleaner rather than a brush?
- Can you substitute one form of timber for something safer, e.g. swap an irritant-type timber such as afrormosia for something less irritating such as beech?
- Can you use a safer form of adhesive, e.g. can you use a liquid adhesive instead of a dry powder to avoid dust?

Health and Safety at Work etc. Act 1974

The Health and Safety at Work etc. Act 1974 (also referred to as HASAWA or HASWA) is the primary piece of legislation covering occupational health and safety in Great Britain. The HSE, with local authorities (and other enforcing authorities), is responsible for enforcing the act. It is covered in more detail in Chapter 1.

Manufacturers' literature and maintenance schedules

All woodworking machinery will be accompanied by a manufacturer's information booklet. This will provide information about such things as:

- the safe setting up of the machine
- the safe use of the machine
- the use of ancillary equipment and delivery aids
- the type and limitation of the tooling
- the maintenance requirements of the machine.

It is important to fully familiarise yourself with the information contained within the manufacturer's information booklet, as this is provided not only to help prevent injury to operators and others but also to help prevent unnecessary damage to the machine. By following the manufacturer's information on maintenance and the required maintenance schedule, you can reduce the machine's downtime.

Maintenance

Maintenance of woodworking machinery not only reduces the risks of injury to operators, and damage to the machine, but is also a legal requirement under PUWER. PUWER requires a written record of the maintenance carried out on woodworking machinery. Maintenance schedules, fault reporting and accidents contribute to the production of risk assessments. Where maintenance logs show repeated damage to a machine or recurring faults, the maintenance schedule should be adjusted in terms of frequency and depth to help eliminate these issues. If there is repeated damage to a machine, such as to guards and tooling, specific extra training may be required for operatives.

Maintenance should be carried out only by people who are trained and competent to do so. Maintenance is not only a legal requirement, it also makes good economic sense. As with a car or any other machine, if you do not change the oil or replace the drive belts, after some time they will break down or fail to perform correctly. Damage, DIY repairs, missing or damaged guards and poor wiring must be addressed immediately, even if this means taking the machine out of use. You must use only suitable replacement parts and tooling.

Name of equipment:		Manufacturer's contact details:	
Label:		Date of purchase:	15/10/2019
Serial number:		Person responsible for equipment:	
Manufacturer:		Date put into service:	23/10/2019

Date:	Maintenance description:	Maintenance performed by:	Date of validation before put into service:	Validation performed by:	Next maintenance planned on (date):	Remarks:

▲ Figure 3.1 Example maintenance log

ACTIVITY

Use the HSE website (www.hse.gov.uk) to research and compare accident statistics for different types of woodworking machine. Choose a machine you will use and present these statistics to your tutor, along with your thoughts on the following three areas:

1 What type of action is being taken to reduce accidents on that machine?

2 In general, why do you think accidents are still happening?

3 Offer any additional recommendations you think could help reduce accidents.

Supervision and training records

Training must be provided to everyone involved in setting up, operating and carrying out maintenance operations on woodworking machines. A list of authorised operators should be kept and displayed in the workplace, stating which machines they have been authorised to use. Poor supervision, and inadequate training and information are the main causes of accidents involving woodworking machinery.

HEALTH AND SAFETY

Most woodworking machinery is designed and manufactured to be used by right-handed operatives. When left-handed operatives are required to use it, they may find the layout and positioning of a machine and its controls slightly more difficult to get used to at first than right-handed operatives. Particular attention and time should be given during the training programme to ensure that left-handed operatives have the right training, help and advice in dealing with such machinery.

E. G. Redfern Ltd
List of authorised machine operators

The authorised trainer of _____ *is* _____
(the company)   *(name of trainer)*

Date _____

I certify that:

(a) *I have carried out training, as indicated on the machines listed*

(b) *I am satisfied that the people named below have demonstrated competence in the operation of the machines listed and have met all the training objectives for those machines, including:*

(i) *correct selection of machine for type of work to be done;*

(ii) *purpose and adjustment of guards and safeguards;*

(iii) *correct selection and use of safety devices – push sticks, push spike, jigs and work-holders;*

(iv) *practical understanding and application of legal requirements;*

(v) *safe working practices to include feeding, setting, cleaning and taking off.*

Signed _____ *(Trainer)*

Machine												
Operator's name	Circular saw	Cross-cut saw	Dimension saw	Surface planing machine	Thickness planing machine	Single-ended tenoner	Single-moulder	High-speed tenoner	Four-sided planer/moulder	Narrow band saw	Band re-saw	etc
J. Brown	✓	✓	✓	✓	✓	✓						
J. Bahri	✓	✓	✓	✓	✓					✓		
T. Yeung	✓	✓										

(Adapted from HSE)

▲ Figure 3.2 Example of an employer's machinery training record

Nobody should use any woodworking machine unless they have received sufficient training and information, and acknowledged that they have understood that training and information. The training should include instructions concerning general aspects of woodworking machinery, as well as machine-specific instructions.

When training is in its early stages, it is expected that supervision levels will be high. As the user begins to demonstrate safe working practices and shows confidence in using the machines, a gradual reduction in supervision levels can be allowed. Poor supervision and inadequate training are major contributors to accidents involving woodworking machinery.

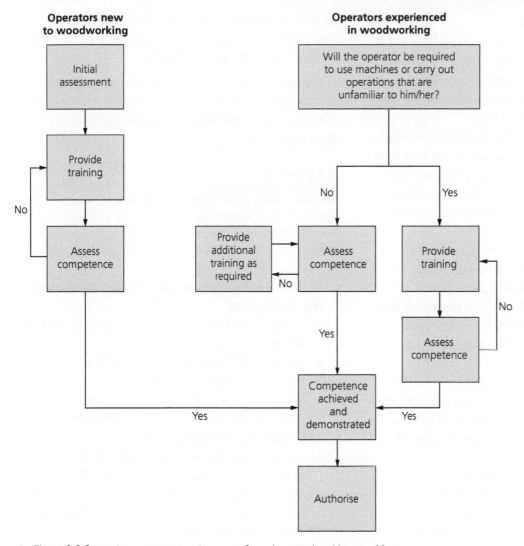

▲ Figure 3.3 Competence assessment process for using woodworking machinery

All woodworking machining training schemes, including those that form part of a joinery qualification, should include the following elements:

- general skills
- machine-specific skills
- machine familiarisation
- demonstrating competence
- competence checklists
- record keeping.

These topics are closely linked to risk assessments.

Risk assessment

Before using woodworking machinery, a risk assessment should be carried out to help identify the user's suitability, maturity, training requirements and supervision ratios.

As well as training, general considerations within risk assessments should include:

- good housekeeping
- noise control
- manual handling, listing aids and operational support
- PPE
- guarding
- tooling
- braking.

Good housekeeping

All workshops should be kept clean and tidy, free from excessive build-up of dust and debris. Bins should be provided for the safe storage of offcuts and trimmings. Sufficient time or staff should be allocated for the safe removal of any offcuts and trimmings.

- Floors need to be in good order, level and non-slip. The area around the machine needs to be kept clear of chippings and offcuts. Any cabling or ducting should either be at high level or below ground, or routed in such a way as not to interfere with the safe operation of the machine.
- A clear working area needs to be established around the machine to allow safe operation without the operator being knocked while operating it. Offcuts should go into dedicated bins. There should also be sufficient room to handle the material being machined and to store the machined components safely.
- Lighting: all areas around the machine must be adequately lit. The light can be natural, artificial or a combination of both. Any artificial lighting that starts to flicker must be replaced straight away.
- Heating: a minimum temperature of 16°C is required for the safe operation of woodworking machines. If the temperature falls below this level then the operator can lose the ability to safely control the work piece, and may lose concentration.
- Dust extraction: before cutting any material, the operator should ensure that an adequate dust extraction system is connected and working correctly. Correctly designed and fitted dust extraction systems should remove the need for dust masks.

Noise control

All employees must be protected from the loud noise produced by woodworking machines and portable power tools. Exposure to loud noise will permanently damage hearing, resulting in deafness and tinnitus. Under the Control of Noise at Work Regulations 2005, duties are placed on manufacturers, employers and employees to reduce the risk of damage to hearing as far as is reasonably practicable.

Manufacturers and suppliers are required to provide low-noise machinery and information about the noise levels produced by any machinery.

Employers must reduce noise levels to a reasonable level to protect their employees' hearing, with specific noise levels of 80 **decibels (dB(A))** and 85dB(A) requiring action.

> **KEY TERM**
>
> **Decibel (dB(A)):** the unit that measures sound levels. The A weighting is applied to instrument-measured sound levels to try to account for the relative loudness perceived by the human ear.

- At 80dB(A) the Noise at Work Regulations require the employer to:
  - make a suitable assessment of the noise workers are exposed to
  - provide suitable ear protection to those employees who request it
  - provide information to employees about the risks to hearing and about the legislation.
- At 85dB(A) the employer must:
  - reduce the amount of noise exposure as far as is reasonably practicable by means other than ear protection, e.g. by erecting physical barriers such as acoustic enclosures or by changing to newer, quieter tooling
  - designate the work area as an 'ear protection zone', and establish a safe system of work to protect visitors and other workers, providing suitable signage stating, 'Ear protection must be worn in this area'
  - provide suitable ear protection and ensure all who enter the designated area wear it.

Employees have a duty to wear and use all ear protection provided, as well as to comply with all other noise control procedures and to report any defects.

Manual handling, lifting aids and operational support

Manual handling must be avoided as far as possible. Where this is not practicable, suitable training and equipment must be provided; this is typically identified as a result of a risk assessment. Large and/or heavy materials require attention not only for lifting them but also for presenting them to the machine and supporting them through the machining process. Some machines, such as crosscut saws, have large roller beds that are specifically designed to handle long and heavy materials, while other machines, such as the circular ripsaw, have an outfeed table to provide support. In other situations, temporary support may be required in the form of adjustable temporary rollers that can be positioned at one or both ends of the machine.

For example, long door jambs may require additional support on a morticer, or the infeed end of a ripsaw and both sides of a narrow bandsaw. When additional manual support is required, it should be used on all occasions and identified in the risk assessment.

Clear verbal communication is vital when team lifting. Each member must fully understand their role in the lifting and moving operation, and know what to do in an emergency.

▲ Figure 3.4 Temporary support roller

Personal protective equipment

The personal protective equipment (PPE) that is needed for a particular activity will be identified by the risk assessment and usually involves such items as:

- ear protection
- eye protection
- head protection
- gloves
- suitable footwear
- suitable clothing
- a suitable dust mask or respirator.

Any PPE identified in the risk assessment must be used. If it is damaged or malfunctioning in any way, it should be reported and a replacement obtained.

HEALTH AND SAFETY

It is recommended that eye protection is worn during cutting operations in order to reduce the risk from flying particles. Gloves can help prevent injury from splinters and are particularly useful if working with toxic timbers such as afrormosia and Douglas fir, whose splinters can easily cause a wound to become septic (infected).

ACTIVITY

Locate woodworking information sheet number 30, 'Toxic woods', on the HSE website and see whether you are likely to machine any of the timbers mentioned. Record the likely effects that could result from using these timbers if the correct precautions are not taken. Display your findings in your workshop.

Braking

All the woodworking machines covered in this chapter require braking of some form. All new machines are fitted with automatic braking systems, but some older machines could have manually operated braking systems. Whichever system is incorporated into the design of the machine, it must be capable of bringing the machine tooling to a controlled safe stop in a maximum time of 10 seconds.

IMPROVE YOUR MATHS

Record the time it takes for each of the following machines to stop:

- surface planer
- narrow bandsaw
- ripsaw
- crosscut saw
- thicknesser
- spindle moulder.

Produce a graph/pie chart or similar that represents your findings on the time required to bring the tooling to a safe standstill. Identify any machines that are within 10% of the maximum stopping time.

The other elements to consider in a risk assessment – guarding and tooling – are covered in the section that follows.

2 CARRYING OUT THE INSPECTION AND MAINTENANCE OF FIXED AND TRANSPORTABLE MACHINERY

Inspection, fault diagnosis and maintenance of woodworking machinery

Inspections

The purpose of inspecting woodworking machinery is to identify whether the work equipment can be operated, adjusted and maintained safely. If any deterioration is detected in the machine, or its guards or tooling, these defects should be rectified before they become a safety risk and cause injury or ill health.

Not all work equipment needs a full formal inspection before use. In many cases, a quick visual check will be enough. The need for full formal inspections and inspection frequencies should be determined through risk assessment and using the manufacturer's information booklet.

The details of checks and inspections required for particular types of machine will be covered throughout this chapter. Below are some general points to consider.

When woodworking machinery should be inspected

Woodworking machinery should undergo a full formal inspection when a risk assessment identifies it is necessary, as well as after any injury to operators or others caused by the equipment's installation or use. The outcome of the inspection should be recorded, and this record should be kept at least until the next inspection. Records do not have to be made in hard copy (such as written in a notebook) but, if they are kept on a computer for example, these records should be held securely and made available upon request by any enforcing authority. Any work equipment that is awaiting an inspection should not be used, until the inspection has taken place and the work equipment has been approved as safe to use.

PUWER requires work equipment to be inspected:
- after installation and before first use, and after reassembly at any new site/location
- at suitable intervals determined by the risk assessment
- where work equipment is exposed to conditions causing deterioration liable to result in dangerous situations
- after exceptional circumstances such as:
  - major modifications to the equipment
  - known or suspected serious damage to the equipment
  - substantial change in the use of the equipment.

All these factors are likely to have jeopardised the safety of the work equipment and have the potential to cause harm or injury.

What should be inspected

The areas covered will depend on the type of woodworking equipment being inspected, its use and the conditions to which it is exposed. The areas to be covered should be determined through risk assessment and take full account of any manufacturer's recommendations. The advice of others, such as trade associations and consultants, as well as other sources such as published advice on health and safety, may also be helpful. Records do not need to be kept for the simplest pre-use checks but must be kept for the full inspections and any maintenance carried out due to the inspection.

Any inspection should concentrate on safety-related parts and activities that are necessary for the safe operation of the work equipment. Inspections can vary in their length and coverage, as outlined in the risk assessment, but the following lists can be used as a guide.

Quick inspection checks before use:
- stop/start switches
- dust extraction
- guards
- tooling
- safety aids such as push sticks.

More detailed or full inspections, undertaken every month or less frequently, as established by the risk assessment:

- all the topics covered under the inspection checks before use
- close examination of the electrical wiring and switch gear
- brake testing
- drive belt adjustment
- general safety of the machine frame.

Any faults found should be reported and the machine should immediately be taken out of use. The machine should be locked off at the isolator with a sign stating that it should not be used until repairs have been made. All repairs need to be entered into the maintenance log, recording what was done by whom and when.

Upon completion of any maintenance work or inspection, the woodworking machine should be left in as safe and isolated condition as possible, with all guards etc. in position. The ACoP also requires that any operator who uses a woodworking machine should leave the machine in a safe condition, e.g. stopped, isolated and with the guards set so that they fully enclose the tooling or any other moving part of the machine.

INDUSTRY TIP

Checklists can aid an inspection, but they should be tailored to the type of woodworking machine. It is essential that inspections do not become a 'tick box' exercise or cease altogether. Remember you only need to inspect what is necessary for safety.

Woodworking machinery can be inspected by anyone who has enough knowledge and experience of the machine to enable them to know:

- what to look at
- what to look for
- what to do if they find a problem.

ACTIVITY

Produce a pre-start inspection checklist for each of the following woodworking machines:

- circular ripsaw
- narrow bandsaw
- crosscut saw
- surface planer/thicknesser
- spindle moulder.

Fix each of the checklists to the relevant machine in your place of work.

Maintenance

PUWER requires all work equipment, including woodworking machinery, to be maintained in an efficient state, in efficient order and in good repair. Where any machinery has a maintenance log, this must be kept up to date, and maintenance operations on work equipment should be carried out safely.

Maintenance work should be undertaken only by those who are competent, and have been provided with sufficient information, instruction and training (PUWER regulations 8 and 9). With high-risk or complex equipment, these demands may be significant and, in some cases, may best undertaken by the manufacturer or specialist contractors. But, in many cases, maintenance can be done in-house by suitably trained, competent staff.

Steps should be taken to manage any risks arising from maintenance activity. Manufacturers' instructions should make recommendations about how to safely maintain their work equipment and, unless there are good reasons to do otherwise, these should always be followed.

Prior to maintenance operations, woodworking equipment should always be shut down and isolated from any power supply, with steps taken to prevent inadvertent reconnection, e.g. by locking off the power supply. Any residual/stored energy must be safely released, e.g. rotating blades stopped, pneumatic pressure released, air reservoirs emptied. Formal systems of work, such as a permit to work, are required in some cases to safely manage high-risk maintenance operations.

It is important that maintenance operations are properly assessed, and staff due to carry out maintenance may need to undertake significant on-the-job risk assessment (essentially considering what could go wrong and how to avoid injury), as the situation may change in ways that could not be foreseen at the outset.

Typical items that may be covered under a maintenance schedule include:

- repairing guards
- changing tooling
- repairing, adjusting or making delivery aids such as push sticks, jigs or saddles
- repairing extraction pipework
- cleaning and oiling moving parts such as shafts and pulleys.

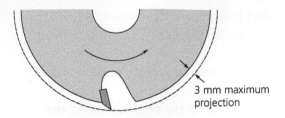

3 mm maximum projection

▲ Figure 3.5 Solid profile block

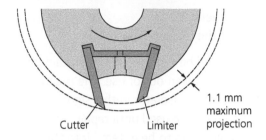

Cutter   Limiter

1.1 mm maximum projection

▲ Figure 3.6 Chip limited cutter block

ACTIVITY

Produce a maintenance log sheet for each of the following woodworking machines:

- circular ripsaw
- crosscut saw
- surface planer
- narrow bandsaw
- spindle moulder.

Refer to Figure 3.1 earlier in this chapter for an example of a log template.

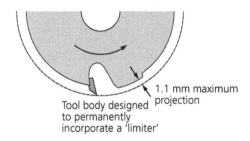

1.1 mm maximum projection

Tool body designed to permanently incorporate a 'limiter'

▲ Figure 3.7 Cylindrical block

Changing tooling

When changing tooling, you must follow the risk assessment and consult the manufacturer's literature. Pre-start checks must be carried out and the machine left in an isolated and safe condition. Details of changing tooling on specific machines will be given throughout this chapter but here is some general information.

The choice of tooling is vital to the safe operation of the equipment. Cutters must be of the correct type, kept sharp and in good condition, and suitable for the task in hand. Tooling should be the correct size and must not exceed the correct safe operation speed as specified on the tooling or by the manufacturer. Only cylindrical blocks can be used on hand-fed machines with limited cutter projection.

ACTIVITY

Go to the HSE website and download wood information sheet number 37: 'Selection of tooling for use with hand-fed woodworking machines'. How does the tooling within your place of work or training establishment compare? Do you need to make changes? Produce a report for your tutor outlining your findings.

3 USING SAWING MACHINES

Circular saw benches are among the most common types of woodworking machines in general use and account for around one-third of all accidents at woodworking machines. Circular saw benches can be divided into three broad types:

1 ripsaws
2 dimension saws
3 crosscut saws.

Ripsaws

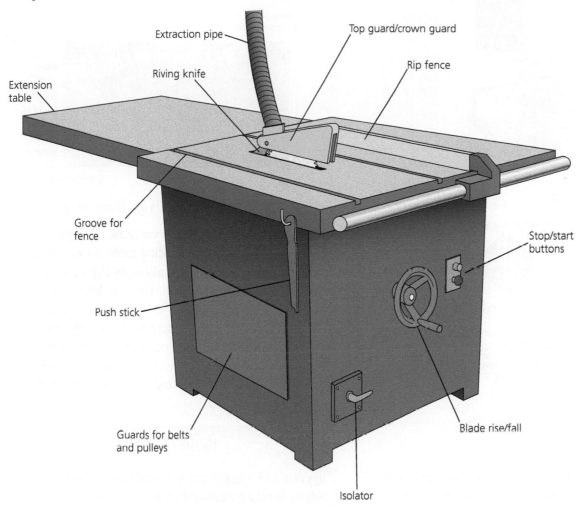

▲ Figure 3.8 Major component parts of a hand-fed circular ripsaw

Circular ripsaws look very much like dimension saw benches (see below) but carry out only three types of basic operation.

1 **Flatting:** the timber is ripped down its length through its thinnest section (thickness) to the required width.

2 **Deeping:** the timber is ripped down its length through its thickest section (width) to the required thickness.

3 **Angled cutting:** material is placed on a jig or saddle that holds it stable at the required angle during the cutting operation. This ensures an accurate angle is produced but, more importantly, aids the safe delivery of the material and safe control of the offcut.

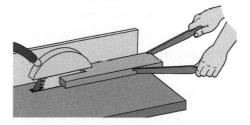

▲ Figure 3.9 Flatting

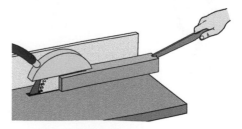

▲ Figure 3.10 Deeping

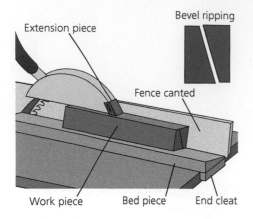

▲ Figure 3.11 Bevel ripping

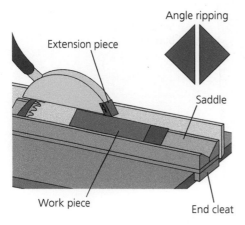

▲ Figure 3.12 45° angle ripping

Ripsaws can only be used for ripping down the grain of the timber and are used mainly to rip timber to the required sizes ready for the surface planer and thicknesser. The processes used for selecting tooling, setting up the saw to comply with current regulations and the safe operation of the saw are almost the same as those required when using the dimension saw.

▲ Figure 3.13 A typical ripsaw found on site

Dimension saws

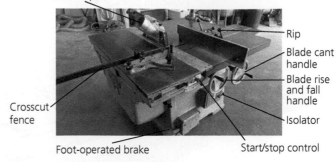

▲ Figure 3.14 Major component parts of a dimension saw bench

The dimension saw is a more versatile machine than a ripsaw. It incorporates a sliding table on which a crosscut fence can be fitted, allowing the dimension saw to crosscut as well as rip. The saw blade can 'cant' or 'tilt' to an angle of 45°, allowing for angle cuts, and, by using the crosscut fence, **compound cuts** can be performed without the aid of saddles or jigs. The dimension saw is usually used for fine, accurate work, and for cutting sheet materials such as MDF and plywood.

INDUSTRY TIP

Jigs help with repeated accurate production of items, helping to reduce machining risks.

ACTIVITY

Design and produce a saddle suitable for cutting 45° angle glue blocks (to be used during staircase production), which can be used on your circular ripsaw.

KEY TERM

Compound cut: a cut that consists of two angles – the bevelled angle from the canted saw blade and the mitre angle (or crosscut angle) from the fence.

Crosscut saws

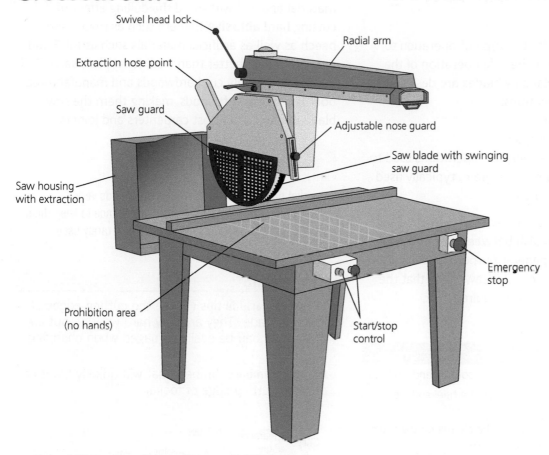

▲ Figure 3.15 Components of a radial arm crosscut bench

Crosscut saws are used to cut across the grain of the timber, usually to a predetermined length from the adjustable end stops. Other operations carried out on the crosscut saw can include:

- cutting tenons
- square cuts
- angled cuts
- compound cuts
- cutting birdsmouths
- cutting housings
- cutting halving joints
- notching
- kerfing.

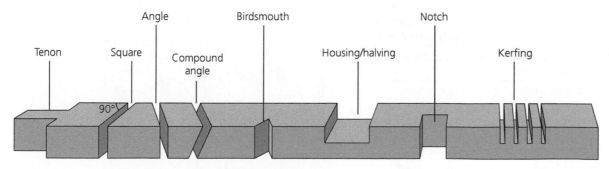

▲ Figure 3.16 Typical types of operation that can be carried out using a crosscut saw

Tooling for circular sawing machines

The choice of saw blade for the type of operation to be carried out is critical for the safe operation of the circular saw bench. Circular saw blades are designed for three main machining functions:

1 ripping down the grain
2 crosscutting across the grain
3 combination or general-purpose operations used to cut both across and down the grain, typically used on a dimension saw bench.

The terms listed in Table 3.1 associated with circular saw blades help to distinguish between ripsaw blades, crosscut saw blades and general-purpose saw blades. It is vital when selecting a circular saw blade that the correct design is chosen for the required task.

▼ Table 3.1 Circular saw blade terms

Term	Description
Pitch	The distance between tooth tip and tooth tip
Clearance angle	The distance between the heel and the cutting circle
Sharpness angle	The angle between the clearance angle and the angle of hook or rake
Hook or rake angle	The angle of the tooth in relation to a line taken from the tooth tip to the centre of the blade
Gullet	The area between the teeth that contains the sawdust during cutting
Point	The top of the tooth, which makes first contact with the material being cut
Heel	The back of the tooth
Face	The leading edge of the tooth
Set	The clearance given to either side of the tooth for traditional plate saws; the tooth tip is slightly bent over alternately to each side
Tangential side clearance	The clearance on either side of the tooth for TCT saw blades
Kerf	The total width of the saw cut, which comprises the thickness of the blade and the set on either side of the blade

Most saw blades in general use today consist of a parallel plate steel blade onto which are brazed **tungsten carbide tips (TCTs)**. It is these TCT teeth that do the cutting, having almost totally replaced parallel plate steel blades alone. TCT is a very hard material and will withstand the dulling effect of cutting hard **abrasive timbers** such as maple and beech as well as artificial materials such as MDF and plywood much better than traditional plate saws. TCT teeth will not only cut hardwoods and manufactured boards but also softwoods, making them the saw blade of choice for most carpenters and joiners.

INDUSTRY TIP

Parallel plate steel blades will cut softwoods very satisfactorily but require regular maintenance to keep them working safely – as a result, they are now rarely used.

KEY TERMS

Tungsten carbide tips (TCTs): the cutting edges of the saw blade. They are very hard wearing but are brittle and can be easily damaged when changing the blade.

Abrasive timbers: timbers that will quickly blunt or dull the cutting edge of tooling.

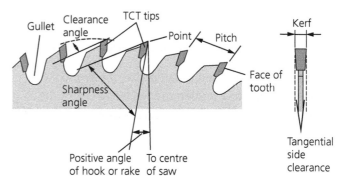

▲ Figure 3.17 Parts of a circular ripsaw saw blade (positive hook angle)

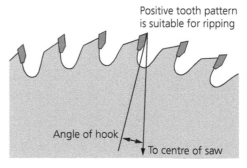

▲ Figure 3.18 Example of a TCT blade tooth

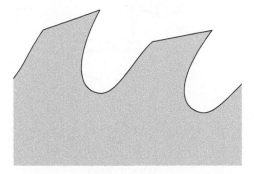

▲ Figure 3.19 Example of a plate saw tooth

The main difference between saw blades designed to rip down the grain and blades designed to cut across the grain is the **angle of hook or rake**.

This angle of hook or rake is referred to as follows.
- Positive angle of hook or rake: to be used when ripping down the grain.
- Negative angle of hook or rake: to be used for cutting across the grain.

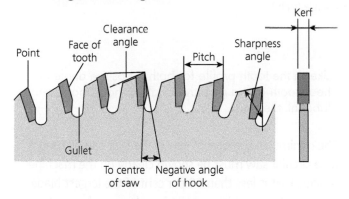

▲ Figure 3.20 Parts of a crosscut saw blade (negative hook angle)

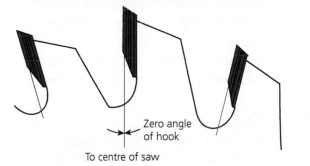

▲ Figure 3.21 General-purpose saw blade teeth

It is vitally important that you do not use a blade with a positive hook angle for crosscutting operations. The tooth design tends to grab at the material, causing the crosscut to snatch and try to run out into the timber, potentially risking injury and causing damage.

A typical ripsaw blade has a hook or rake angle of 20–25° positive for softwood, and 10–15° for cutting hardwood. In contrast, a crosscut blade will have a negative hook angle of 5° for softwood and 10° for hardwood.

Combination or general-purpose saw blades have a zero-degree angle (also known as a 'neutral angle') of hook or rake and are used on dimension sawing machines that continually change from ripping to crosscutting operations. A zero angle of hook only allows for a slow ripping operation.

All saw blades require clearance during cutting. This is to prevent friction and binding during the cutting process and is provided by the tangential side clearance with TCT blades. Traditional parallel plate saw blades achieve this clearance by applying **spring set**; this is the bending of alternating teeth in opposite directions around the saw blade.

> **KEY TERMS**
>
> **Angle of hook or rake:** the angle at which the face of the saw tooth slopes from the tooth tip, either down and forward from the tip, as in the case of negative tooth profiles for crosscutting, or down and backwards from the tooth tip, as in the case of positive tooth profiles for ripping.
>
> **Spring set:** the bending of the top third of alternating teeth in opposite directions around the saw blade.

Another difference between ripsaw and crosscut saw teeth is the way the tooth face is ground. The crosscut tooth design needs to sever the timber fibres across their length and therefore requires teeth with a more needlepoint design. Teeth belonging to the ripsaw blade have a chisel edge to them, allowing the blade to pull the fibres out as opposed to trying to cut them.

Most TCT saw blades, particularly the larger blades, have expansion slots cut into them. This is to allow for the slight expansion of the saw blade material when the cutting edges become hot during the cutting process. If the blade did not have an expansion slot, the saw blade could become distorted, and the cutting edge could start to wobble. This wobbling could mean the blade is unable to cut a true straight line, increasing the risks of saw blade sticking or causing

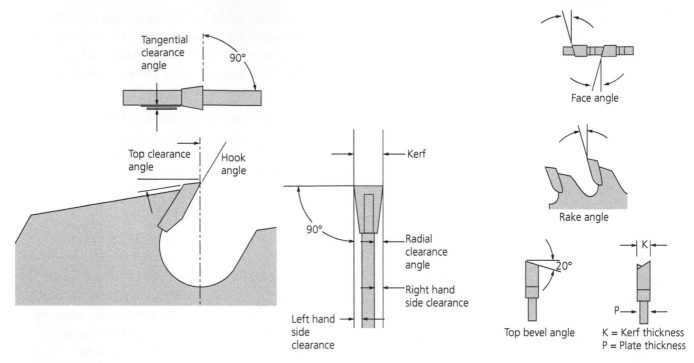

▲ Figure 3.22 Ripsaw tooth face angle (left) and crosscut saw tooth face angle (right)

kickback and increasing the risk of injury. As an added safety feature, anti-**kickback** shoulders can be built into the saw tooth design, and are now widely available for both rip and crosscut tooth profiles.

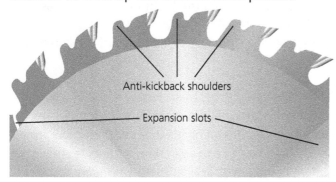

▲ Figure 3.23 Ripsaw blade with expansion slots and anti-kickback shoulders

INDUSTRY TIP

The more teeth a saw blade has, the finer the finished cut will usually be. Larger gullet areas allow for a faster feed speed of the material.

The minimum diameter of saw blade that can be used on a circular saw must be shown clearly on the machine. A blade that is less than 60 per cent of the largest blade for which the machine was designed cannot be used. This is because smaller-diameter blades produce lower **peripheral speeds**, meaning the cutting speed of the blade will not be high enough to cut the timber efficiently. To efficiently and safely cut timber, it is recommended that a peripheral speed of 50 m per second is reached.

KEY TERMS

Kickback: when the material is thrown forcefully back towards the operator, potentially causing serious injury.

Peripheral speed: on a circular saw, the speed at the periphery of the blade – how fast the outer edge of the blade is travelling.

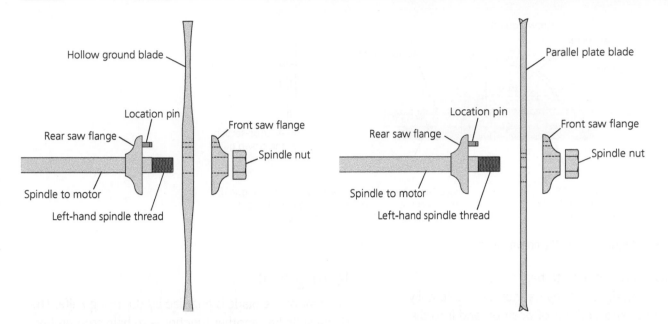

▲ Figure 3.24 Hollow ground saw blade (left) and traditional plate saw with spring set (right)

There are also other special types of blade, of which the hollow ground blade is the most common. It is typically used on crosscut and dimension saws. It gives a very fine finish and is therefore typically used for finish cutting.

Hollow ground saw blades do not have set applied as it is not needed to achieve clearance; instead, clearance is achieved by the hollow ground shape of the blade.

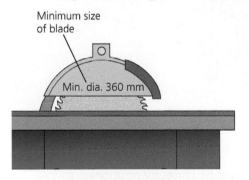

▲ Figure 3.25 Notice on machine showing minimum size to be used

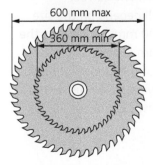

▲ Figure 3.26 The smallest blade must not be less than 60 per cent of the largest size for which the machine was designed

IMPROVE YOUR MATHS

Locate the minimum-sized saw blade label on the machine you will be using. Compare this size with the largest the machine can take. Work out the difference between the two. Does it fall within the 60 per cent rule?

ACTIVITY

Sketch the different profiles of a hollow ground saw, spring set saw and TCT saw blade, showing how the set appears in relation to the kerf.

Safe operation of circular ripsaws and dimension saws

The principles for safe operation of circular ripsaws and dimension saws are the same and can be dealt with together. Where there are differences, they will also be described in this section.

To comply with PUWER, all hand-fed circular saws must be guarded. This is achieved in three ways, as follows.

1 A guard must be fitted to the top and front of the saw blade above the table. This is known as the 'crown guard' and can have adjustable nose pieces fitted at the front.
2 Guarding from the rear is achieved using a riving knife and extension table.
3 Guarding from below is provided by the frame of the machine bed and framework.

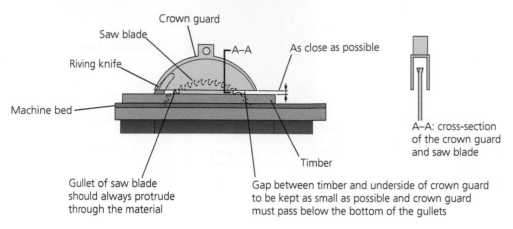

▲ Figure 3.27 Positioning of the crown guard

Saw guards are required to be:
- strong and rigid so they cannot be accidentally moved or knocked out of position and into the saw blade
- easy to adjust and correctly set in position
- set to the current requirements of PUWER and the ACoP for woodworking machinery
- checked and maintained on a regular basis.

Crown guard

The upper part of the saw blade is guarded by the crown guard. Traditionally, crown guards were made from steel and usually incorporated an extra adjustable nose section; more contemporary designs are made from rigid see-through plastic and are easily adjusted via a spring-loaded mechanism.

The crown guard should cover both sides of the blade and be set as close as possible to the top of the material being machined. The saw blade should always protrude through the material so that the lowest part of the gullet is always covered by the crown guard.

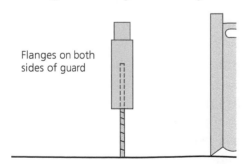

▲ Figure 3.28 Correct guard positioning in relation to the centre of the saw blade

Riving knife

The rear of the blade is guarded by the riving knife. The riving knife has another function – to help stop timber from binding on the saw blade. This could forcefully throw the timber back towards the operative, risking serious injury and damage to the machine.

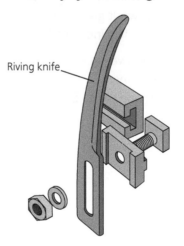

▲ Figure 3.29 Riving knife and holder

The riving knife must be correctly positioned, to comply with the ACoP, and to aid the correct and smooth operation of the machine. The correct positioning of the riving knife will depend on the size of the blade. For blades up to 600 mm in diameter, the riving knife must be within 25 mm of the top of the blade and 8 mm from the back of the blade. For blades over 600 mm in diameter the riving knife must extend above the machine bed by at least 225 mm.

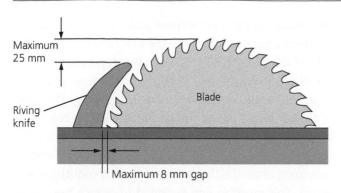

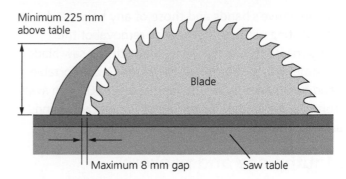

▲ Figure 3.30 Riving knife position for blades less than 600 mm in diameter (left) and for blades more than 600 mm in diameter (right)

Riving knives should have a leading edge that has a slight taper or rounded edge, and should be thinner than the kerf but approximately 10 per cent thicker than the gauge of the blade.

Extension table

When an assistant is used to remove material at the outfeed end of the saw bench, an extension table must be fitted so that the distance between the saw spindle and the end of the outfeed table measures at least 1200 mm.

It is also good practice to have support rollers or a support table if long lengths are to be cut, to provide support for the material during the cutting process.

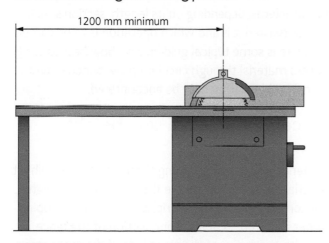

▲ Figure 3.31 Correct positioning of extension table in line with PUWER

INDUSTRY TIP

Always have a selection of wedges behind the saw blade. They can be easily and quickly inserted into the saw cut as soon as there are any signs of the material closing up behind the blade.

Push sticks

Every circular sawing machine should have at least one push stick to aid delivery and removal of components. Push sticks should be around 450 mm long with a birdsmouth shape cut into one end to aid a positive location on the end of the timber. They must be used to push the last 300 mm of material through the saw.

▲ Figure 3.32 Push sticks in use

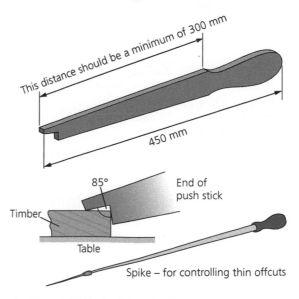

▲ Figure 3.33 Push stick and spike

The operative's hands and those of any helpers who are assisting with the delivery or removal of the components should not be in line with the saw blade or come within 300 mm of the blade. A comfortable hand hold needs to be provided at least 300 mm away from the birdsmouth to ensure the operative's hands are kept at least 300 mm away from the blade.

Mouthpiece and packers

Packers are inserted either side of large-diameter saw blades (typically those over 600 mm in diameter). These are usually made from dense felt and help to provide stability to the saw blade during cutting. They can be used as a means of lubrication to help keep the blade cool and clean, because they are impregnated with thin oil or an oil/paraffin mix.

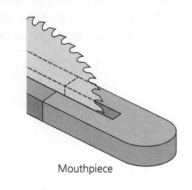

Mouthpiece

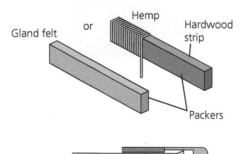

Gland felt or Hemp Hardwood strip

Packers

Timber mouth

Packer should only go as far back as the centre of the saw, but for a dimension saw the timbers should extend all the way along the bed

▲ Figure 3.34 Timber mouthpiece and side packers for circular ripsaw and dimension saw

Close the dust extraction system gates to the machine before gaining access to the saw blade. This will reduce the likelihood of the packers being lost down the extraction system.

On smaller-capacity machines and dimension saw machines, timber is fixed to the machine bed on either side of the blade. This is to keep the gap between the blade and the metal part of the machine bed as small as possible, in order to aid safe cutting. The use of timber removes the risk of damage to the blade if it develops a slight wobble (possibly due to loss of tension). Immediately in front of the saw teeth, and wrapping around either side, is a timber mouthpiece insert. This insert is also designed to reduce the gap around the teeth and the machine bed to the smallest amount possible, in order to give support to the material being cut and to reduce the chances of **spelching** or breakout on the underside.

> ### KEY TERM
>
> **Spelching:** when, during cutting on saw machinery, part of the material breaks away in an uncontrolled manner, damaging the components being cut.

Setting up and operation of circular ripsaws and dimension saws

Each piece of timber that is offered up to be cut poses its own problems, depending on its length, section size, how bent or twisted it is and which operation is to be carried out. Here is some typical guidance on how best to safely process material through circular ripsaw benches, and typical problems that may be encountered.

Fence position and saw height settings

The fence positions for ripping timber and ripping sheet materials are different, as are the required saw heights. The correct position for the ripsaw fence when ripping timber is with the curved front section of the fence in line with the radius of the saw blade. If the fence stops too far back from the saw blade then the material will not remain stable at its end and could jump about, giving a poor finish and making the push stick slip. If the fence is pushed too far forwards, the timber could become trapped and start rubbing on the saw blade, risking damage and kickbacks.

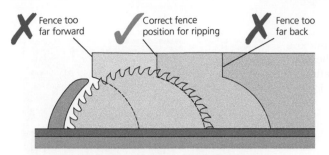

▲ Figure 3.35 Correct fence position when ripping timber

When cutting artificial sheet materials, the fence can be pushed further forward, towards the centre of the blade. This will give extra guidance for the material during and after cutting. Because these materials are much more stable than timber, there is very little chance of them bending and twisting during cutting, unlike with timber.

Dimension saws usually have a smaller capacity than ripsaws and so the blade usually has a smaller diameter than a circular ripsaw blade. During the cutting of artificial materials, the saw blade can be lowered so that it only slightly protrudes above the material being cut – but the saw teeth must still be covered by the crown guard. Keeping the saw blade lower when cutting sheet materials, particularly those with face veneers, will help produce a cleaner cut with less breakout on the underside of the material. Because manufactured boards are less likely to distort than natural timber, there is less likelihood of kickback even though the material is being cut near the top of the saw blade; with natural timber, the kickback risks would be greatly increased.

To help reduce breakout further, a running board can be fitted to the machine bed. This can be a thin piece of plywood or similar material that is part-passed through the saw and fixed in position with a cleat at the infeed end to prevent any further forward movement. The running board is then fixed at the outfeed end onto an extension table. This running board helps to reduce breakout of the material by keeping the gap between the saw kerf and the machine bed (in this case the running board) as small as possible.

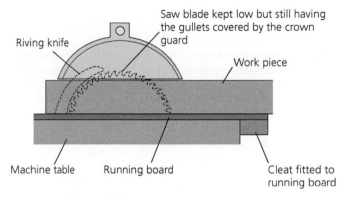

▲ Figure 3.36 Running board position

Changing circular saw blades

The sequence for changing circular saw blades on ripsaws and dimension saws is similar for all types of design (see Table 3.2). However, you should always refer to the manufacturer's booklet for information on any specific requirements for that machine.

▼ Table 3.2 Changing circular saw blades: step by step

Step 1	Isolate the machine and close the dust extraction gates. If any packers fall into the extraction ducting, this will prevent them from being drawn away and requiring replacement.	 ▲ Figure 3.37 Isolate the machine

→

▼ Table 3.2 Changing circular saw blades: step by step (continued)

Step 2	Raise or remove the crown guard to expose the saw blade. Remove the finger plate and any timber packers to allow access to the saw blade locking nut. Remove the saw blade from the spindle after releasing the saw nut. For most standard machines, tighten in an anti-clockwise direction and loosen in a clockwise direction.	 ▲ Figure 3.38 Remove the blade using suitable hand protection
Step 3	Remove the riving knife. This step may not be needed if the riving knife is clean and the correct size and shape to suit the replacement blade.	 ▲ Figure 3.39 Remove the riving knife
Step 4	Select the correct size and type of blade, e.g. positive TCT and tooth design. Fit the replacement saw blade, ensuring the teeth are facing in the correct direction. Make sure the blade is fully tightened using the correct spanners and collars for the machine; the tightening direction will be away from the saw blade's cutting rotation direction (i.e. for most standard machines tighten in an anti-clockwise direction and loosen in a clockwise direction).	 ▲ Figure 3.40 Refit replacement blade ensuring the correct hook or rake angle is chosen
Step 5	Reposition the riving knife, ensuring its height and the gap to the saw blade are correct.	 ▲ Figure 3.41 Refit correct riving knife and position to comply with PUWER

▼ Table 3.2 Changing circular saw blades: step by step (continued)

Step 6	Refit the finger plate, ensuring it is flush with the machine bed. Refit timber packers, ensuring a good close fit to the size of the saw blade to help reduce spelching on the underside of the material. Reset the fence position to line up with the saw gullets at table height for cutting timber.	 ▲ Figure 3.42 Set fence to correct position
Step 7	Set all guards, including the extension table; ensure the extension table is at least 1200 mm long from the centre of the saw blade.	 ▲ Figure 3.43 Refit and position all guards and extraction in line with PUWER
Step 8	Ensure a suitable push stick is available that is at least 450 mm long. Conduct a trial run, listening for any unusual sounds or vibrations. Carry out a test cut.	 ▲ Figure 3.44 Conduct a test cut

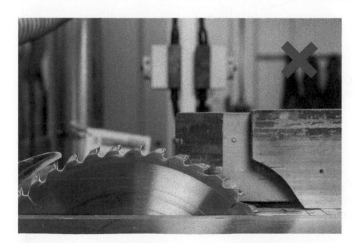

▲ Figure 3.45 Incorrect ripping of timber – set the fence as in Figure 3.42

INDUSTRY TIP

When releasing saw blade nuts, rotate the spanner in the direction of the saw rotation while keeping the saw blade stationary. To tighten the saw blade nut, keep the blade stationary while rotating the spanner in the opposite direction to the saw's normal cutting direction.

Standard operating position

While cutting material, do not stand directly in line with the saw cut but slightly off to the side. This reduces the risk that any timber that is kicked back will be forcefully propelled at you, resulting in possible injury.

▲ Figure 3.46 Stand slightly to one side when using a circular saw bench

Cutting material

The way timber is presented and fed through the saw bench is vital to its safe operation. Timber is subject to movement as a result of seasoning or atmospheric changes resulting from storage conditions. This movement can take several forms, the most common being:

- cupping – where the timber 'cups' over the face of the board; wide boards that are cut tangentially will usually cup over time
- bowing – a curvature along the board's face (its widest section) from one end to the other
- springing – a curvature along the board's edge from one end to the other
- twisting – a curvature along both edges of the board's length, producing a propeller-shaped twist.

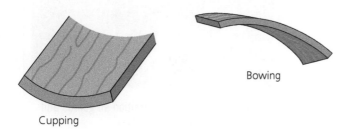
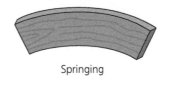
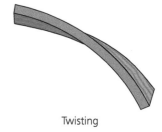

Cupping

Bowing

Springing

Twisting

▲ Figure 3.47 Examples of timber defects in relation to feeding through the circular saw bench

Make sure you position your timber as shown in Table 3.3.

▼ Table 3.3 Correct timber positioning

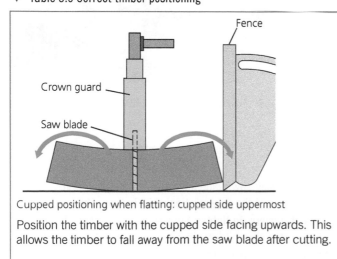

Cupped positioning when flatting: cupped side uppermost	Cupped positioning when deeping
Position the timber with the cupped side facing upwards. This allows the timber to fall away from the saw blade after cutting.	Position the timber with its cupped side against the fence. This allows for safer feeding through the machine and allows the timber to fall away from the saw blade.

▼ Table 3.3 Correct timber positioning (continued)

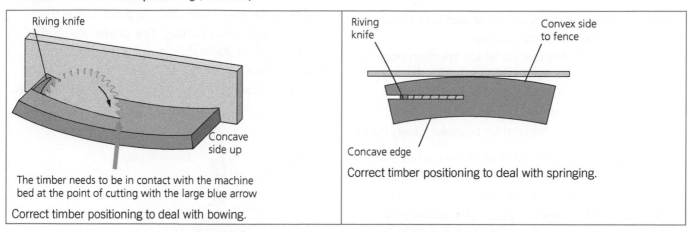

Riving knife

Concave side up

The timber needs to be in contact with the machine bed at the point of cutting with the large blue arrow

Correct timber positioning to deal with bowing.

Riving knife

Convex side to fence

Concave edge

Correct timber positioning to deal with springing.

Machining twisted timber can be problematic, so use your best judgment when following the instructions in Table 3.3. In the case of excessive twisting or doubt, do not use the material.

During the cutting operation timber can, from time to time, close up on the saw blade, causing friction; if this happens, it may be forcefully pushed back at the operator. This is usually a result of **case hardening** from poor seasoning. To help prevent this effect, drive a wedge into the saw cut behind the riving knife, to stop the timber closing in any further.

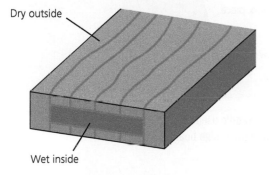

Dry outside

Wet inside

▲ Figure 3.48 Case hardening

ACTIVITY

Research the moisture content for different types of timber and their intended uses, then present the information in an easily understood format. Present your findings to your peer group and discuss the issues that may occur if the moisture content is incorrect.

Describe the main considerations when selecting tooling for cutting down the grain of timbers such as afrormosia on circular saw benches. Analyse and present your findings to your tutor.

KEY TERM

Case hardening: a defect caused by the timber being dried too rapidly, leaving the outside dry but the centre still wet. It typically causes the material to bend and twist during cutting, resulting in binding on the saw blade and kickback.

Jigs and saddles used during cutting operations at circular saws

Circular saw benches can be used for more than the straight cutting of timber and manufactured materials. For materials such as plywood and MDF, the dimension saw is the best type of saw bench, because of its versatility in being able to use the sliding crosscut table and fence, its accuracy, its ability to produce compound cuts and, in general, its finer sawn finish. These reasons make this the machine of choice for fine, accurate working and, in particular, when cutting manufactured materials.

Most types of circular ripsaw do not allow the blade to tilt or cant, but the dimension saw does, which allows the production of bevelled cuts. With both types of machine, angled or bevelled cuts can be produced with the aid of jigs and saddles.

PUWER requires that the most suitable and safest production method should be used, to reduce the risk of injury. Correctly made and positioned jigs and saddles offer the safest way to produce angled components on circular sawing machinery. They also enable consistent accuracy and continued safe operation of the machine.

The following are typical examples of jigs and saddles that can be used in manufacturing joinery items.

Saddles

Saddles are used for quick, accurate and safe cutting of angled or bevelled timbers, such as:

- long lengths of glue blocks, which are then crosscut to the required length
- arris rails used with fencing
- feather-edge boarding using in cladding
- any angled lengths required for bespoke joinery items.

The saddle is constructed so that its base or running board sits on the machine bed and passes through the saw and onto the outfeed table, where it can be fixed in place. The angled support pieces that hold and support the timber during cutting should be long enough to fully support the maximum length of timber likely to be cut. This ensures that the material being cut does not lift off the saddle at the infeed side of the saw cut during cutting, when the outfeed end of the material outweighs the infeed part. The use of saddles also provides a consistent cutting angle for repeat jobs.

▲ Figure 3.49 Saddle in use to produce long lengths of glue blocks for staircases

The timber to be cut sits on the saddle, which holds and controls the timber safely in place and at the required angle while cutting. The saddle design also supports all cut items during and after the cutting process, reducing the risk of any kickback. This is not only a vital safety feature but also adheres to PUWER.

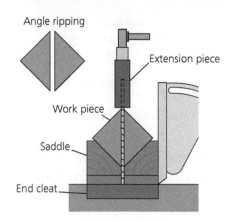

▲ Figure 3.50 Saddle for 45° angled ripping of glue blocks

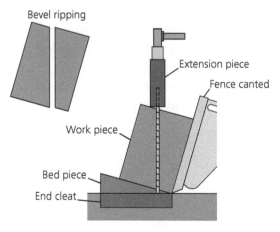

▲ Figure 3.51 Saddle used for bevelled ripping, such as to produce a feather edge board

Jigs

Jigs enable the safe, accurate and fast cutting of identical square or tapered components such as wedges and tapered firring strips. All jigs should include a suitable hand hold unless the cutting process allows the operator's hands to be placed at least 300 mm away from the cutting process, and the use of push sticks for the last 300 mm of the cut. In most cases, hand holds will be required and should be positioned in a convenient location that is not directly in line with the cut or within 300 mm of the cutting process or finishing position. Hand holds should be constructed of good-quality material, avoiding any short grain as far as possible.

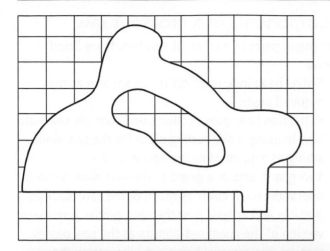

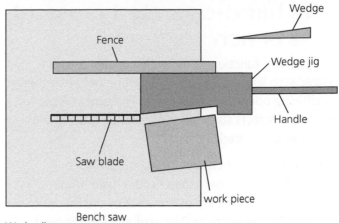

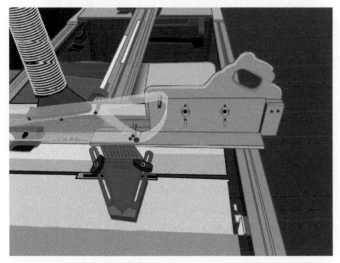

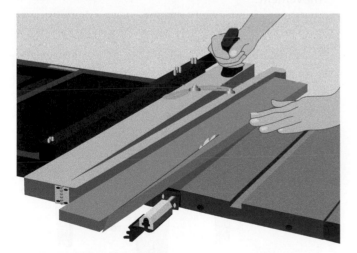

▲ Figure 3.52 Example hand hold that could be incorporated into a jig or push block (top); handle built into a push block (bottom)

At least 300 mm clearance between hands and saw cut

▲ Figure 3.54 Wedge jig (top) and jig in use (bottom)

▲ Figure 3.55 Tapering jig in use

▲ Figure 3.53 Example of jig used to produce long wedges or firring strips on a circular saw bench (note: crown guard removed for clarity)

119

The hand-operated crosscut saw bench

The primary function of the crosscut saw is to cut material to length. This is usually done before further machining operations take place. The timber is cut to minimise defects such as knots, shakes, bowing and twists in the material.

Two main types of crosscut machine are used:
1   The travelling head crosscut: the head of the saw and the arm are at the back of the machine, requiring more room. The arm and cutting head travel out towards the operator.
2   The radial arm crosscut: the arm on which the cutting head travels is positioned over the table at the front. This saves a lot of space and makes it easier to carry out some operations.

ACTIVITY

Go to the HSE website and download wood information sheet number 36, 'Safe use of manually operated crosscut machines', and display it next to your machine. Use the information from the sheet to create a ten-question safety quiz that can be used as a knowledge checker before use.

All crosscut saws must be guarded in three ways:
1   Guards are fitted around the back, top and front of the saw, often with an adjustable frontal extension piece.
2   There is no access to the saw blade when it is in the rest (behind the fence) position; on older machines, the saw returns automatically into a saw housing.
3   Guards are fitted below the machine table, usually formed by the main frame of the machine table.

All guards must be:
●   strong and rigid so they cannot be accidentally moved or knocked out of position and into the saw blade
●   easy to adjust and correctly set in position
●   set to the current requirements of PUWER and the ACoP for woodworking machinery
●   checked and maintained regularly.

Components of a crosscut saw

The major parts of a crosscut saw bench are listed below.
●   Swivel head lock: locks off the radial arm at the required angle.
●   Extraction hose point: collects and controls sawdust.
●   Saw housing: a safe resting place for the saw blade, acting as a guard while the saw is at rest.
●   Saw guard: acts as a guard to the saw blade while in operation; must cover as much of the saw blade as possible and at least past the saw spindle. A modern version of the guard self-adjusts as the saw passes through the material being cut, eliminating the need to set the guard height.
●   Saw blade: negative tooth patterned blade for cutting across the grain.
●   Start/stop switch: starts and stops the machine; must be within easy reach of the normal operating position.
●   Prohibition area: the ACoP requires an area to be marked on the machine bed measuring 300 mm each side of the saw's travel. This is a 'no hands' zone – while cutting material, your hands should be kept out of it. For short lengths of timber, a holding device is to be used.
●   Emergency stop: allows you to stop the saw with your knee in an emergency, while your hands are holding the material and controlling the saw movement.

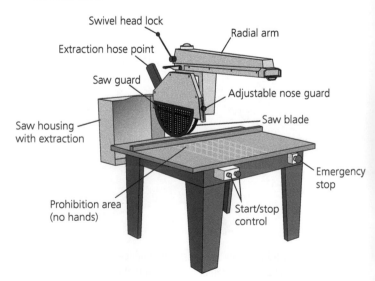

▲ Figure 3.56 Major component parts of a radial arm crosscut bench

- Adjustable nose guard: reduces the gap between the guard and the timber to be as small as possible.
- Radial arm: enables the saw to rotate to the left or right to make a cut from 45–90° in either direction.
- Back fence: prevents material moving into the blade and is of sufficient height to support the material being cut. The gap for the saw cut should be kept as small as possible.
- Machine bed: a 'sacrificial' bed should be incorporated into the machine, enabling the saw teeth to pass fully through the material being cut, reducing spelching on the underside.
- Machine end stops: determine the cut length of the material. These are either the standard lift-up end stops or spring-loaded end stops. In the case of spring-loaded end stops, the end stop becomes flush with the fence when the material is pressed against the stop.
- Self-return system for the cutting head: for crosscut saws that have a traditional guard with an adjustable nosepiece, PUWER requires the fitting of a self-return system for the saw head. This automatically returns the saw head to its resting place, reducing any risk of contact with the saw blade. It is good practice to have a spring-loaded return system fitted to the saw head, which is fitted with a self-adjusting guard – although not a legal requirement, this means that increased safety measures have been considered and acted upon.

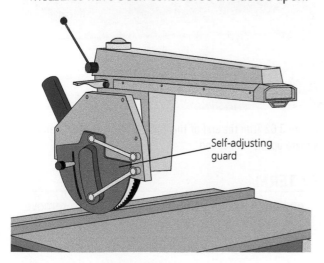

▲ Figure 3.57 Self-adjusting guard

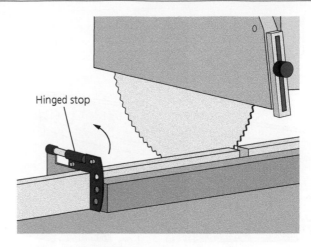

▲ Figure 3.58 Machine end stops

Material location and dealing with distorted material

The correct way to deal with timber that is bowed, cupped, sprung or twisted is to ensure it is positioned with the **convex** side down and to the fence. This ensures stability during cutting, which allows the rotational thrust of the saw blade to keep the material down on the machine bed and against the fence during the cutting operation. As an added safety feature, the outer ends of the material fall away from the cutting edges of the blade, reducing any risks of binding and kickback.

KEY TERM

Convex: in timber, a face that bulges out in the middle.

Where short lengths of material require cutting it may be preferable to cut these in multiple lengths, e.g. if the required finished length is 200 mm it may be preferable to cut to 410 mm and then trim to the finished length on a dimension saw or crosscut saw after other machining operations have been completed. Where this is not possible, short lengths can be held safely using a pivoted handle fixed to the back fence. Alternatively, a push stick can be used.

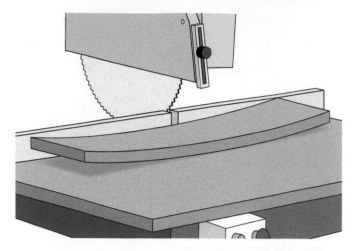

▲ Figure 3.59 Bowed board cut with convex side facing down: the ends of the board fall towards the machine bed, opening up the saw cut after cutting; this reduces the risks of the material binding and prevents kickback after cutting while the saw blade travels back into its guarded resting place

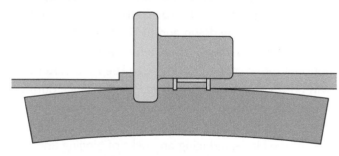

▲ Figure 3.60 Sprung boards: outer ends can move back towards the fence, resulting in an opening of the saw cut after cutting, reducing the risks of snatching and kickback

Push stick attached to hinge on back fence

▲ Figure 3.61 Using a holding device while cutting short lengths on a radial arm crosscut saw

Safe procedure for crosscutting material

As well as correctly offering up the material to the crosscut, the operator needs to consider how to position themselves and how to operate the crosscut. Here is an example of a safe crosscutting procedure.

1   Bring up the material to square up and trim the end of any defects. This is called **fair ending**. The correct stance and hand positioning are vital to the safe operation of the crosscut. Do not cross your arms when using the machine. Instead, keep your arms parallel to the machine, and ensure the hand not pulling the crosscut is not resting on the material within the marked danger zone, 300 mm each side of the saw cutting line. This greatly reduces the chance of accidental contact with the saw blade.

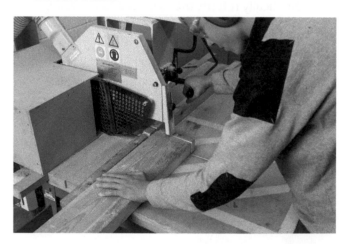

▲ Figure 3.62 Trim the end of the material of any defects and square up the timber end

KEY TERM

Fair ending: trimming the end of a board to remove defects such as splits and to square it up.

2 Pull out the cutting head at a steady, controlled rate and do not allow the cutting head to rush out towards you. If the blade begins to slow down during the cut, stop and pull it towards you but push slightly away, allowing the saw's speed to normalise. If the blade becomes jammed in the cut and stalls, stop the machine straight away and then carefully return the cutting head to its home position. Set any end stops to the required lengths as written on the cutting list.

3 If the blade constantly becomes jammed or stalls, either the blade has become blunt or the incorrect tooth design is being used. In both cases, you need to change the blade for a sharp blade of the correct design.

Push the trimmed length up to the required length position. If using an end stop, ensure you do not trap any debris between the stop and the material, and do not hit the end stop too hard or you risk moving the stop.

▲ Figure 3.63 Setting the end stop for the required length of cut

ACTIVITY

Design and manufacture a suitable clamping system to comply with the woodworking machinery ACoP, for holding short lengths of timber in place while they are being crosscut to length on the crosscut machine you will use.

Using the narrow bandsaw

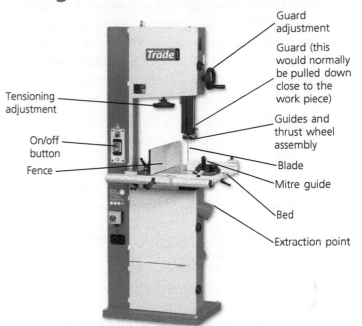

▲ Figure 3.64 Major parts of a narrow bandsaw

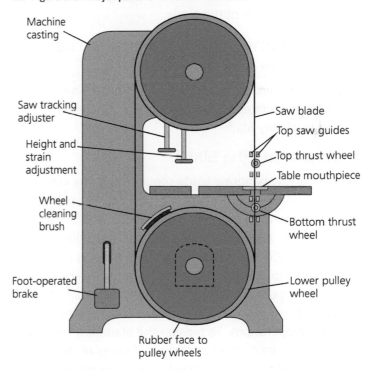

▲ Figure 3.65 Guards removed showing internal components of a narrow bandsaw

The narrow bandsaw is primarily designed for cutting curved shapes and joints such as tenons. It can cut down the grain and across the grain, and cut manufactured materials such as plywood and MDF. Narrow bandsaws are defined as 'narrow' if the blade is less than 50 mm wide; 'narrow' does not refer to the size of the **throat** or the maximum cutting capacity. All narrow bandsaws cut with a continuous band of teeth running around two pulley wheels and cutting in a vertical straight line, rather than a circular band of teeth as with the ripsaw and crosscut saw blades.

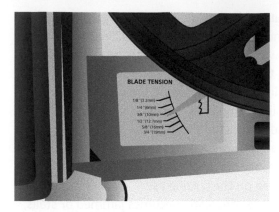

▲ Figure 3.66 Narrow bandsaw tension gauge

> ## ACTIVITY
>
> Go to the HSE website and download wood information sheet number 31, 'Safety in the use of narrow bandsaws', and display it next to your narrow bandsaw. Use it to create a multiple-choice safety quiz that can be used as a knowledge checker before use.

The bottom pulley wheel of the narrow bandsaw is driven by the machine motor, while the top wheel is driven by the blade. Both wheels are rubber coated to help prevent the blade from slipping and damaging the cutting edge of the blade.

Tensioning and tracking the narrow bandsaw blade

Always ensure that the power to the machine has been isolated before attempting to tension and track any blade.

Tensioning the blade

The narrow bandsaw blade should run in the centre of the pulley when correctly tensioned. The purpose of tensioning or straining a blade is to keep the blade running straight and in the centre of the pulley. If too little tension is applied to the blade, there is a danger it could slip during cutting operations or even fall from the pulleys. If too much tension is applied to the blade, there is an increased risk of it snapping during cutting operations. To tension the blade, the top pulley wheel is raised via a hand wheel. To ensure the correct amount of tension is applied, all machines have a tensioning gauge fitted with pre-marked positions for different widths of blade. The marked positions help the operator to apply the correct amount of tension each time the blade is used.

Tracking the blade

To allow for different-width blades and to ensure the blade runs in the centre of the top pulley wheel, the top

pulley wheel can be tilted slightly (**tracking**). Tilting or canting the top wheel slightly makes the blade run up to the highest part of the pulley wheel and it will stop there if it is correctly tensioned. When tracking the blade, the thrust wheels and guide assemblies should be moved clear to allow the blade to move freely on the pulley wheels, both backwards and forwards, until the desired position is achieved.

Guide wheel assembly

With the blade correctly tensioned and tracked, the saw guide assemblies can be positioned. The saw guides consist of side guides on each side of the blade and a thrust wheel or thrust bearing at the back of the blade.

The way the guide assembly is formatted or arranged and adjusted will depend on the type of narrow bandsaw. Most guides and thrust wheels are made from hardened steel roller bearings, which run more smoothly and last longer. The positioning of the guides and thrust wheel will be the same for all makes and models.

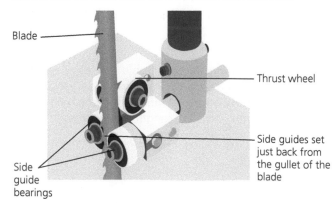

Blade

Thrust wheel

Side guides set just back from the gullet of the blade

Side guide bearings

▲ Figure 3.67 Guide assembly with side guides and thrust wheel

The correct positioning of the side guides and **thrust wheel** are as follows.

The side guides should be set:
- as close as possible to the blade without touching or distorting the blade in any way
- with the front edge of the guide set just back from the bottom of the gullets.

The thrust wheel should be set:
- so that there is a gap between the thrust wheel and the back of the blade of 1–2 mm.

Narrow bandsaw blades

Narrow bandsaw blades are a continuous length of steel blade that has been cut and welded to the required length for the type and model of narrow bandsaw being used. The teeth cut at an inclined triangular angle, and are set and sharpened ready for cutting. Blades are made from carbon steel, which requires regular sharpening and setting, usually by a specialist saw doctor (someone who maintains machine tooling such as narrow bandsaw blades).

Narrow bandsaw blades can be heat treated on their cutting edge – these types of blade are called hard point blades. The heat treatment preserves the cutting life of the blade, but when they finally lose their cutting edge they become scrap as they cannot easily be sharpened. Hard point blades are particularly good for cutting hardwoods, chipboards or MDF because the cutting edge is more durable than standard steel blades.

Various designs of blade cut different types of material. Some have been hardened and are discarded when they become blunt; these are particularly useful for cutting hard, abrasive materials.

Selecting the correct size and type of saw blade is vital for the safe operation of the machine and safe, accurate cutting of the material. The smaller the width of the blade, the smaller the radii that can be cut. As a rule, larger-width blades are more suitable for cutting straight lines using a fence, such as when forming tenons and using a power feed unit.

The pitch of the teeth is another vital consideration when selecting a suitable blade. The tooth pitch must always be smaller than the thickness of the material being cut. As a general rule, coarse teeth or a blade with a large pitch are used for softwoods and fine teeth or a small pitch are used for hardwoods.

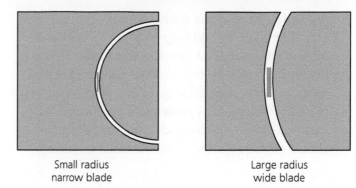

Small radius narrow blade

Large radius wide blade

▲ Figure 3.68 Use a small-width blade for tight, small-radius curves and a larger-width blade for large-radius curves

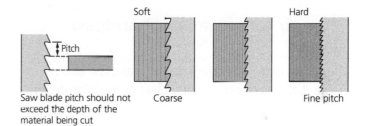

Soft Hard

Pitch

Saw blade pitch should not exceed the depth of the material being cut

Coarse

Fine pitch

▲ Figure 3.69 Saw blade pitch

Guarding narrow bandsaws

The narrow bandsaw is guarded in two main ways:
1 The blade is fully encased in guards, apart from the section that is doing the cutting. The top and bottom pulley wheels have **interlocked doors** that prevent access to the blade while it is moving, and stop the machine starting if the doors are open. The top pulley door also locks into place the side access guard covering access to the side of the framework, which allows the blade to be removed and fitted.

> **KEY TERMS**
>
> **Throat:** the horizontal distance from the cutting edge of the blade to the machine casing.
>
> **Tracking:** tilting the top pulley wheel of a narrow bandsaw to make the blade run in the centre of the pulley.
>
> **Thrust wheel:** the small wheel behind the blade of a narrow bandsaw, which prevents the blade from being pushed backwards during cutting operations.
>
> **Interlocked doors:** on a narrow bandsaw, doors that are impossible to open while the machine is in motion and which prevent the machine from starting if the doors are open.

2 A guard is fitted to the guide assembly rise and fall arm. This guard should fully enclose the front and sides of the blade. The operator must adjust this guard as required by the ACoP to suit the type of operation. This adjustable guard needs to be set as close to the top of the material as practicable. As with adjustment of the guide assembly, the guard should not be adjusted while the blade is moving.

ACTIVITY

Draw and label the correct positioning of the saw guides and thrust wheel in relation to the saw blade.

Changing the narrow bandsaw blade

The sequence for changing narrow bandsaw blades (Table 3.4) is very similar regardless of machine design.

However, you should always refer to the manufacturer's booklet for information about any requirements for that specific machine.

Whenever the blade is changed, always check that the **timber mouthpiece** is in good condition, with as small a gap as possible around the blade. If the timber mouthpiece becomes damaged and worn, you must replace it. This ensures bits of cut material cannot become trapped between the blade and the mouthpiece, which would risk breaking the blade, potentially injuring the operator and damaging the machine.

KEY TERM

Timber mouthpiece: used to prevent material becoming trapped between the machine bed and the blade; this should be replaced as often as required.

▼ Table 3.4 The sequence for changing narrow bandsaw blades

1 Isolate the machine from the power supply and remove the keys.

2 Open the interlocked door guards and remove the adjustable front guard, exposing the saw blade.

3 Move the saw guide assemblies back away from the saw blade.

4 Loosen the tension on the saw blade by lowering the top pulley wheel.

▼ Table 3.4 The sequence for changing narrow bandsaw blades (continued)

5 Fold the saw blade and store in a safe place.

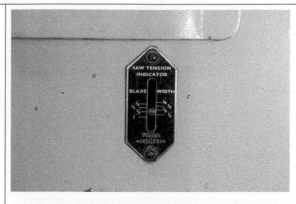

6 Clean down the saw guides and ensure all parts move smoothly. Fit the new saw blade and re-tension the blade, ensuring it is tensioned correctly.

7 Track the saw blade and ensure it is running in the centre of the top pulley.

8 Adjust the guide assemblies so the guides are just back from the bottom of the gullets and set as close to the blade as possible without making contact.

9 Adjust the thrust wheel so it is just clearing the back of the blade by about 1–2 mm. Repeat this process for the lower guide assembly.

10 Re-fit the timber mouthpiece, ensuring it is as close as possible to the side and the back of the blade to reduce the risk of thin pieces of cut material becoming trapped against the saw blade and timber mouthpiece. Set all guards and close both doors, ensuring the interlocks are located correctly. Conduct a trial run, listening for any unusual sounds or vibrations. Carry out a test cut.

Folding a narrow bandsaw blade

There are several ways to fold a narrow bandsaw blade (see step 5 in Table 3.4). The steps presented in Table 3.5 outline one simple and safe way. This method not only keeps the blade away from the face but also keeps it under a greater degree of control than other methods.

▼ Table 3.5 Folding a narrow bandsaw blade: step by step

Step 1	Remove the blade using suitable protective gloves.
Step 2	Hold the blade approximately one third of the way down, with the teeth facing away from you. Your thumbs will need to be on the outside of the blade, with one of your feet holding the blade down on the floor; this prevents unwanted movement from the lower part of the blade.
Step 3	Keeping your elbows still, rotate your wrists so your thumb now faces you and your hands have moved from the vertical position to the horizontal position; the top part of the blade will loop over away from you. Next, keeping your elbows and arms still, rotate your wrists outwards; this will try to bring the top part of the blade back towards you. While the blade is trying to come back towards you, push down with your arm and the blade will form into a circle on the floor.
Step 4	Push the blade down onto the floor after folding, forming a small band which can be transported safely.

Practise the steps first without a saw blade until you are familiar with the required sequence.

Using the narrow bandsaw safely

The narrow bandsaw is an extremely versatile machine. In general terms, though, the type of cutting done is:

- straight cutting with a fence and power feed
- straight cutting using a fence with push block
- cutting tenons
- curved or freehand cutting.

Straight cutting using the fence and power feed

It is best to cut straight long lengths of timber using a wide blade and fence. This not only helps to keep the cutting line parallel to the edge of the material but also helps to stop the blade from wandering in the cut. The extra strength of a wide blade will also help to prevent any blade breakages. When using the narrow bandsaw to rip long lengths of material, a **demountable power feed unit** should be used; this helps feed the material in a safe, controlled manner, keeping the operator's hands away from

the blade, reducing risks and following the advice given in the ACoP.

> **KEY TERM**
>
> **Demountable power feed unit:** an automatic feed system that can be moved out of the way. It is used mainly for continuous feeding of material during cutting or profiling operations.

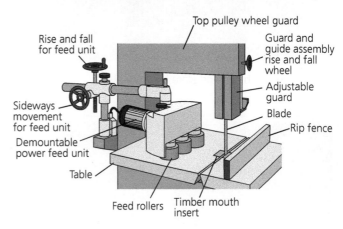

▲ Figure 3.70 Narrow bandsaw set up with a demountable power feed unit for ripping long lengths of timber

Straight cutting with fence and push block

When there is no demountable power feed unit available, or when ripping short lengths of material against the fence, a guide block should be used. The guide block helps to keep the material up against the fence at the point of cutting, without the need to place your hand close to the blade. This method of cutting on the narrow bandsaw reduces the risk of your hand coming into contact with the blade, helping to comply with the ACoP as well as improving the quality of the finished component.

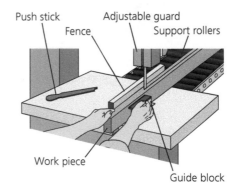

▲ Figure 3.71 Straight cutting on narrow bandsaw using a guide block

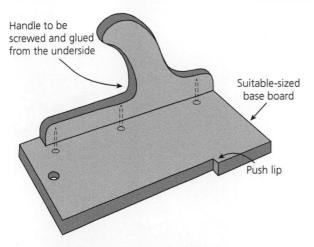

Handle to be screwed and glued from the underside

Suitable-sized base board

Push lip

Cutting tenons

The narrow bandsaw is useful for producing tenon and open mortice joints. With a combination of the fence, push block and end stop, most straight shouldered tenon joints can be produced.

▲ Figure 3.72 Example push block used when cutting short lengths

▼ Table 3.6 Sequence for cutting tenons on a narrow bandsaw

1	Set the fence and end stop to the required locations and cut down the length of the tenon on one side. Readjust the fence position when all components have been cut, and cut down the other side of the tenon length.	
2	Re-set the fence and end stop so the tenon cheeks can be formed. Using a crosscut fence for the bandsaw or using a push block worked from the fence, cut the tenon cheeks.	
3	Finished tenon.	

→

▼ Table 3.6 Sequence for cutting tenons on a narrow bandsaw (continued)

4 Set fence and end stop, and form haunches.

Curved or freehand cutting

Often, it is not practicable to use a fence – for example, when cutting irregular shapes or curves. In these situations, the material should be fed through the cutting process with a steady and even forward motion, without applying any undue pressure to the blade. Hands should be kept out of the line of the cutting path and as far away from the blade as practicable.

Where possible, use a jig and guide when carrying out repetitive work. The guide is fixed in front of the blade and runs against the jig or **templet**; it is set so the finished product is 2–3 mm bigger than the finish cutting position. This extra 2–3 mm is to allow for final finishing to the required size and profile on a spindle moulder or router. Using jigs improves the overall safety, productivity and accuracy of the operation.

KEY TERM

Templet: also known as a 'template', this is a thin piece of hardboard, MDF or plywood that is cut to the shape required and is then used to help reproduce the shape.

INDUSTRY TIP

You can safely cut wedges using a jig and fence. This helps to increase the production and continuity of wedges.

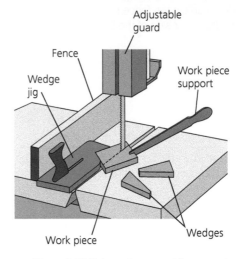

▲ Figure 3.73 Using a fence and jig to produce wedges

When cutting complex shapes, choose your starting and stopping points with care. Generally, you do not want to be pulling the material out of the cut. Sometimes, however, this will be unavoidable and extra care should be taken in this case: there is a risk that if you do not follow your original saw path you will pull the blade from the pulley wheels or cause the blade to break. Always cut the shortest length first; this will ensure the risks are reduced to the minimum.

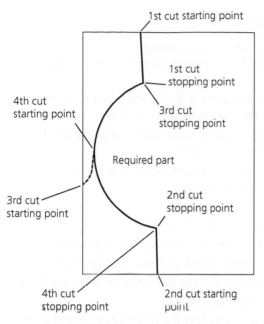

▲ Figure 3.74 Cutting stages for shapes having tight internal corners

Always ensure that the material being cut is sufficiently supported both at the infeed end and also at the back of the bandsaw. If no fixed extension table is available, use mobile support rollers.

INDUSTRY TIP

Always have a push stick or spike available at a narrow bandsaw to remove any short lengths and trimmings from the danger area.

ACTIVITY

Create a templet for a push stick on suitable material such as 18 mm MDF. Using a suitably sized blade, cut out the push stick for your own use.

4 USING PLANING MACHINES

There are three main types of woodworking machine that plane timber:
1 surface planer
2 thicknesser
3 combination planer.

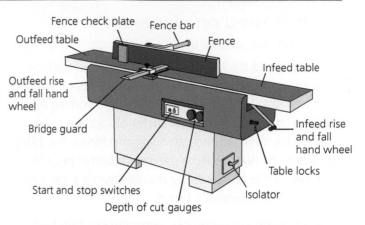

▲ Figure 3.75 Typical example of hand-fed surface planer and components

▲ Figure 3.76 Typical example of a thicknesser

▲ Figure 3.77 Typical example of a combination planer/thicknesser

Tooling used on planers and thicknessers

All planers and thicknessers make the timber surface smooth, flat and planed, but the three types of machine do it in slightly different ways. The woodworking ACoP requires that only cylindrical cutter blocks are to be used on hand-fed surface planning machines. Pre-1995 models must not exceed a maximum cutter projection of 3 mm, while for machines manufactured after 1995 this should not exceed 1.1 mm.

These circular cutter blocks normally contain two cutting **knives**, although they can have more. The cutters revolve at speeds up to 6000 **revolutions per minute (rpm)**, which is 12,000 cuts per minute for a two-knife cutter block. These circular cuts leave ripples on the finished surface called **pitch marks**; the sizes of these pitch marks determine the quality of finish. The faster the timber is fed across the cutters, the larger the pitch marks will be, resulting in a poorer-quality finish than if the material is fed across the cutter knives more slowly. A slower feed produces pitch marks that are closer together, giving the feel and appearance of the surface being totally smooth and flat. A better quality of finish at this stage, with small pitch marks, saves considerable time later in the sanding and finishing process.

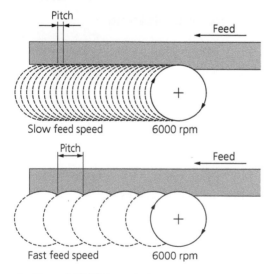

▲ Figure 3.78 Pitch mark sizes

For general joinery, a pitch mark length of 2–2.5 mm is acceptable; for better classes of work and cabinet work, a pitch mark length of 1 mm will be required. To work out the length of the pitch marks, three pieces of information are required:

1 feed speed – how fast the material is passing over the cutters
2 number of cutters – the number of cutters in the cutter block, typically two, three or four
3 how fast the **cutter block** is travelling – usually 4500 or 6000 rpm.

Using the following formula, the pitch of cutter marks can be determined:

$$P = \frac{1000F}{N \times R}$$

where F = feed speed

N = number of cutters cutting

R = speed of block (rpm)

P = pitch of cutter marks

Example

Assuming a two-knife planer block runs at 4500 rpm with a feed speed of 15 m per minute, the pitch of cutter marks would be:

$$\frac{1000 \times 15}{2 \times 4500} = 1.66 \text{ mm}$$

Therefore, the pitch of cutter marks would be 1.66 mm, suitable for general joinery products.

ACTIVITY

Calculate the pitch of cutter marks for a surface planer that has four cutters cutting a feed speed of 20 m per minute and a block revolving at 6000 rpm.

Replacing the tooling in surface planers and thicknessers

All cutter knives should be kept sharp, and should be replaced when they become blunt. To help keep a sharp cutting edge and reduce the number of times the knives need changing, they can be honed in the machine. The honing process involves passing an oilstone or diamond slip stone over the cutting edge. As with the replacement of the cutting knives, always ensure the isolator has been turned and locked off, to eliminate any possibility of accidental start-up of the machine.

The type and age of the machine will affect the type of cutters it uses and how they are changed. There are two basic methods used for changing the cutter knives.

Positive location

This is the simplest method and involves the use of disposable tooling, with each cutter in the set (usually two) able to be turned over as one edge becomes dull. This method requires little in the way of specialist tooling or skill to replace the knives. This type of tooling is used on newer machinery and limits the cutter knife projection to a maximum of 1.1 mm, complying with the ACoP. The cutters are held in place either by a wedge bar that tightens up onto the cutters, thereby holding them in place, or by **centrifugal force**. When released, the cutters are either lifted out or slid out of the side of the block.

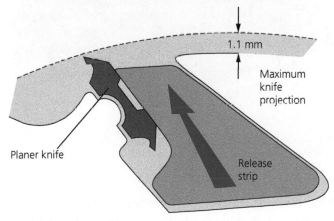

Centrifugal force tightens the tapered release strip into the slot and wedges the planer knife in position

▲ Figure 3.79 Centrifugal force cutter knife holding system

KEY TERM

Centrifugal force: a force created by a rotating body, such as by a rotating cutter block, where objects are forced away from the centre of rotation.

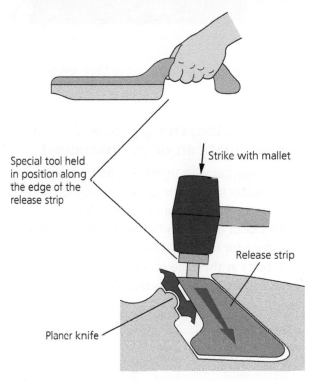

Special tool held in position along the edge of the release strip

Strike with mallet

Release strip

Planer knife

▲ Figure 3.80 Loosening the release strip for knife removal

Adjustable-position cutter knives

This type of tooling allows for the cutter knives to be re-ground after they have become blunt, rather than be thrown away. Because of the reduction in size resulting from re-grinding, the cutting circle position changes each time the cutters are changed. As a result, a great deal more skill and time is required to correctly set this type of tooling.

The cutter knives are normally held in place by a wedge bar that tightens up against the knives, preventing movement. The maximum amount of cutter knife projection allowed past the cutter block for these types of cutter block is 3 mm. The machine usually comes with a specific tooling setting device that enables the correct positioning of the tooling.

133

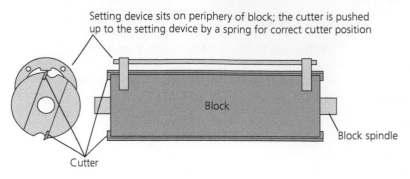

Setting device sits on periphery of block; the cutter is pushed up to the setting device by a spring for correct cutter position

Block

Block spindle

Cutter

▲ Figure 3.81 Setting device located on cutter block with cutter pushed up to setting stand

Spring

Wedge

Single-edge high-speed steel cutter set to a maximum projection of 3 mm

Captive bolt

▲ Figure 3.82 Example of wedge bar and re-grindable cutter positioning

Tightening and loosening procedure for cutter knives that can be re-sharpened

Wherever a wedge bar with tightening bolts is used to secure the cutter knives, the correct tightening and releasing sequence must be followed. This helps to eliminate the risk of distortion to the cutter knife. The correct sequence for releasing the cutters is to start undoing the bolts at the edge of the block, working in from each end alternately towards the centre.

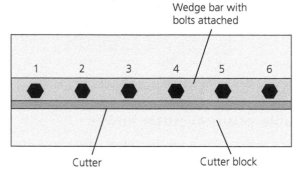

Wedge bar with bolts attached

1 2 3 4 5 6

Cutter

Cutter block

▲ Figure 3.83 Plan view of the cutter block with the cutter knife and numbered wedge bar

1 Starting at the outer edge of the block, undo bolt 1 to release the pressure.
2 Move to bolt 6 and release the pressure.
3 Release the pressure from bolt 2.
4 Release the pressure from bolt 5.
5 Release the pressure from bolt 3.
6 Finally, release the pressure from bolt 4.
7 Remove the cutter and wedge bar.

INDUSTRY TIP

Always use the correct tooling supplied with the machine to change the cutters. Over-tightening by using the incorrect length of spanner can be just as dangerous as under-tightening with a spanner that is too short.

When undoing the cutter, always pull the spanner up and away from the cutting edge. This will reduce the likelihood of it accidentally slipping onto the cutting edge when the bolt becomes loose. Use the same principle when tightening the cutters.

▲ Figure 3.84 Undoing a wedge bar bolt

After cleaning the block and wedge bar with the appropriate cleaning fluids, the tightening sequence is the reverse of the loosening sequence described above.

▲ Figure 3.85 Position setting device, start tightening from the centre of the cutter block and work outwards

▲ Figure 3.86 Check the cutting circle is level with the outfeed table

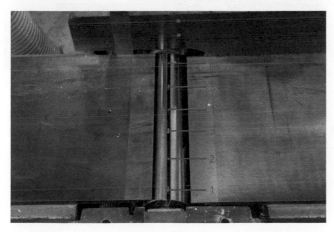

▲ Figure 3.87 Bolt numbers for the undoing and tightening sequence

Extraction to planing and thicknessing machines

All types of planers and thicknessers should have suitably sized extraction fitted and working during all cutting operations. This a legal requirement of COSHH and also a requirement of the woodworking machinery ACoP.

Correctly fitted and working extraction is essential for the safe operation of the surface planer/thicknesser as it not only keeps the atmosphere clean and dust free, but also helps to keep the floor clean and less slippery from the planer chippings and shavings. It also improves the quality of the surface finish: because there are no excess chippings or shavings around the cutter knives, the knives are able to produce a cleaner cut and, in the case of the thicknesser, the excess chippings do not get stuck to the thicknesser feed rollers or machine table bed. This results in a smoother feeding process without any indentations from chippings stuck to the feed rollers or machine table.

Correct positioning of the planer tables

To ensure smooth, safe and accurate planing, correct positioning of the infeed and outfeed tables is vital.

The infeed table controls the amount of material that is removed in one pass. As a general safety rule and to enable safe control of the material, a maximum depth of cut of 3 mm per pass is recommended. If this amount fails to achieve the desired flat surface – for example, due to the timber being twisted, cupped or just to remove some other defect – then send the material over the planer several times.

To ensure a smooth transition from the infeed table, over the cutter block and onto the outfeed table, the outfeed table needs to be set in line with the cutting circle at its maximum height. Failure to correctly set this position will result in a poor finish and, in extreme cases, could be dangerous.

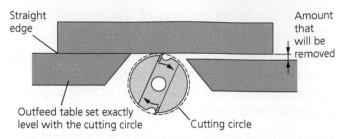

▲ Figure 3.88 Correct setting of the outfeed table in line with the cutting circle

If the outfeed table is set too high above the cutting circle, the timber will not pass onto the outfeed table but will come to a stop at its front edge. However, if the outfeed table is set very slightly too high above the cutting circle, the timber may still be able to pass onto the outfeed table, but it will result in a tapered cut. The infeed end of the timber will receive a planed finish, but the outfeed end of the timber will not.

If the outfeed table is set too low in relation to the cutting circle, the timber will receive a 'drop-in' equal to the difference in height of the outfeed table, as indicated in images a–d in Table 3.7.

Safe use of the surface planer

The hand-fed surface planer carries out two main functions:

1 Facing: producing a smooth, flat surface along the widest face of the material, known as the 'face side'.
2 Edging: producing a smooth, flat surface along the narrowest face of the material, known as the 'face edge'.

The planer can be used to produce bevels, chamfers and rebates.

Guarding the planer

Every planing machine must be correctly guarded. This is primarily achieved with the bridge guard, situated on the operator's side of the machine. It should be:

▼ Table 3.7 Incorrect positioning of the outfeed table resulting in a 'drop-in' at the end of the timber

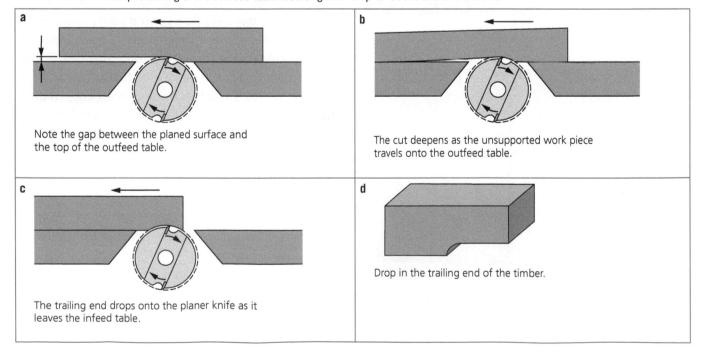

a Note the gap between the planed surface and the top of the outfeed table.	**b** The cut deepens as the unsupported work piece travels onto the outfeed table.
c The trailing end drops onto the planer knife as it leaves the infeed table.	**d** Drop in the trailing end of the timber.

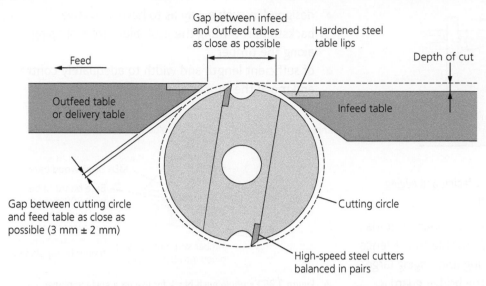

▲ Figure 3.89 Correct positioning of the machine table for hand-fed planing machine

- strong and rigid (to support heavy timbers)
- made from a material such as wood or aluminium, so that in the event of contact with the cutter block neither guard nor cutter will disintegrate
- constructed so that it is not easily deflected, which would expose the cutter block
- long enough to cover the table gap with the fence at maximum adjustment (telescopic guards are available for large machines)
- sufficiently wide (at least equal to the cutter block diameter)
- easily adjustable, both horizontally and vertically, without the use of a tool.

Attached to the back side of the fence there is another guard that is fixed, meeting all of the above criteria.

Facing and edging timber requires the timber to be placed on the infeed table; it is then passed over the cutters and on to the outfeed table. To ensure this is carried out safely, the infeed and outfeed tables should be set as close as possible to the cutting circle of the block.

While material is fed across the cutter block, the bridge guard needs to be positioned correctly. The guard should be set as close as possible to the top and side of the timber being planed. For combined planing

of the face followed by the edge, the guard should be set as shown in the illustrations, but when facing ensure that the timber is fed through on the part of the machine bed that is fully covered by the bridge guard.

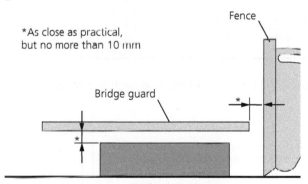

▲ Figure 3.90 Facing or flatting the timber

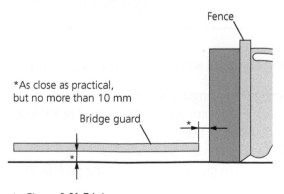

▲ Figure 3.91 Edging

137

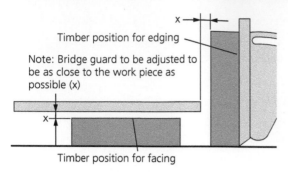

Timber position for edging

Note: Bridge guard to be adjusted to be as close to the work piece as possible (x)

Timber position for facing

▲ Figure 3.92 Bridge guard positioning for facing and edging operations following on from each other

When facing and edging small-sectioned square material, position the bridge guard as close as possible to the fence and the top of the timber. When facing and edging large-sectioned square material, position the bridge guard as close as possible to the bed and the timber side.

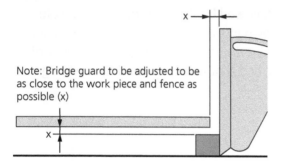

Note: Bridge guard to be adjusted to be as close to the work piece and fence as possible (x)

▲ Figure 3.93 Facing and edging small-sectioned square material

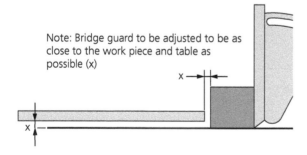

Note: Bridge guard to be adjusted to be as close to the work piece and table as possible (x)

▲ Figure 3.94 Facing and edging large-sectioned square material

When planing short lengths of timber, use a well-designed push block with a suitable method of holding it. Any handles should be:
- suitable in size, number and location
- securely fixed to give a firm grip on the push block

- designed in such a way as to have a positive backstop preventing the push block from slipping along the timber
- of sufficient length and width to adequately control and move the timber being planed.

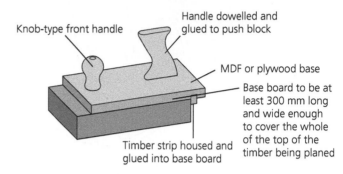

Knob-type front handle

Handle dowelled and glued to push block

MDF or plywood base

Base board to be at least 300 mm long and wide enough to cover the whole of the top of the timber being planed

Timber strip housed and glued into base board

▲ Figure 3.95 Example push block for use on a surface planer enabling safe control of short lengths of timber while being planed

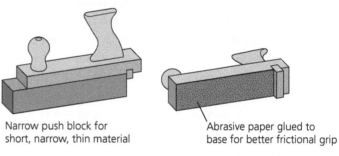

Narrow push block for short, narrow, thin material

Abrasive paper glued to base for better frictional grip

▲ Figure 3.96 Narrow push block for short, thin timber

▲ Figure 3.97 Abrasive paper can help reduce the risk of any sideways slipping

ACTIVITY

Design and construct a suitable push block for planing short lengths of 25 × 50 mm timber.

Safe operational positioning while feeding timber

When operating a hand-fed planer, it is not only important to correctly position the bridge guard but also to correctly position your body and hands while passing the timber across the cutter block. As a rule, your body will be positioned at the side of the planer on the infeed side. Your feet should provide a stable and firm base. Where long and/or heavy lengths of timber are being planed, make use of support rollers at both the infeed and outfeed ends of the machine or additional manual support by another person.

You should be positioned as at point 1 in Figure 3.98, with your feet about 300–400 mm apart. As you work towards the cutter block, slowly move your feet in a steady, unhurried continuous movement to position 2.

Facing timber

Feet should provide a firm base and allow forward movement to position 2

▲ Figure 3.98 Stance and foot movements when facing timber

The work piece should be fed by pressure with the right hand, with the left hand initially holding the material down on the infeed table (this applies to both right-handed and left-handed operatives). As soon as there is enough timber on the outfeed table, the left hand

can pass safely over the top of the bridge guard to apply pressure on the outfeed table, to be followed by the right hand as it becomes close to the end of the material to complete the feeding operation.

It is not necessary to exert feeding pressure directly over the cutter block and your hands should never pass directly over the unguarded part of the cutter block even if they are in contact with the material. This surface is usually known as the **face side**.

While edging the material, you should follow the same principles as for facing material, but this time the main function of your hands is to exert side pressure to the work piece to keep it square to the fence. This surface is usually known as the **face edge**. If the work piece is short, use a push block to feed it through the machine As when fencing, your hands should not pass directly over the cutter block.

KEY TERM

Face side and face edge: the two sides that have been planed square to each other on the surface planer.

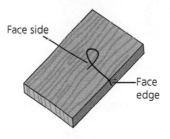

Face side

Face edge

▲ Figure 3.101 Face side and face edge mark

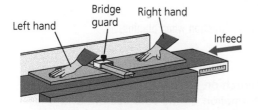

Left hand   Bridge guard   Right hand   Infeed

Left hand to apply pressure onto timber as it passes under bridge guard

Right hand to apply firm forward pressure

▲ Figure 3.99 Hand positioning for facing timber towards the start of the operation

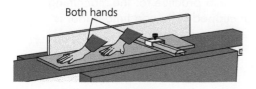

Both hands

Both hands on the outfeed end of the timber towards the end of the cut

▲ Figure 3.100 Hand positioning for facing timber towards the end of the operation

Bevelling and rebating

The woodworking machine ACoP requires that the most suitable machine available (the one posing the smallest risk) is used for every machining operation. For example, cutting a rebate on a correctly guarded vertical spindle moulder is safer than cutting it on a surface planer. New hand-fed planing machines are designed not to allow rebating, but on pre-1995 machines rebating can be carried out provided:

- a tunnel guard is used, usually with two Shaw guards
- the material is properly supported.

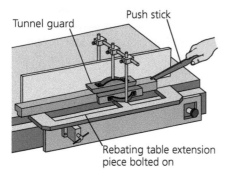

▲ Figure 3.102 Rebating on a hand-fed planing machine using a tunnel guard and push stick

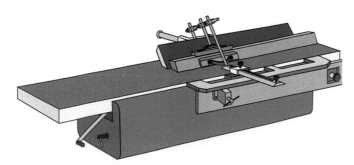

▲ Figure 3.103 Bevelling on a hand-fed planing machine using a tunnel guard

Presenting the timber to the machine

The timber to be planed should be offered to the hand-fed surface planer in such a way as to make the timber as stable as possible while it is passing over the cutter block. This is achieved by placing the timber with its concave side facing down on the machine bed.

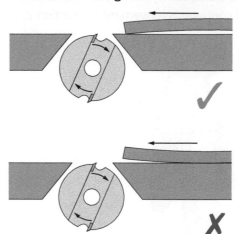

▲ Figure 3.104 Correct positioning of timber when planing

Wherever possible, you should feed the timber through the machine with the grain direction running with the cutter block rotation.

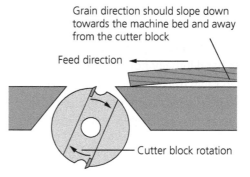

▲ Figure 3.105 Grain direction when planing

The images below show the correct positioning of the bridge guard, material and hands for facing only. Note that the bridge guard is pushed all the way over to the fence and kept as close to the top of the timber as practicable.

The images below show the correct positioning of the bridge guard, material and hands for edging only. Note that the bridge guard is dropped down to the bed and kept as close to the timber as practicable when the timber is against the fence.

The images below show the correct positioning of the bridge guard and machining material for facing and edging. Note the correct positioning of the bridge guard when facing and edging operations are carried out consecutively.

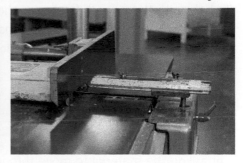

Thicknessers

The thicknesser incorporates an encased cutter block and powered feed roller system. The feed rollers grip the timber and pass it under the rotating cutter block like that of the surface planer. The timber is planed or thicknessed to the required finished size. It is fed through the thicknesser with the face side or edge facing down. Always thickness the edge side first – this will enable greater stability while the timber passes through the machine. Some machines allow for the thicknessing of more than one piece of timber at a time; the thicknesser should have a label stating whether this can be done. Alternatively, the machine can be fitted with split sectional feed rollers to convert it to take more than one piece of timber at a time.

It is extremely important to establish whether your machine can feed more than one item at a time – do not assume that it can. If you feed more than one piece of material at a time into a machine that has one fixed serrated feed roller, instead of split feed rollers, the timber may be forcefully thrown back at the operator, causing serious injury.

To further reduce the risks from kickback, anti-kickback fingers are fitted to the infeed end of the thicknesser. These metal fingers lift as the material passes under them but, if the timber is kicked back, they dig into the timber, preventing any movement towards the operator at the infeed end of the machine.

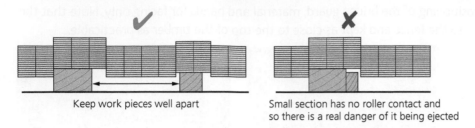

Keep work pieces well apart

Small section has no roller contact and
so there is a real danger of it being ejected

▲ Figure 3.106 Example of split-feed roller and multi-feeding

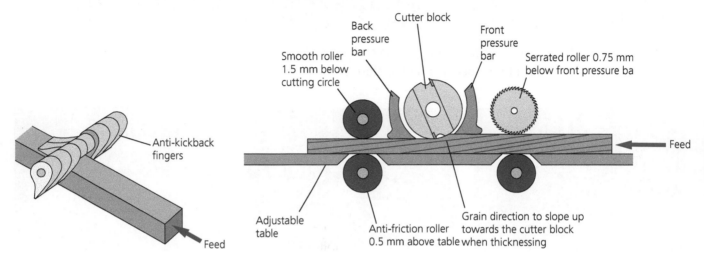

▲ Figure 3.107 Anti-kickback fingers   ▲ Figure 3.108 Typical example of a section through the thicknesser feed system

To help with the smooth feeding of the timber, extra rollers can be incorporated into the machine bed. These **anti-friction rollers** aid the smooth delivery of the timber through the thicknesser.

KEY TERM

Anti-friction rollers: adjustable, free-running rollers set into the machine bed of a thicknesser, enabling the material to move through the machine without sticking.

Guarding while thicknessing

The cutter block and feed mechanism are fully enclosed within the thicknesser casing, presenting little danger if the machine is used correctly. Combination planers need changing from their surfacing position to their thicknessing position; this is covered later in the chapter.

Operating the thicknesser

The timber is fed into the machine with the sides requiring thicknessing uppermost; these are the sides opposite the face edge and face side, in that order. The serrated infeed roller grabs the timber, feeding it past

and under the infeed pressure bar, which helps keep the timber on the machine bed and reduces the size of the chippings produced. Keeping the chippings small helps the extraction mechanism to take away the chippings quickly without blockages and also helps to produce a smoother, defect-free finished surface. The material is fed past the cutter block, where another feed roller continues to feed the timber through the machine. The outfeed roller will have a smooth surface to prevent damage to the finished surface.

It is important not to remove too much material in one pass; a maximum depth of cut of 4 mm for each pass is good practice. The required finished size of the material is obtained by either raising or lowering the machine bed, with either a mechanical hand wheel or an electronically driven motor on modern machines. The feed speed of the timber can be adjusted to cope with different species of timber as well as the required pitch mark finish.

INDUSTRY TIP

Feed the timber into the machine so that the revolving cutters are rotating down, round and towards you.

Tapered components can be produced on a thicknesser if you use well-designed jigs and saddles.

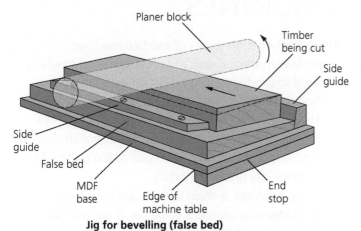

Jig for bevelling (false bed)

▲ Figure 3.109 Tapered cut being produced with the aid of a saddle

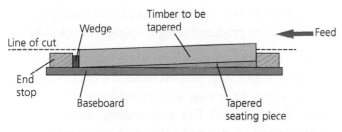

▲ Figure 3.110 Section through a jig suitable for producing long tapers on a thicknesser

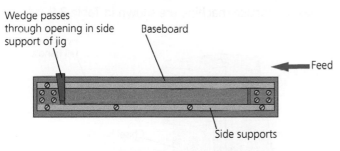

▲ Figure 3.111 Plan view of a jig suitable for long tapers on a thicknesser

---

INDUSTRY TIP

Always ensure that, for both surface planing and thicknessing, you have enough support in place to ensure the safe, controlled delivery and removal of the timber, using either temporary support rollers or additional personnel.

Combination planer/thicknesser

The combination planer/thicknesser, as its name implies, combines all the functions of a surface planer and

thicknesser in one machine. All the regulations, safety principles, setting up and operating techniques described for the individual machines above apply to this machine.

The combination planer/thicknesser is particularly useful in workshops that have limited room because the machine has a similar footprint to a surface planer, thereby saving the space needed for a thicknesser. Other advantages include only having to pay for one machine (both purchase and installation costs), not needing additional ducting, and the fact that there is no need for separate tooling.

Although there are clear advantages to having only a combination planer/thicknesser, there are also some disadvantages to be considered. The major one is that, if the machine breaks down or becomes unusable for some other reason, both planing and thicknessing operations have to stop. Other disadvantages include:

- the possibility of a backlog of jobs with the same machine being used for both planing and thicknessing operations
- the time taken to change guards etc. from planing to thicknessing and back
- increased risk of operatives not correctly changing guarding etc. from one operation to another
- small chips etc. to cutters affecting both planing and thicknessing operations.

Consider carefully all the advantages and disadvantages for your workshop and working practices before deciding which type of machine to purchase.

The combination planer will require changing from its surfacing position to its thicknessing position. This involves removing the bridge guard and fitting an additional guard that incorporates the chippings extraction point, to which the ducting needs to be fitted in order to take away the chippings. The reverse process happens once the thicknessing has been completed.

On some smaller models, the machine beds are folded out of the way and then the additional guard incorporating the extraction point is fitted before thicknessing operations can be carried out. This additional guard must still completely cover the cutter block, restricting access and preventing accidental contact with the cutter block, and meeting other guarding criteria (described previously).

Infeed table lifted out of the way

Outfeed table lifted out of the way

Additional guard added incorporating extraction

▲ Figure 3.112 Example of a small combination planer set up for thicknessing, with tables lifted out of the way

▲ Figure 3.113 Examples of dust extraction fitted to a combination planer, set for thicknessing (top) and surface planing (bottom)

5 USING MORTICE MACHINES

There are two main types of mortice machine:

1 Hollow mortice chisel: the most common machine used to produce a mortice, consisting of a square hollow mortice chisel in which revolves an auger bit. Using this method to produce a mortice can take longer than the chain morticer, as several strokes could be required from each side of the material to produce the finished mortice.

2 Chain morticer: a chain like the one used on a chainsaw runs around a guide bar. It can produce in one go a mortice that penetrates all the way through the material, making it quicker than a hollow chisel morticer. As a result, it is often used for mass production. The finished mortice is not usually as neat as one produced using a hollow mortice chisel.

In both cases, the depth of cut can be controlled by an adjustable depth stop. In some cases, the depth stop can be swung out of the way, allowing for more than one depth of cut to be set. This is particularly useful when morticing and haunching on the same component.

Advantages and disadvantages of the two different types of mortice machine are shown in Table 3.8.

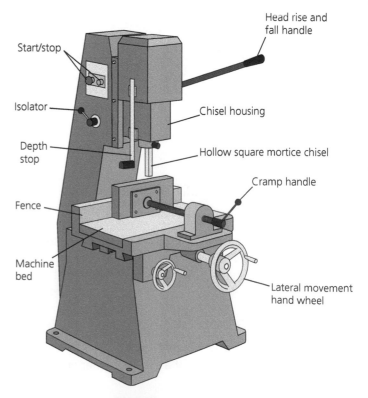

Head rise and fall handle

Start/stop

Isolator

Depth stop

Fence

Machine bed

Chisel housing

Hollow square mortice chisel

Cramp handle

Lateral movement hand wheel

▲ Figure 3.114 Hollow chisel mortice

▼ Table 3.8 Advantages and disadvantages of the two different types of mortice machine

Mortice type	Advantages	Disadvantages
Hollow mortice chisel	Easy to use Easy to set up Safer to use Easy to maintain Easy to sharpen Produces neat mortice holes Bigger range of mortice hole sizes	Takes longer to produce a mortice Small-sized chisel sets can split easily if forced to cut too fast More marking out required unless end stops are available
Chain morticer	Quick to produce a mortice Able to penetrate all the way through the material	More dangerous to use More difficult to sharpen More difficult to maintain in good working order Limited in the size of mortice it can produce Small-pitch chains can break if extra care is not taken

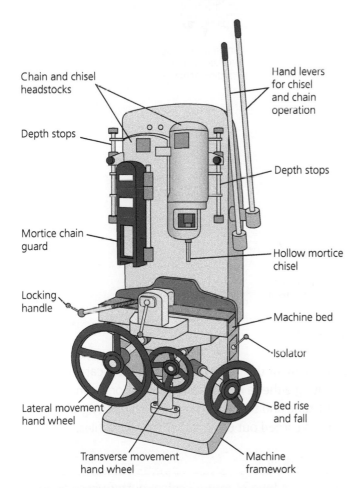

▲ Figure 3.115 Combination chain and chisel morticer

The hollow mortice chisel is generally the preferred method for cutting a mortice for the reasons shown in Table 3.8. The need for higher production output can dictate that a chain morticer is used.

Hollow mortice chisel tooling

Hollow mortice chisel sets are generally available in sizes from 6–25 mm². For sizes below 16 mm, the set usually consists of four parts, while for sizes of 16 mm and over only three parts are required. The component parts that make up the hollow chisel set are:

- hollow mortice chisel: the part that cuts the outer shape of the mortice.
- auger: the part that cuts out the bulk of the mortice.
- chisel bush: this part has a variable internal diameter to suit the size of the chisel, but a constant external diameter to suit the **collet** size of the machine.
- auger bush: this part has a variable internal diameter to suit the size of the chisel set and a fixed external diameter to fit the machine. For chisel sets of 16 mm and above, an auger bush is not required because the shaft size of the auger is the correct size to fit into the machine housing.

KEY TERM

Collet: a means of centralising the auger in a chisel headstock.

When correctly set up, the bottom of the auger will protrude out of the bottom of the hollow mortice chisel by about 0.5 mm for chisel set sizes up to 12 mm and about 1.5 mm for sizes larger than this.

The side window in the chisel is to allow the chippings to escape and should be positioned so it is on either the right or left side as you look at the machine. This means, as you work away from the window, the

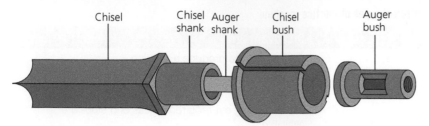

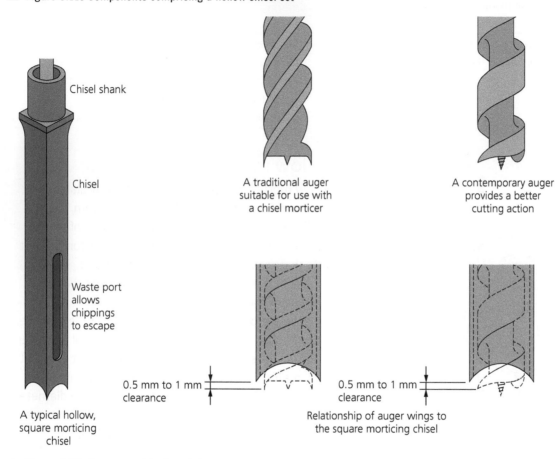

▲ Figure 3.116 Components comprising a hollow chisel set

▲ Figure 3.117 Correct positioning of the auger in relation to the bottom of the chisel

chippings flow into the previously cut mortice hole, reducing the chance of blockages, overheating or masking of the line you want to cut to.

Procedure for sharpening the hollow mortice chisel set

Regular sharpening is required to ensure the hollow mortice chisel produces a clean mortice, without risk of splitting the chisel or the breaking away of one of the small 'wing spurs' at the bottom of the auger. Sharpening can be divided into hollow mortice chisel sharpening and auger sharpening.

Sharpening of the hollow mortice chisel is carried out using either a **reamer** and pilot set or a conical grindstone and triangular files. The sharpening of the auger is carried out using flat or triangular files.

KEY TERM

Reamer: a type of rotary conical cutter used for cutting metal.

The step-by-step guide in Table 3.9 outlines the procedure for sharpening a hollow mortice chisel and auger.

▼ Table 3.9 Sharpening a hollow mortice chisel and auger: step by step

Step 1	Select the correct-sized reamer and pilot to suit the size of the hollow chisel and fit into a power drill.	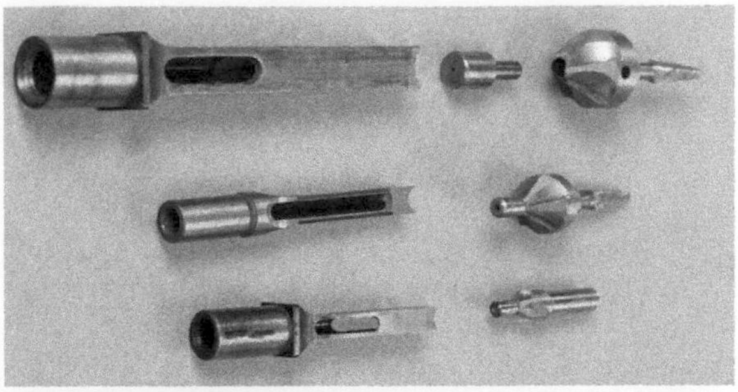
Step 2	Tighten the chisel in a vice, being careful not to over-tighten (which risks breaking the hollow chisel). Insert the pilot into the opening, ensuring you keep the drill parallel to the chisel, and re-cut the bottom of the chisel, cutting away enough material to develop a sharp edge with corners of equal length.	
Step 3	File the relief angles in all four corners.	
Step 4	File the relief angle of 25° to the four sides.	

→

▼ Table 3.9 Sharpening a hollow mortice chisel and auger: step by step (continued)

| Step 5 | Remove the burr from the outside of the hollow chisel by placing it flat on a fine oilstone and rubbing it up and down the stone. Apply a small amount of grease to the inner part of the chisel, and the chisel is ready for use. |  |

The step-by-step guide in Table 3.10 outlines the procedure for sharpening the auger.

▼ Table 3.10 Procedure for sharpening an auger

Step 1	Using a fine triangular file to suit the auger, file the end face of the auger, maintaining the original grinding angle.	
Step 2	File the wing spurs lightly, just enough to ensure they are sharp.	
Step 3	The wing spurs must always project beyond the bottom of the auger so that they cut in advance of the chisel.	Auger wing spur

Installation of a hollow mortice chisel

Before changing the tooling in any machine, always refer to the risk assessment. This will ensure you follow the correct safety procedures required for that specific machine. The steps presented in Table 3.11 assume you have referred to the risk assessment and are following the requirements as specified. They are a guide to the correct procedure for the installation of a 10 mm hollow mortice chisel set (local machine variations may need to be applied).

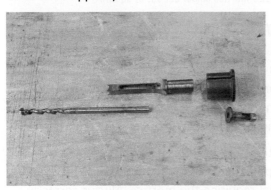

▲ Figure 3.118 Auger and chisel set

Always be careful to square up the chisel properly, as not doing so will result in a series of holes that are out of square, leaving an untidy mortice. Accurate positioning of material in the machine is vital to ensure mortice holes are cut accurately. Always ensure no chippings are trapped between the material and the machine fence and bed.

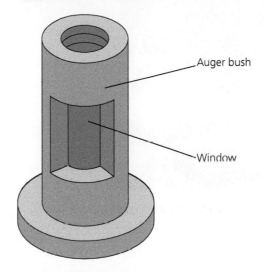

▲ Figure 3.119 Auger bush

▼ Table 3.11 Installing a hollow mortice chisel: step by step

Step 1	Isolate the machine from the power supply.
Step 2	Select the required-size chisel set, which will include the hollow mortice chisel with its matching auger, and auger and chisel bushes. Ensure the mortice set is in good condition with the chisel having sharp edges and corners. The chisel should not have any splits or other defects. The auger should have wings that are sharp and intact. Both the chisel and auger bushes should be of the correct size.
Step 3	Insert the auger bush into the mortice head, ensuring the window of the bush lines up with the Allen key. Tighten the Allen key only sufficiently to stop the auger bush from falling out.
Step 4	Insert the assembled chisel set into the headstock, ensuring that the split in the chisel bush lines up with the split jaws of the headstock. This will allow for easier tightening around the chisel without putting undue force onto the headstock. Tighten only sufficiently to stop it dropping out; there is no need at this stage to square up the chisel.
Step 5	The end of the auger will have a flat ground on it to ensure that, when tightened, the auger will not slip during the cutting operation. It is this flat that should face the window in the auger bush, which allows the Allen key to tighten a grub screw firmly onto the flat of the auger. The auger should be set so the wings of the auger protrude out of the bottom of the chisel. In this case, the correct amount of projection is about 0.5 to 1 mm. Tighten the grub screw with the Allen key.
Step 6	Slightly slacken the chisel cramp and square up the chisel with a set square against the fence. Make sure the window in the side of the chisel is facing either to the right or to the left. Tighten the chisel cramp, keeping the chisel fully pushed home.
Step 7	Make a final check of the auger position in relation to the bottom of the chisel and tighten the chisel bush. Remove all tools.
Step 8	Turn the power back on and do a quick start and stop of the machine. The auger should run quietly even though it is steel running inside steel. Any loud squeaking sound from the auger will probably be due to it being set too high within the chisel. If the auger is not set too high in relation to the chisel, excessive noise could be due to the lack of an auger bush or because you have used the incorrect size of auger bush.
Step 9	A trial cut will allow you to check that the chisel is cutting correctly and that it is square to, and the correct distance away from, the fence.

▲ Figure 3.120 The assembled chisel set ready for inserting into the headstock

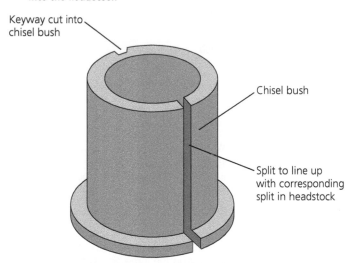

Keyway cut into chisel bush

Chisel bush

Split to line up with corresponding split in headstock

▲ Figure 3.121 Chisel bush

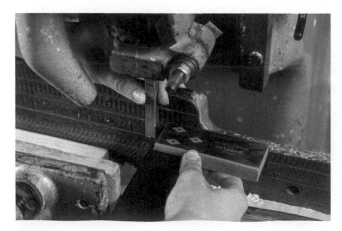

▲ Figure 3.122 Square up the chisel

▲ Figure 3.123 Mortice chisel set out of square

Forming mortices using a hollow mortice chisel

Four types of mortice are generally formed using a hollow mortice chisel:

1. through mortice
2. open mortice or bridle
3. stub/blind mortice
4. haunched mortice.

As a rule, the size of the mortice chisel should be equal to approximately one third of the width of the material requiring the mortice. Always correctly position and cut your mortices before cutting your tenons to ensure the correct tenon thickness is formed to suit the mortice hole.

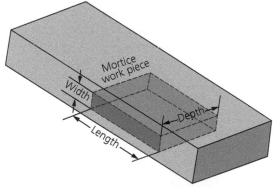

Mortice work piece

Width

Depth

Length

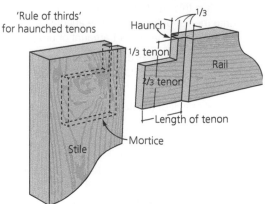

'Rule of thirds' for haunched tenons

Haunch

1/3

1/3 tenon

2/3 tenon

Rail

Length of tenon

Mortice

Stile

▲ Figure 3.124 Terms used with mortices

Through mortice

The through mortice is perhaps the most common mortice and, as the name suggests, the mortice hole passes all the way through the material. They tend to be positioned away from the ends of the material, with the tenon held tightly in place with wedges.

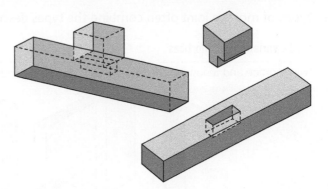

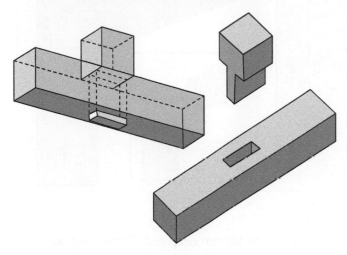

▲ Figure 3.125 Through mortice and tenon joint

Open mortice or bridle

The open mortice is formed at the end of the material and is normally a through mortice with the tenon held in place with a dowel, although it can be a stub mortice. Open mortices can be problematic because holding the tenon tight to the mortice edge can lead to misalignment and possible movement compared to other mortice joints.

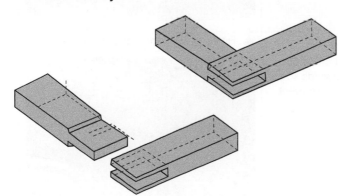

▲ Figure 3.126 Open or bridle mortice joint

Stub or blind mortice

The stub or blind mortice passes only part way through the material requiring the mortice and is commonly used as a decorative feature as no end grain is visible at the joint.

▲ Figure 3.127 Stub or blind mortice joint

Haunched mortice joint

The haunched mortice is used at the end of the material, typically with door stiles. Haunches are used to provide extra tenon strength and fixing area without having an open mortice. The haunch is cut as a stub partway through the material and out of the material end. Again, the edges of the tenon are held tightly in place with wedges.

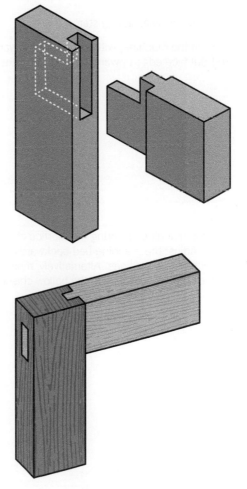

▲ Figure 3.128 Haunched mortice joint

Other types of mortice joint often combine the types described above, as depicted in Table 3.12.

▼ Table 3.12 Variations on mortices

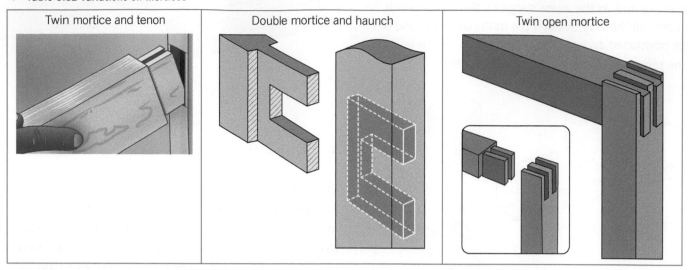

Twin mortice and tenon	Double mortice and haunch	Twin open mortice

Sequence for cutting a stub mortice hole

The step-by-step guide shown in Table 3.13 assumes that the mortice will be 50 mm deep and 50 mm long, and is being cut using a 10 mm hollow mortice chisel.

▼ Table 3.13 Cutting a stub mortice hole: step by step

Step 1	Place the timber in the machine, with its face side towards the fence and its marked-out face edge upwards, and clamp tight.	▲ Figure 3.129 Clamp material into machine
Step 2	Without turning the machine on, bring the mortice chisel down to the top of the timber. Adjust the machine bed backwards or forwards so the chisel lines up with the gauge lines. Alternatively, measure the correct distance from the fence to the back of the hollow chisel.	▲ Figure 3.130 Aligning the mortice chisel

→

▼ Table 3.13 Cutting a stub mortice hole: step by step (continued)

Step 3	Mark the required depth by drawing a line on the end of the timber to be morticed, then bring the chisel down to that point and set the depth stop. When producing a through mortice, the depth should be set at just past half the material's depth so, when the material is turned over to mortice from the other side, a clean through mortice is produced.	 ▲ Figure 3.131 Set the required depth
Step 4	After checking that everything is set up correctly and safely, turn on the machine. Bring down the chisel carefully at the end of the mortice and push it into the timber, taking care to go only halfway through.	 ▲ Figure 3.132 Positioning the mortice chisel for the first outer edge cut
Step 5	Cut the other end of the mortice. This ensures that the ends of the mortice are straight.	 ▲ Figure 3.133 Cutting the second outer edge cut

→

▼ Table 3.13 Cutting a stub mortice hole: step by step (continued)

Step 6	Remove the middle of the mortice, always working away from the window of the chisel so the chippings fall into the mortice that has already been cut. Always overlap the mortice hole by about a quarter of the mortice chisel size.	 ▲ Figure 3.134 Removing the middle portion of the mortice

INDUSTRY TIP

Never produce a through mortice by morticing from just one side, as this will result in spelching and splintering on the underside. Always mortice from both sides.

If a through mortice is required, turn over the material and repeat the above steps, except when setting the depth in step 3. For a through mortice, this depth should be set about 5 mm past the middle of the material.

When producing deep mortices or mortices in hard materials such as oak and ash, the first cut should be done in stages to reduce the risk of the chisel splitting. The correct sequence for these stages is shown in Figure 3.135. This is the procedure to use when the chisel window is on the right as you look at it. If the window is on the left, reverse the procedure.

1 Position the mortice chisel with its window on the right-hand side of the required mortice (position 1).
2 Mortice to about half the required mortice depth.
3 Move to position 2 and mortice to full depth.
4 Move back to position 1 and mortice to full depth (position 3).
5 Move to position 4 and mortice to full depth.
6 Repeat for the remainder of the mortice.

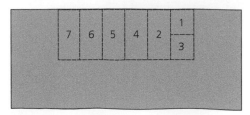

▲ Figure 3.135 The mortice sequence for deep mortices

INDUSTRY TIP

Never force the chisel when forming a mortice – use steady downward pressure. If hard materials such as knots are met when morticing, take several cuts at the mortice to help remove the chippings.

INDUSTRY TIP

Never drag the chisel across the bottom of the mortice to make a smooth mortice bottom, as this can break the chisel. Instead, make multiple small step cuts across the bottom of the mortice.

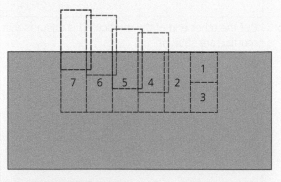

▲ Figure 3.136 Overlap of mortice strokes to help create a clean bottom to the mortice and reduce the risk of the chisel splitting

HEALTH AND SAFETY

Make use of support rollers when working with long lengths of timber to prevent accidental movement of the material when tightening and loosening the work clamp to the morticer.

Fault diagnosis with hollow chisel sets

Typical examples of faults found when using and setting up hollow mortice chisels are shown in Table 3.14.

▼ Table 3.14 Typical examples of faults found when using and setting up hollow mortice chisels

Faults with hollow mortice chisel tooling	Possible reason	Corrective action
Mortice is not at right angles (90°) to the fence 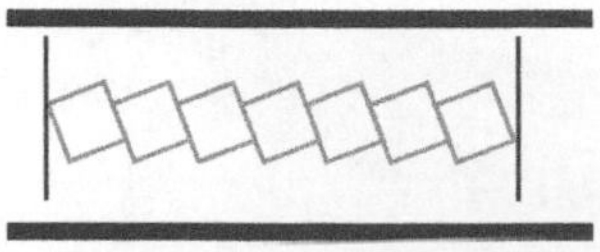	The back face of the chisel is not set parallel to the fence.	Adjust the chisel so the back edge is parallel to the fence.
The bottom of the chisel set leaves the corners of the chisel, producing an uneven bottom to the mortice 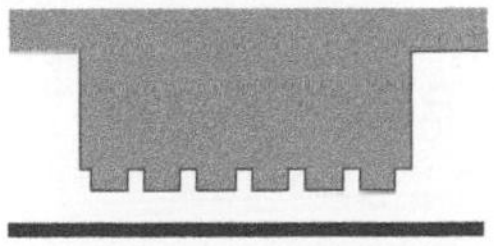	The auger is protruding too far from the bottom of the chisel.	Reposition the auger at the correct distance below the chisel to suit the size of the chisel set.
The chisel shows signs of burning (blueing) around the bottom cutting edge 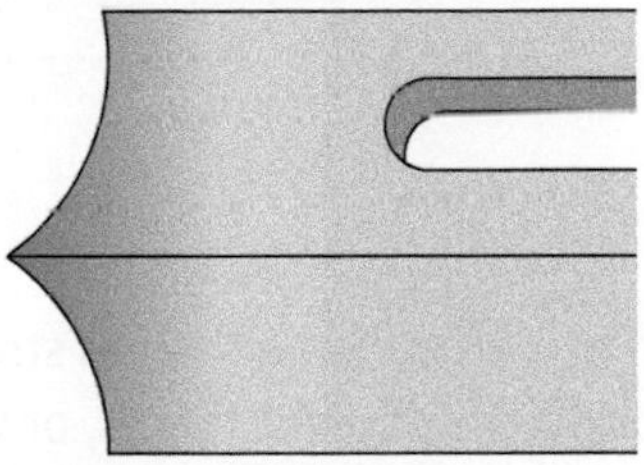	The auger is set too deep within the chisel, causing the auger to rub against the chisel.	Reposition the auger at the correct distance below the chisel to suit the size of the chisel set.
The chisel develops a crack 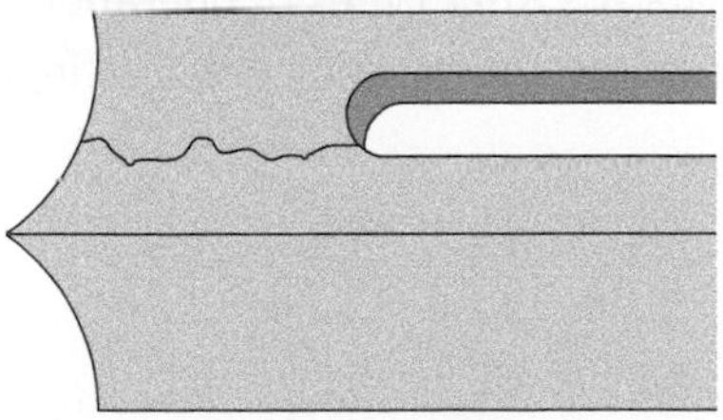	The auger may be positioned too close to the bottom of the chisel, so the chippings cannot escape properly. This problem may also be caused by poor sharpening of the chisel.	The chisel has become dangerous and should not be used. Always ensure that the chisel is regularly and correctly sharpened and the auger is set correctly before use, particularly with small chisel sets.
The bottom of the auger has a broken spur wing 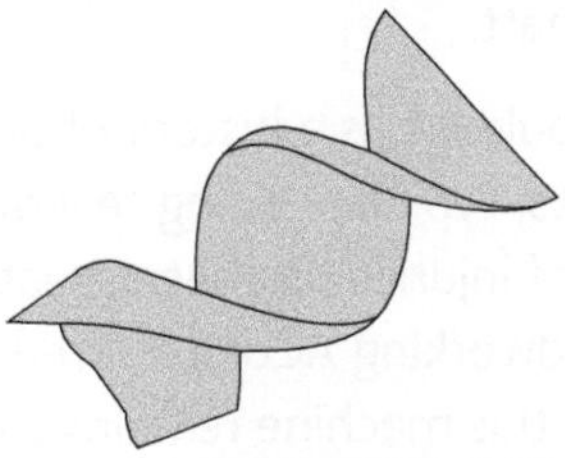	Poor or lack of sharpening of the auger.	Dispose of the auger and ensure regular and correct sharpening of the replacement auger.

Spindle moulder/router table

The spindle moulder (also known as the vertical spindle moulder, or simply the spindle) and a router table perform very similar tasks and incur very similar risks. The spindle is basically a larger and heavier version of a router table, more able to perform larger, neater cuts. The router table incorporates a hand-held portable router fixed to a removable bed, while the spindle's motor and cutter block shaft are permanently fixed within the machine.

The woodworking machinery ACoP defines a portable router fixed into a router table as a spindle moulder; therefore all the safe systems of work, guarding processes and machining techniques outlined below should be followed for both types of machine. Where specific differences do apply, they will be highlighted.

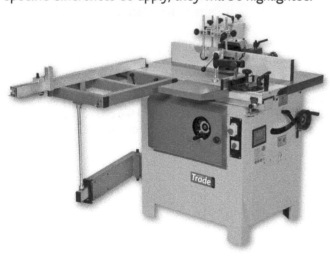

▲ Figure 3.137 Spindle moulder

A spindle moulder consists of a main framework incorporating the machine bed, fence, motor, spindle or arbour upon which a cutter block is mounted, and the machine guards. The spindle can be raised or lowered with an adjusting wheel, and on some models the spindle can tilt. The fence is also adjustable. If required, usually for curved work, a special fence known as a ring fence and bonnet guard can be fitted to the machine. An independent motor with rubber wheels (known as an automatic feed or demountable self-feed) can be used to feed the timber through the spindle, resulting in a higher-quality finish and improved safety.

The router table is usually a smaller, lightweight machine, capable only of producing smaller cuts on smaller-section material. Due to their light weight, router tables are not generally suitable for heavy or long sections of timber, as there is a risk of moving and overbalancing the router table during processing.

▲ Figure 3.138 Router table

Spindle moulder and router table uses

The spindle and router table are primarily used to perform the following tasks:

- rebating
- moulding
- grooving
- bevelling.

These operations can be performed either on straight lengths of material, using the straight fences, or when making or working on curved materials using the ring fence and bonnet guard. Other types of operation carried out on these types of machine are creating:

- dovetails
- corner locks
- stair trenching
- tenons.

These types of operation require the use of ancillary attachments and, for spindles, will usually involve changing the tooling shaft.

The vertical spindle moulder has a history of operators having serious accidents, typically losing several fingers. Although the severity of injuries has been greatly reduced since the woodworking ACoP required the use of chip limited tooling, this machine remains a common cause of accidents in the woodworking industry.

Accidents typically occur as a result of:

- lack of training
- incorrect setting up of the machine and poor positioning of the guards
- cutters snatching
- work piece kickback
- working with short-grain materials
- not using a false fence for straight work
- not using pressure pads (Shaw guards) to adequately enclose the cutters
- not using backstops, jigs or work piece holders when performing stopped work
- poorly designed jigs.

The woodworking machinery ACoP requires the spindle moulder (and all woodworking machinery) to be capable of coming to a controlled stop within 10 seconds. This is usually achieved by an electronic control box that automatically applies the brake as soon as the stop button is pressed. On older machines, the braking mechanism was a hand brake that applied pressure to the spindle shaft to stop it.

Demountable power feed unit

Whenever possible, a power feed unit should be used when working with straight lengths of timber. This not only keeps the operator's hands well away from the cutting area, but also maintains a stable, controlled feed speed of the material, resulting in a better quality of finish. Whenever a power feed unit is used on a spindle moulder, the spindle is still classed as a hand-feed machine and all the appropriate safety measures outlined by PUWER and the ACoP for woodworking machinery must be followed.

The correct positioning of the demountable power feed unit is vital for the safe, controlled feeding of the material. In most cases, the feed unit has three sprung rubber-faced feed rollers. The middle feed roller is usually positioned in line with the centre of the cutter shaft.

When the feed rollers are pushing the material down on to the machine bed, the feed unit should be set so the outfeed roller is slightly leading or pointing into the machine fences. This positioning helps to ensure the material is kept pushed up against the fence.

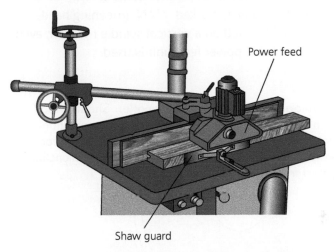

▲ Figure 3.139 Demountable power feed unit positioned for feeding above the material

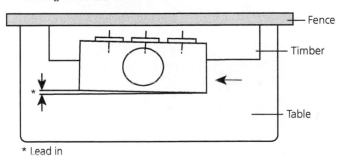

▲ Figure 3.140 Plan view of demountable power feed unit positioning (note the lead-in towards the outfeed fence)

When wide, thin lengths need to be machined, the edge of the demountable feed unit can be adjusted so its feed rollers are on the wide surface, pushing the material against the fence and down onto the machine bed.

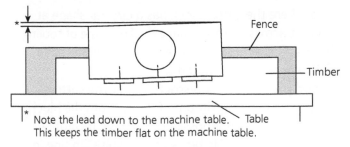

▲ Figure 3.141 Demountable feed unit positioned against the side of the material for feeding wide, thin material up against the fence

Tooling used on the spindle

Always use limited cutter projection tooling on hand-fed machines such as a vertical spindle moulder. This reduces the risk of kickback and the severity of injury if the operator's hand comes into contact with the tooling. Only tooling marked 'MAN' (meaning hand-feed) should be used on a vertical spindle moulder, even if a demountable power feed unit is used.

ACTIVITY

Download Woodworking Information Sheet WIS37 from the HSE website and produce a presentation that outlines the advantages of using limited cutter projection tooling, which can be used in conjunction with a training programme.

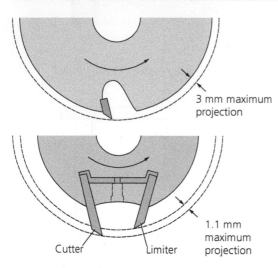

3 mm maximum projection

1.1 mm maximum projection

Cutter   Limiter

▲ Figure 3.142 Examples of chip limited tooling for use on the spindle moulder

Limited projection tooling is usually defined as either:

- round form tooling
- non-round form tooling.

Round form tooling

This is where the cutter block has a circular shape at any point across its rotational axis. This type of tooling achieves its chip limiting protection with either solid profile blocks or chip limited tooling blocks.

- **Solid profile blocks** restrict or limit the cutter projection to a maximum of 3 mm from the block surface. As it is not possible to change its profile shape, a solid profile block can be quite limiting in its versatility.
- **Chip limited tooling blocks** have balanced pairs of removable cutters with a reverse profile set

of limiters to match. The limiters and cutters are produced and fixed so that the cutter projection past the limiters is restricted to a maximum of 1.1 mm.

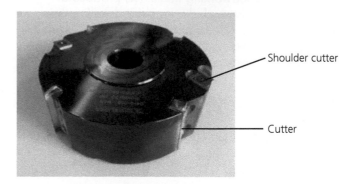

Shoulder cutter

Cutter

▲ Figure 3.143 Rebating block incorporating disposable cutters with a cutter projection of 3 mm

Cutter

Limiter

Limiter

Cutter

▲ Figure 3.144 Chip limited tooling with a limited cutter projection of 1.1 mm

Non-round form tooling

This type of tooling should be designed in such a way that cutters project a maximum of 1.1 mm beyond the edge of the tool body or limiter built in to the cutter block.

Circular portion of block restricting cutter projection to 1.1 mm maximum and acting as chip limiter

Cutters not exceeding 1.1 mm from block

▲ Figure 3.145 Non-round form limiter projection grooving cutter block

Chip limited tooling not only helps reduce the risk of injury but also has other benefits, such as:

- limited cutter projection tooling is simpler and quicker to set up, reducing the time required for a change of profile
- limited projection tooling is more likely to be right first time – less time and materials are used in checking the profile is correct
- chip limited cutter blocks can be used for many different profile cutter sets
- chip limited tooling is well balanced, limiting vibration
- reduction in noise levels
- lighter aluminium cutter blocks reduce forces on the motor during braking and make the tools easier and safer to handle
- improved finish and less wear on the shaft and bearings of the machine.

Safe operation of the spindle moulder and router table

The spindle and router table should be effectively guarded. This is achieved in four ways.

1 guarding to the rear of the fence and to the back of the cutter bock
2 guarding to the top of the cutter block
3 guarding in front of the fence and the front of the cutter block
4 guarding below the machine table.

> **HEALTH AND SAFETY**
>
> The woodworking machine ACoP applies to both router tables and vertical spindle moulders so you are required to follow the same guarding and safe operation principles for both types of machine.

Guarding to the rear of the fence

All machines should be adequately guarded behind the fences. The guarding usually incorporates a means of attaching the dust extraction hosing. The rear guarding must not become easily detached and must be robust enough to withstand the rigours of use. The rear guard often incorporates a lift-over top portion which acts as the top guard; this also allows easy access for changing the tooling, without requiring the complete removal of the rear guard.

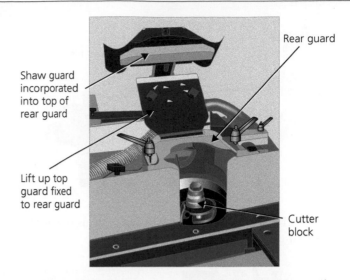

Shaw guard incorporated into top of rear guard

Rear guard

Lift up top guard fixed to rear guard

Cutter block

▲ Figure 3.146 Rear guard incorporating swing-over top portion and top Shaw guard

Guarding to the top of the cutter block

Most spindle moulders have the top guarding cover incorporated into the rear guarding assembly. This type of top guarding format often incorporates an adjustable top pressure pad, particularly on new machines.

Guarding to the front of the fence

A demountable power feed unit should be used, if possible, for straight working. This not only acts as a guard but also helps to produce a smooth, regular finish to the component. When it is not possible to use a demountable power feed unit, the cutting area should be enclosed by vertical and horizontal spring-loaded pressure pads (Shaw guards). These guards should form a tunnel that is long enough to prevent the operator's hands from reaching the cutters. Traditional Shaw guards had timber pads fixed to a sprung steel mount that was bolted or clamped to the machine. Modern versions incorporate flexible clear Perspex guards bolted and clamped to the machine casing.

The guarding of either design should be of such a width that it is only slightly smaller than the material being machined, so that there are no gaps between the machine bed or machine fence.

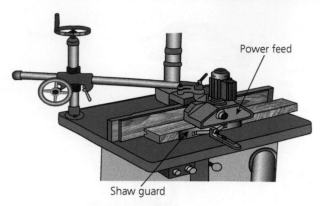

▲ Figure 3.147 Demountable power feed unit being used to feed material

▲ Figure 3.148 Guarding in use on a router table

When machining wide boards, a side pressure guard can be omitted provided that a wide top pressure pad is used.

▲ Figure 3.149 Machining wide material with a wide top pressure pad

Guarding to the underside of the machine bed

Any pulleys and drive belts below the table should be guarded, as should access to the spindle shaft and the lower part of the cutter block. For router tables, the router and lower part of the cutter should be enclosed to eliminate any risk of accidental contact.

ACTIVITY

Research and identify the different types of feed unit and guarding system that could be used on the spindle moulder or router table that you will use. Suggest a preferred option along with reasons for your choice.

Setting up and operating the vertical spindle moulder and router table

Spindle moulders and router tables have a hole in the machine bed that the router cutter or spindle cutter block shaft penetrates through. To allow for different-diameter tooling, removable inserts are used to keep this aperture as small as possible.

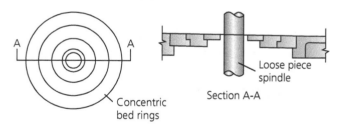

▲ Figure 3.150 Inserts used to reduce the aperture in the spindle bed

All vertical spindle moulders and router tables will have a fence for the material to run against, ensuring a straight, constant position for the material and finished profile. Fences on older spindles can slide closer to or further away from the cutter block, allowing different-diameter cutting circles to be used, either reducing or increasing the gap between the infeed fence and the outfeed fence. This gap between the cutting circle and the fences should always be kept as small as possible, but even with the fences set as close to one another as practicable without the risk of the fences catching the cutters, the woodworking ACoP considers the gap left above and below the cutting circle too large. To further reduce the gap between the fences and the cutting circle, and in particular to eliminate the gap left above and below the cutters, a timber false fence should be fitted. This timber false fence not only helps to enclose any gap around the cutters, making the machine safer to use, but also helps to improve the quality of the cutters' finish.

Some vertical spindle moulders may have the newer version of fence fitted, which incorporates adjustable fingers. These bridge the gap between the infeed and outfeed fences above and below the cutters as required, reducing the gap, providing support to the material around the cutting area and eliminating the need for a timber false fence.

Always set the height of the cutter block before fitting the false fence or positioning the fingers on the fence. This avoids the risk of the cutters catching the fence fingers, or the top of the block rubbing on the false fence, when the cutter block is raised or lowered.

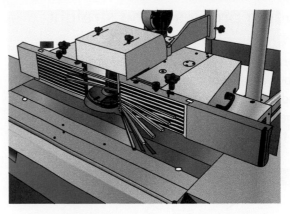

▲ Figure 3.152 Adjustable fingers on the vertical spindle moulder's fence

Any timber false fence will have to be pushed back into the cutters to create an opening of the required size. It is vital that the operator is still protected by guards during this process. One method of ensuring this is to drop the top Shaw guard to the machine bed. You must make sure there will be enough clearance between the Shaw guard and the cutting circle after the fence has been pushed back the required distance.

Alternatively, another heavy-duty false fence can be positioned over the fences in front of the cutters to act as the guard. You must ensure that it is both large and strong enough to cope with accidental contact with the cutters.

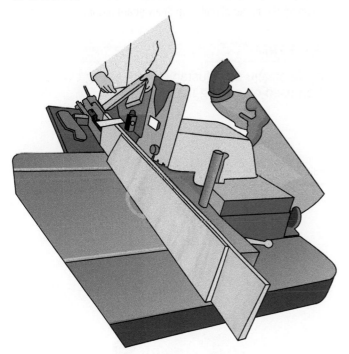

▲ Figure 3.151 Timber false fence on a vertical spindle moulder (front guards removed for clarity)

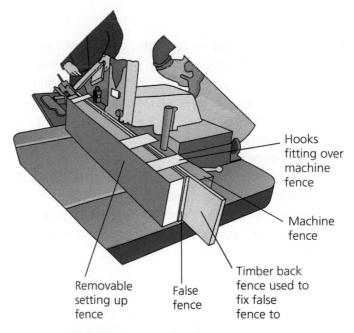

Hooks fitting over machine fence

Machine fence

Timber back fence used to fix false fence to

Removable setting up fence

False fence

▲ Figure 3.153 Example of heavy-duty removable guard hooked over the machine fences and to be used when breaking through false fence

Producing profiles using straight fences

When forming rebates, mouldings and grooves that do not remove a full-face cut from the material, the rear fence and infeed fence should be set parallel to each other. The cutters should project through the false fence to the desired cutting depth.

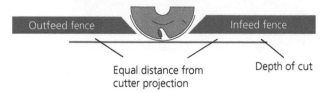

▲ Figure 3.154 Both infeed and outfeed fences parallel for part face cut, e.g. when cutting a rebate

When a full-face cut is required, the outfeed fence should be set in line with the cutting circle and the infeed fence slightly behind the cutting circle, giving the required material removal of 1–2 mm.

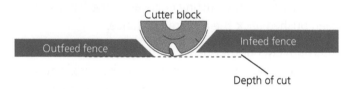

▲ Figure 3.155 Infeed fence set in front of cutting circle with outfeed fence in line with cutting circle for full-face cut

Stop work

Spindle moulders and router tables can be used to form cuts that go only part way along the material – just the beginning part, the end part or any portion in between. If any of these operations is required, the machine needs to be guarded in a slightly different way. The woodworking ACoP requires the use of end stops to reduce the risk of kickback and potential injury. These must be positioned at both the infeed and outfeed ends at the required starting and finishing profile points. The stops must be fastened in such a way that any potential kickback resulting from **dropping on** will not move the end stops.

KEY TERM

Dropping on: on a cutting machine, the pushing of the material into the revolving cutters part way along.

Wherever possible, the material should be held in a suitably constructed jig that is able to withstand the rigours of use and deliver the material in a safe, controlled manner. The woodworking machine ACoP requires all jigs to have suitable handholds as well as suitable holding methods for the material. Larger sections of material become impracticable to position in jigs. They still require end stops but it is necessary to leave enough room for good hand positioning away from the cutters.

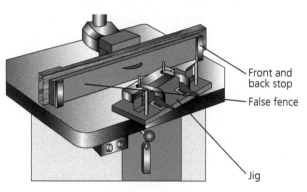

▲ Figure 3.156 End stops fitted to both fences for use with stopped work, jig with hand holds and toggle clamps for holding the material (note top Shaw guard missing for clarity)

KEY TERM

Toggle clamp: an adjustable quick-acting clamp often used on cutting machines.

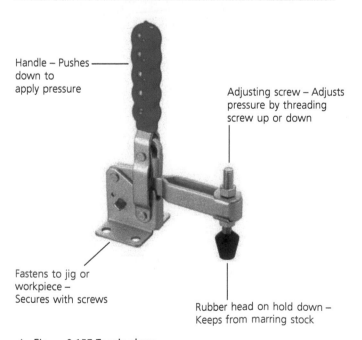

▲ Figure 3.157 Toggle clamp

Safe procedure for dropping on while using a jig

1 When dropping on, the jig containing the material is placed with its corner against the infeed fence and up against the end stop that is fixed to the infeed fence.
2 The jig is slowly pushed round into the cutters, so it is entirely against the fences.
3 The jig is then passed along the fences until it reaches the stop on the outfeed fence.
4 The jig is pulled away from the cutters, starting with the part of the jig next to the cutters, leaving the other end of the jig against the fence and end stop. When the end nearest the cutters is clear of the cutters, the whole jig can be removed.

Safe procedure for dropping on with large or long lengths of timber

When dropping on without a jig (for example, when cutting large or long pieces of material) the same process is followed, except there is no jig or hand holds (as provided by the jig). You must always ensure your leading hand is kept on the part of the material that will rest on the infeed fence (i.e. before the cutter block). Your leading hand should move to the outfeed fence only when the timber has come to rest fully against the fences. Your hand should not pass directly in front of the cutters; you should remove it and place it on the material away from the cutter block. If the side Shaw guard is not used when dropping on (e.g. because it will prevent the use of the jig), a larger top Shaw guard should be used to cover a larger area of the machine bed.

Producing curved sections on a vertical spindle moulder or router table

The production of curved sections on a vertical spindle moulder is one of the most dangerous operations carried out on woodworking machinery. A thorough risk assessment must be carried out for each operation to establish the suitability of the operator and to identify safe systems of work.

To produce curved sections, straight fences are removed and replaced with a ring fence and bonnet or cage guard.

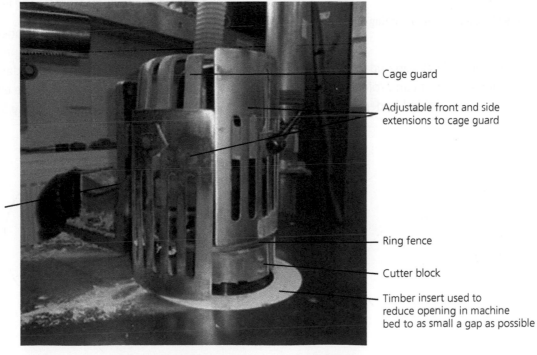

Cage guard

Adjustable front and side extensions to cage guard

Ring fence and cage guard support and fixing arm

Ring fence

Cutter block

Timber insert used to reduce opening in machine bed to as small a gap as possible

▲ Figure 3.158 Traditional type ring fence and cage guard fitted to a vertical spindle moulder

Setting up the ring fence and cage/bonnet guard

For the correct production of curved work, the ring fence should be positioned with its maximum outer curvature in line with the centre of the cutting circle. The ring fence can be used and positioned below the cutter block close to the machine bed, but preferably it should be positioned as close as possible above the cutter block, when and where the machining operation allows. The jig should be designed to suit the safest machining method.

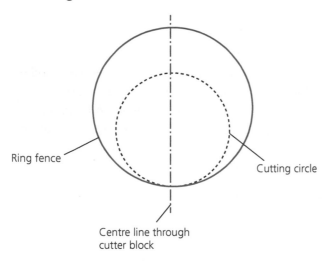

▲ Figure 3.159 Maximum curvature of ring fence in line with cutting circle

The ring fence should always line up with the centre line of the cutting circle but can project past the cutting circle or remain behind it, depending on the depth of cut required for the profile and how the jig is constructed.

The cage guard is positioned as close to the top of the cutter block or ring as practicable. The front adjustable extension should be lowered so it is as close as practicable to the top of the material or jig being used. The side adjustable extensions should be positioned down on the machine bed where possible or as close as possible to the top of the material or jig being used.

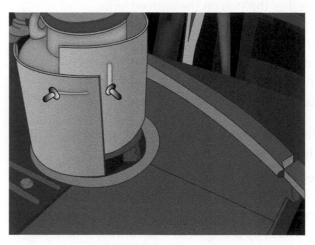

▲ Figure 3.161 Cage/bonnet guard correctly positioned with side extensions positioned on machine bed, while front extension is positioned as close to the top of the ring fence as practicable

An adjustable lead-in arm is positioned to enable easy and smooth location of the jig against the ring fence. Side brushes, which help to deflect any debris away from the operator and into the extraction, are an additional safety feature that may be used.

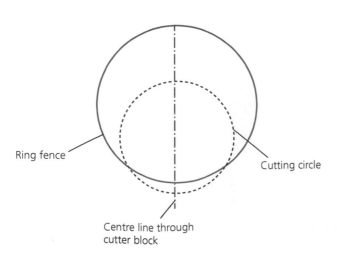

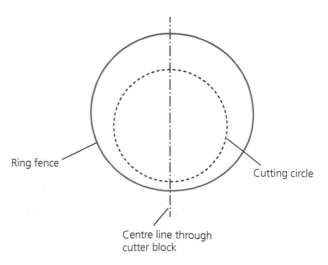

▲ Figure 3.160 Ring fence positioned behind the cutting circle (left) and in front of the cutting circle (right); in both cases, the ring fence is in line with the centre line of the cutter block

ACTIVITY

Using the internet and any manufacturers' information you have, source a new ring fence and bonnet guard for the vertical spindle moulder you will be using and include reasons for your choice.

Jig design

Jigs for spindle moulders and router tables are usually used for dropping on, or curved working with ring fences. The jig profile or template will determine the shape of the finished components. It is this part of the jig or template that runs against the ring fence while passing across the cutter block. Care should be taken to maintain a high-quality finish to the edge of the jig or template that is to run against the ring fence, as any defects will be transferred to the finished component.

In all cases, the construction of the jig must comply with strict manufacturing and safety rules, even if the jig is going to be used only for limited production runs.

Jigs must:

- be constructed of sturdy material of a suitable size and weight to allow for safe presentation of the material and to withstand the rigours of continued use
- have suitable hand holds, allowing them to be handled safely and easily during machining operations
- have means of fixing and holding the material being machined in the jig
- have a lead-in and lead-out to the jig, allowing safe presentation of the material to the cutter block.

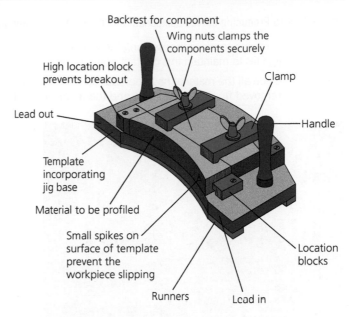

▲ Figure 3.162 Example spindle jig for producing curved components

HEALTH AND SAFETY

Always limit the amount of short grain any component contains when producing curved components on a vertical spindle moulder with a jig and ring fence. Any short-grained material is likely to break up during the cutting operations, which may cause injury to the operator.

Producing curved components using a ring fence and jig

The steps in Table 3.15 outline a safe procedure when producing curved sections on a vertical spindle moulder using a ring fence and jig.

▼ Table 3.15 Producing curved components using a ring fence and jig: step by step

Step 1	Fit the required cutter block, and position it at the required height. Use timber or machine inserts to close the hole in the machine bed so it is as small as possible.
Step 2	Position the ring fence at the required depth of cut and line up with the running surface of the jig.
Step 3	Position the lead-in arm to help guide the material onto the ring fence but away from the cutting circle.
Step 4	Position and set the cage/bonnet guard as close to the material and machine bed as practicable.
Step 5	Fit extraction.
Step 6	Use suitable PPE as outlined in the risk assessment, e.g. ear defenders and eye protection.
Step 7	Start extraction and machine.
Step 8	Position the jig and material against the lead-in arm and gently feed the jig onto the ring fence. The lead-in portion of the jig should be in contact with the ring fence and the cutters should not at this stage be in contact with the material.
Step 9	Move the jig round the ring fence and into the cutters so it is in contact with the ring fence at the centre point of the cutter projection.

→

▼ Table 3.15 Producing curved components using a ring fence and jig: step by step (continued)

Step 10	Continue to move the jig along the ring fence, always maintaining contact with the ring fence at this centre point. If the jig fails to maintain this contact point, the finished profile depth will vary.
Step 11	Once all the material has been cut and the lead-out portion of the jig is positioned on the ring fence, the jig can be removed from the ring fence and the machine stopped to remove the material from the jig.

▲ Figure 3.163 Correctly used jig and cage/bonnet guard to produce curved components

ACTIVITY

Produce a risk assessment for producing on a vertical spindle moulder curved sections that are suitable for the top rail of a standard-sized door. Then design and produce a jig suitable for producing a curved top rail for a standard door.

Maintaining machines

Maintenance is a legal requirement, not an optional activity. Maintenance carried out with the correct parts and at the correct time intervals will benefit the user in the long term.

Other advantages of carrying out maintenance are:
- prolonged machine life
- fewer breakdowns, resulting in a more productive machine that is cheaper to run
- less risk of injury to operatives
- better-quality work produced when all parts of the machine function correctly.

All the information required to carry out maintenance should be available in the manufacturer's handbook for that type of machine and model number. Examples of the type of information available include:
- types of lubricant required, e.g. oil, grease, water
- how much lubricant to use each time and when, e.g. 500 ml every 200 working hours, 'three depressions of the grease gun' daily
- where lubricants should be used, e.g. 'points A and B'
- suitable replacement parts, e.g. belts, bearings
- how often parts should be replaced, e.g. weekly, monthly, yearly
- how to check conditions and tensions of belts, e.g. amount of free movement, signs of cracking or fraying, stopping times.

Remember that, while carrying out maintenance operations, attention should be paid to the personal protective equipment (PPE) you will need. This could include:
- barrier creams – to protect the skin from oils etc.
- clothing – to keep contaminants away from the skin
- safety glasses or goggles – to keep dust and chippings etc. away from the eyes
- gloves – to keep contaminants away from the hands and for use when handling tooling
- a face mask – to prevent dust and fumes from being breathed in
- ear defenders – to protect against loud noise.

ACTIVITY

Produce a maintenance schedule using the manufacturer's information for a machine that has been covered in this chapter. Include in your plan all the equipment, materials and PPE required.

The typical contents of a maintenance schedule are shown in Table 3.16.

▼ Table 3.16 The typical contents of a maintenance schedule

Step	Notes
1 Inspect and follow the risk assessment	Always locate the risk assessment and identify all significant risks along with relevant control measures. Ensure you are authorised to carry out the maintenance.
2 Isolate the machine	–
3 Clean down the machine using extraction or vacuum cleaner	Do not blow down the machine as this puts harmful dust into the atmosphere.
4 Follow the manufacturer's instructions on maintenance required	Always dispose of waste oils, liquids and parts in the approved way. Do not dispose of waste oils etc. down drains or in skips – use only authorised methods of disposal.
5 Replace tooling with suitable and legally compliable replacements	Replacements should conform with PUWER and the ACoP for woodworking machines.
6 Check all safety devices, e.g. guards, push sticks	These should conform with all requirements discussed earlier in the chapter.
7 Check extraction hosing for splits etc.	As a temporary fix, heavy-duty tape could be used to repair holes or splits, but this will soon fail, so suitable replacements need to be arranged straight away.
8 Replace any damaged timber lippings or mouthpieces, e.g. bandsaw or rip/ dimension saw	–
9 Set guards and carry out test run	Always carry out a test run after any maintenance. Listen carefully for any unusual noises or vibrations; these are usually signs of a fault developing.

Test your knowledge

1 PUWER is short for which of the following?

   A Provision and Use of Work Equipment Requirements

   B Provision and Use of Work Equipment Regulations

   C Provision and Understanding of Work Equipment Requirements

   D Provision and Understanding of Work Equipment Regulations

2 Before you change tooling on a circular saw bench, the first action should be to:

   A isolate the power supply

   B remove the guard

   C check the riving knife is set correctly

   D turn on the extraction.

3 What is the guard covering the top of a ripsaw blade called?

   A Bridge guard

   B Landing guard

   C Shaw guard

   D Crown guard

4 ACoP is short for what?

   A Approved Code of Provision

   B Approved Control of Practice

   C Approved Code of Principle

   D Approved Code of Practice

5 What three pieces of information should be obtained before carrying out maintenance on fixed and transportable machinery?

   A Manufacturer's maintenance schedule, risk assessment, authorisation

   B Manufacturer's maintenance schedule, authorisation, time sheet

   C Risk assessment, authorisation, HSE permission

   D Manufacturer's maintenance schedule, risk assessment, HSE permission

6 What is the maximum stopping time for the cutter blocks on a surface planer?

   A 12 seconds

   B 10 seconds

   C 6 seconds

   D 3 seconds

7 A saddle used on a circular ripsaw bench will assist with:

   A deep cuts

   B angled cuts

   C using negative tooth pattern saw blades

   D using positive tooth pattern saw blades.

8 Guards on woodworking machines should be:

   A strong, easily adjusted and always used

   B flexible, fixed position and used when required

   C adjustable and used if required

   D set to the largest capacity of the machine.

9 What is the maximum permitted distance from the leading edge of the riving knife to the uprising teeth of the saw blade at table height?

   A 12 mm

   B 10 mm

   C 8 mm

   D 6 mm

10 What is the maximum projection of the knives on a circular cutter block fitted to a surface planer dating after 1995?

   A 1.1 mm

   B 2.1 mm

   C 3 mm

   D 4 mm

11 During a pre-start-up inspection, faulty equipment has been identified. What is the correct order of action?

  A Isolate the machine, put a warning sign on the machine, inform your supervisor

  B Put a warning sign on the machine, inform your supervisor, start the repair

  C Isolate the machine, put a warning sign on the machine, start the repair

  D Carry on with the job, inform your supervisor, start next job

12 What is the minimum cutter projection for chip limited tooling fitted to a spindle moulder?

  A 5.5 mm

  B 2.1 mm

  C 1.5 mm

  D 1.1 mm

13 Where would you find a thrust wheel?

  A Behind the circular saw blade

  B Fitted above the planer block

  C Fitted at the back of the narrow bandsaw blade

  D At the top of the hollow mortice chisel

14 Which one of the following lists the four component parts of a 10 mm standard hollow mortice chisel?

  A Chisel, chisel bush, gouge, gouge bush

  B Chisel, chisel spacer, gouge, gouge spacer

  C Chisel, chisel bush, auger, auger bush

  D Chisel, borer bush, auger, auger bush

15 What is the minimum distance from the back of the outfeed table to the centre of the ripsaw spindle?

  A 1200 mm

  B 1100 mm

  C 12,000 mm

  D 10,000 mm

16 The purpose of a riving knife is to help:

  A stop the outfeed operative from touching the saw blade and to stop the timber from closing up

  B stop the outfeed operative from touching the saw blade and help the timber to close up

  C speed up the feeding of timber into the saw and enable the timber to close up

  D enable the outfeed operative to touch the saw blade and help stop the timber from closing.

17 On what type of machining operation would you use a bonnet guard?

  A Morticing door stiles on a morticer

  B Producing a curved section on a spindle moulder

  C Thicknessing timber to size on a thicknesser

  D Cutting MDF to size on a dimension saw bench

18 Which type of tooth hook should a crosscut saw have?

  A Single angle hook

  B Negative angled hook

  C Sharpened angled hook

  D Positive angled hook

19 What is the purpose of a 'lead-in' area in jig design for curved working with a ring fence?

  A To allow the jig to move freely across the fence

  B To allow sufficient room to locate the jig against the ring fence before making contact with the cutters

  C To allow for sufficient room to clamp the material in the jig

  D To allow for room to test the setting up of the cutter block in relation to the ring fence

20 Too little tension applied by the top pulley on a narrow bandsaw is likely to result in:

  A the blade being under too much strain

  B the blade being unable to move

  C increased risk of the blade coming off the pulley

  D the saw guides being in the way during cutting operations.

CHAPTER 4

CONSTRUCTING CUT ROOFING

INTRODUCTION

Traditional roofing methods require the carpenter to construct a roof by hand using timber components. This is carried out on site using relatively simple tools and jointing methods applied to individual timber components. These components are then erected and fixed on site, in contrast to modern roof design and construction using trusses. All roofing work must comply with the Building Regulations Part A.

LEARNING OBJECTIVES

By reading this chapter you will learn how to:
1 understand roofing types and component terminology
2 construct a traditional cut roof
3 lengthen roofing timbers.

1 ROOFING TYPES AND COMPONENT TERMINOLOGY

Types of roof construction

Traditional cut timber pitched roofs can be divided into three types, the choice of which usually depends on the span required:
- single roof
- double roof
- triple roof.

Cut roofs can be further categorised by their shape. The common ones are:
- flat
- lean-to
- mono-pitch
- gable-end
- hipped-end
- mansard
- valley
- gambrel
- jerkin-head.

KEY TERM

Cut roof: a construction technique that builds up the roof without the use of trusses.

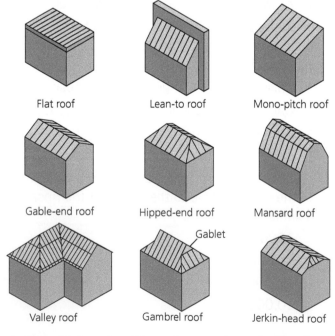

▲ Figure 4.1 Examples of the most common types of roof profile

Single roof

Single pitch roof designs rely on rafters that are supported only at either end. Single roof designs are not suitable for spans of more than 5.5 m because of the exceptional section sizes the timber will require – in its length, width and thickness – in order to support the weight of the roof. The design of the roof should enable the **live** and **imposed loads** of the roof to be transmitted through the rafters so that they do not push the wall out, causing collapse.

> ### KEY TERMS
>
> **Live loads:** the weight of the structure – for example, the timber and tiles or slates on the roof.
> **Imposed loads:** the live loads and any additional loads a roof may incur, such as the weight of snow (so it is sometimes called the snow load).

Single roofs can be divided into three types.

Couple roof

This type of roof has a pair of rafters, each of which typically sits at its lower end on a wall plate that sits on top of the inner wall, and at its highest point in the centre of the roof against another piece of timber called a ridge board. (See the next section of this chapter for more information about the components used within traditional cut roofs.)

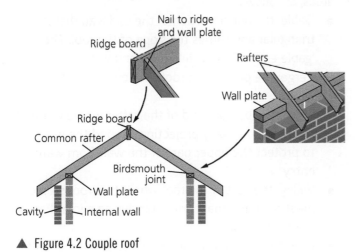

▲ Figure 4.2 Couple roof

Close-couple roof

This type of roof is very similar to the couple roof, except the ceiling joists are fixed to the ends of the rafters at wall plate height. Vertical timbers called 'hangers' can be fixed from the common rafters to the ceiling joists to help stop the ceiling joists from sagging. This type of roof construction can help to reduce the risk of the roof **spreading**.

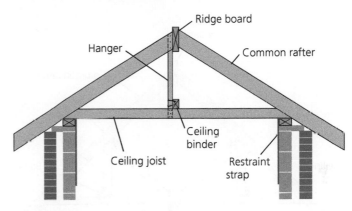

▲ Figure 4.3 Close-couple roof

> ### KEY TERM
>
> **Spreading:** where the roof structure starts to move and pull outwards from the wall plate, causing the ridge to sink and possibly destabilising the whole structure.

Collar-tie roof

This design of roof has a tie part way up the rafters, allowing an increased span. The tie should not be positioned more than one-third of the rise up the rafters.

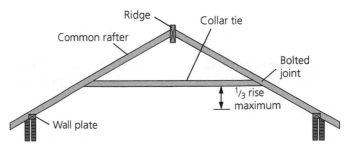

▲ Figure 4.4 Collar-tie roof

Double roof

Double-roof designs allow greater spans to be covered. They require longer rafters but do not have to use timbers with excessively large sectional sizes. This is achieved by using **purlins**, intermediate supports along the rafters which are usually positioned midway down the rafter. Purlins are usually fixed vertically but can be fixed with their face edge square to the common rafter.

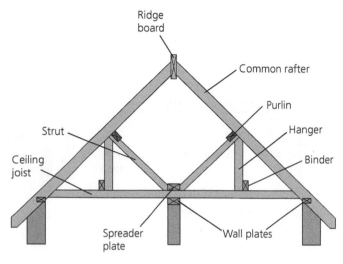

▲ Figure 4.5 Double roof

KEY TERM

Purlin: a large horizontal structural timber used to support rafters.

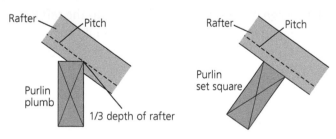

▲ Figure 4.6 Purlin plumb to roof pitch

▲ Figure 4.7 Purlin set at an angle to match pitch line

Triple roof

This type of roof was traditionally used for very large spans. These days, it is mainly found in renovation work, heritage buildings, barn conversions

and oak-framed buildings. This roof design includes a truss, typically a 'king post truss', which is fixed at intervals along the roof to provide intermediate support for the purlins, which in turn support the common rafters.

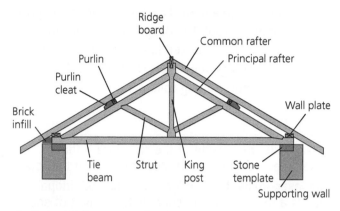

▲ Figure 4.8 King post roof truss

ACTIVITY

Without looking at the illustrations, list and describe as many components as you can from the different roof structures described above.

Components within a traditional cut roof

The traditional cut roof can be divided into several clear areas, as follows.

- Gable: the upper portion of the end wall that is triangular and carries the slope of the roof. The 'gable end' refers to the whole end wall.
- Verge: the part of the roof that overhangs the gable.
- Eaves: the bottom end of the rafter where it meets the walls, normally projecting beyond the walls to protect the upper part of the wall from water entry.
- Valley: the part of the roof where two roof surfaces meet at an internal corner (producing a join shaped like a letter V).
- Hips: the part of the roof where two external roof surfaces meet on an external corner.
- Gablet: a small gable at the top of a hipped roof.

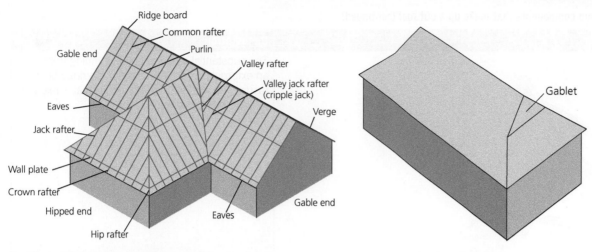

▲ Figure 4.9 Components within a traditional cut roof

▲ Figure 4.10 Gablet on roof

The main components that make up a cut roof are shown in Table 4.1.

▼ Table 4.1 The main components that make up a cut roof

Component	Description
Wall plate 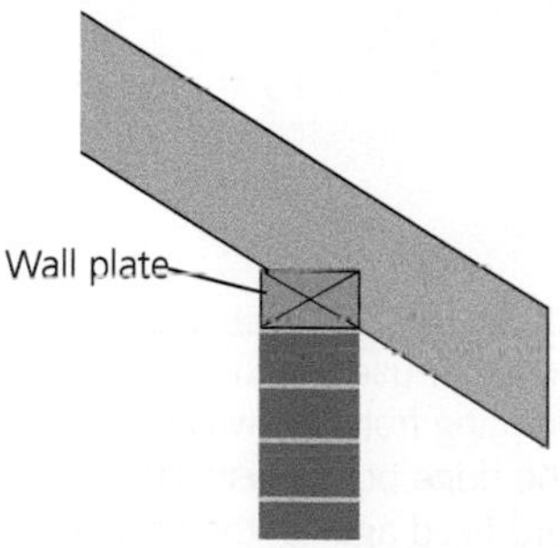	The piece of timber bedded on top of the internal walls, held in place with metal restraining straps to prevent movement. The wall plate helps to spread the load of the roof and provides a fixing for the bottom of the rafter.
Ridge board 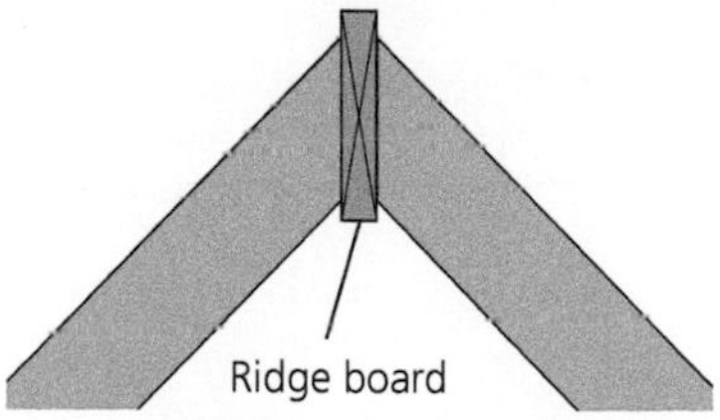	The horizontal board at the apex (top) of the roof against which common rafters are fixed, providing longitudinal support for the roof.
Common rafters 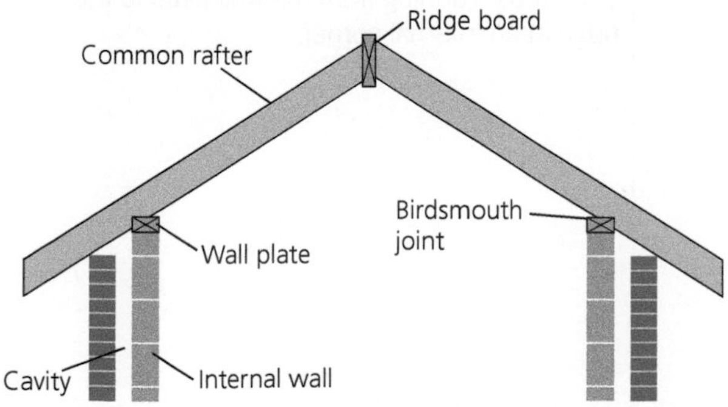	The main load-bearing components of the roof, these are fixed at the lower end by a birdsmouth joint to the wall plate, pass over a purlin (if fitted), and are then fixed at the top of the roof to the ridge board.

→

▼ Table 4.1 The main components that make up a cut roof (continued)

Component	Description
Hip rafter 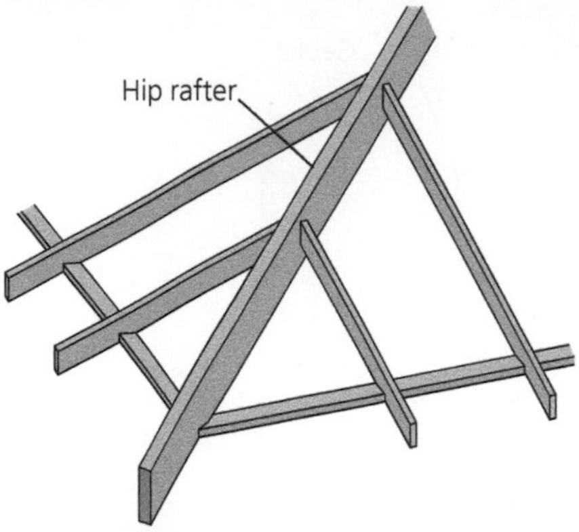 	The substantial piece of timber running from an external corner of the roof to the ridge board (usually up against a saddle board that is fixed up against the end of the ridge).
Saddle board  	A piece of timber or plywood fixed to the end of the ridge board where the first set of common rafters starts; used as a surface to fix the crown rafter and hip rafters to.
Crown rafter 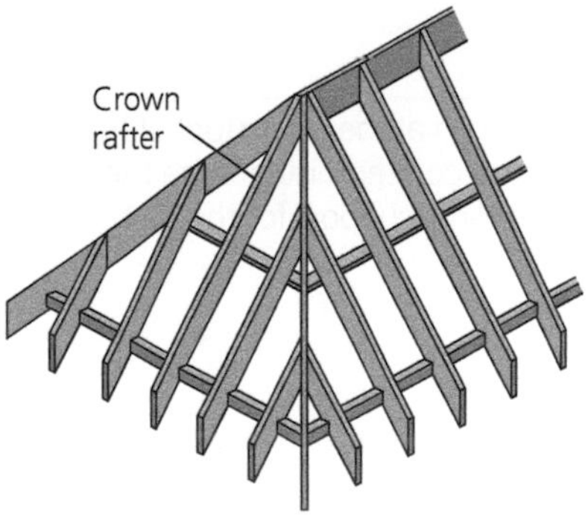 	The longest common rafter used at the centre of the hipped end, running from the wall plate up to the centre of the ridge board (usually up against a saddle board fixed against the end of the ridge).
Valley rafter 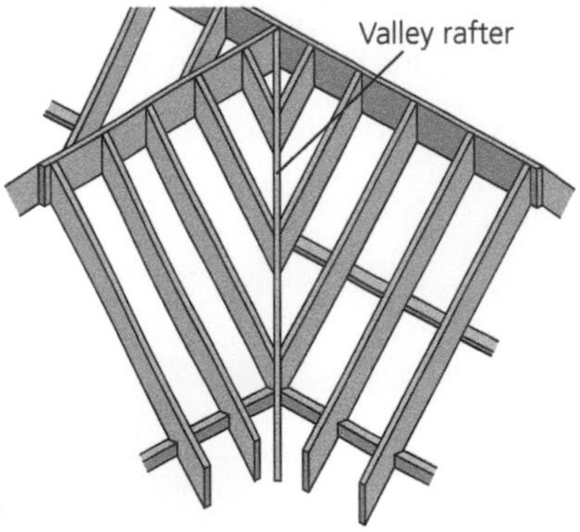 	The timber running from the wall plate to the ridge at an internal corner.

▼ Table 4.1 The main components that make up a cut roof (continued)

Component	Description
Hip jack rafters 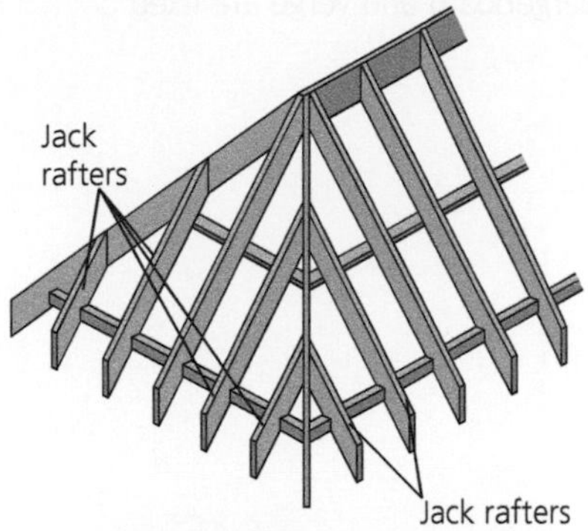	The timbers running from the wall plate to the hip rafter at an external corner.
Valley jack rafters 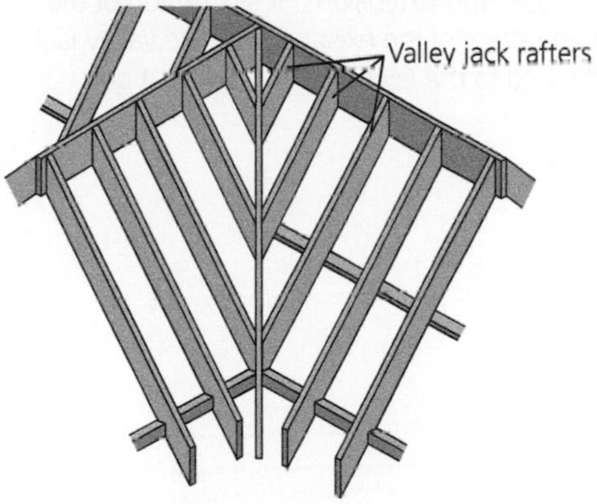	The timbers running from the ridge board to the valley rafter at an internal corner.
Cripple jack rafter 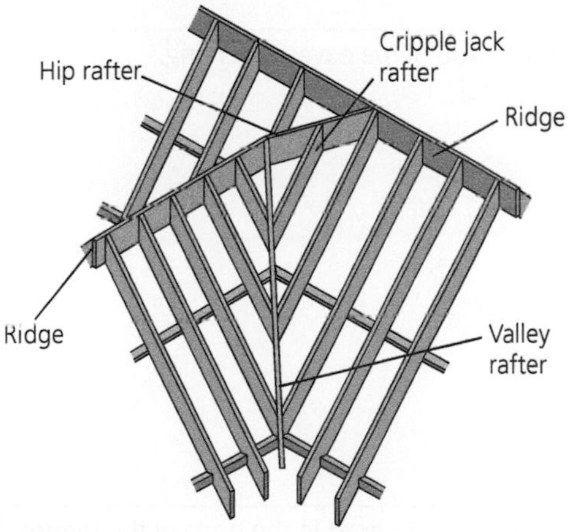	Cripple jack rafters touch neither the ridge board nor the wall plate – they are positioned between the valley rafter and a hip rafter; on this type of roof the hip is often referred to as a broken hip.
Purlin 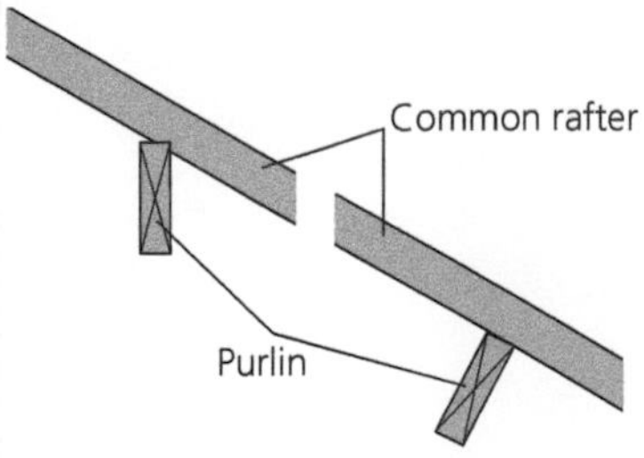	The timbers carrying part of the load of the roof rafters, traditionally placed at right angles to the rafters at the mid-span point but now usually fitted vertically.

▼ Table 4.1 The main components that make up a cut roof (continued)

Component	Description
Gable ladder 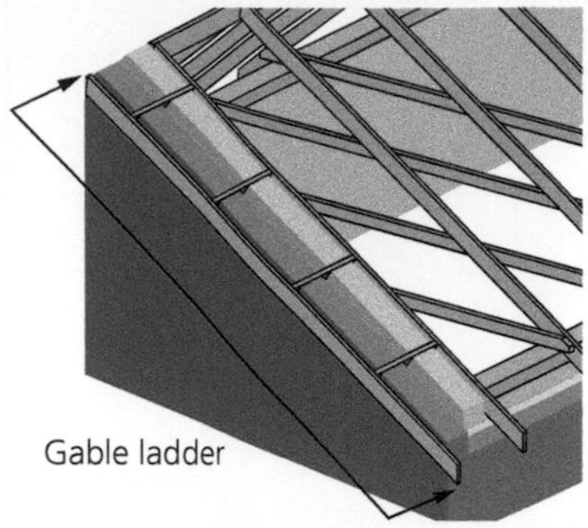 Gable ladder	The framework that overhangs the gable end, to which the bargeboard and verge are fixed.
Lay board Lay board 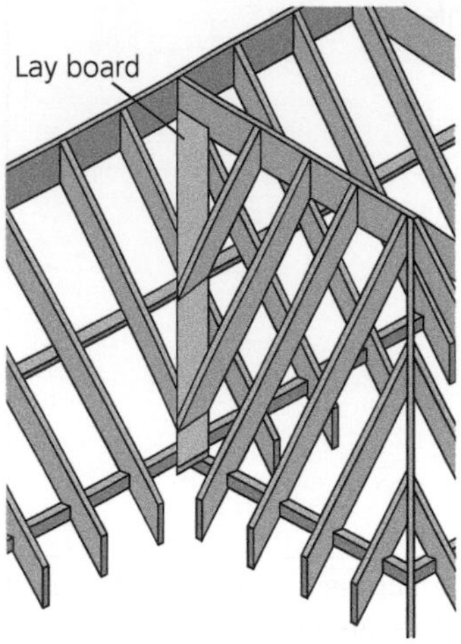	The timber running down the valley, typically for dormer roofs and extensions. It sits on top of the common rafters of the existing roof; the valley jacks are attached to the lay board at their seat cut.
Soffit 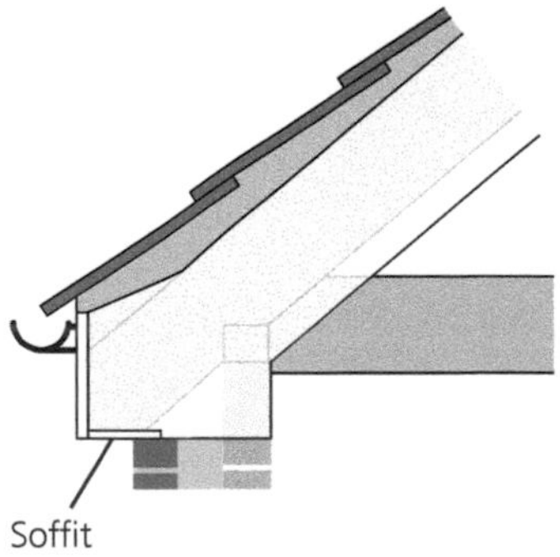 Soffit	The underside of the eaves overhang.
Fascia Fascia 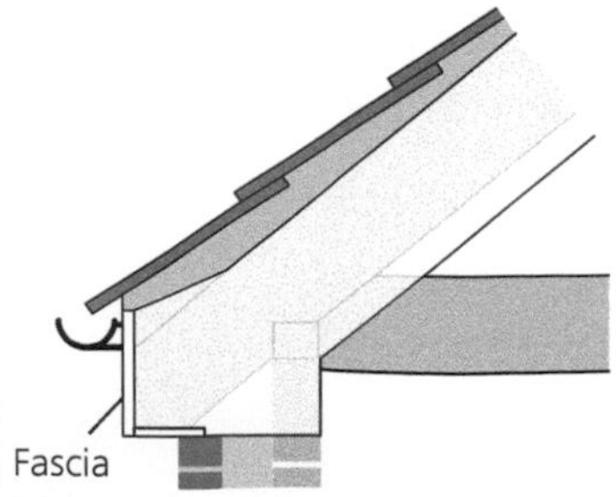	The vertical board fixed to the ends of the rafters, carrying the guttering.

➜

▼ Table 4.1 The main components that make up a cut roof (continued)

Component	Description
Bargeboard 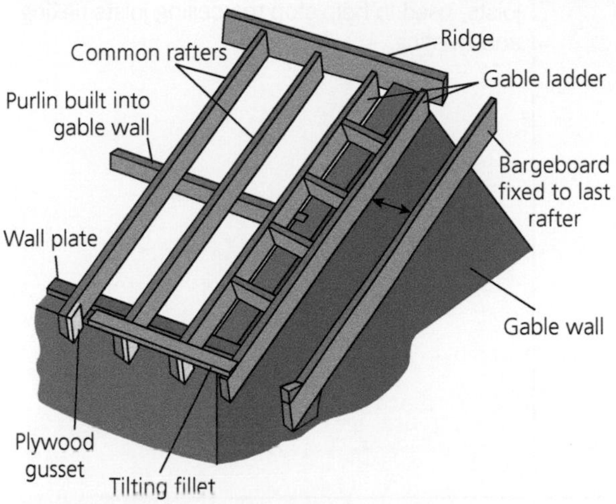	A board fixed parallel to the side of the gable ladder following the pitch line, often showing decorative features, particularly in older and exclusive properties.
Ceiling joist 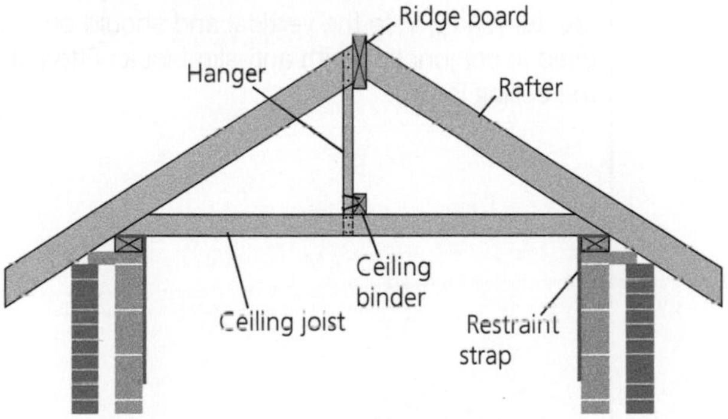	Joist usually fixed at wall plate height to both the wall plate and the common rafter. The underside of the ceiling joist provides the fixing for the ceiling plasterboard.
Collar ties 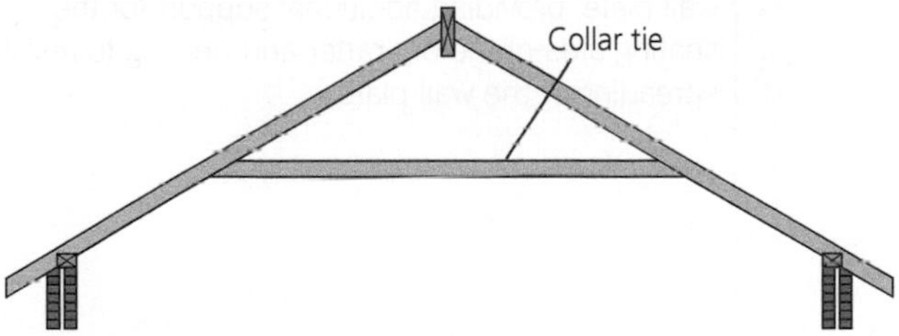	Timbers attached to the common rafters to help reduce the spread of the roof.
Restraining straps 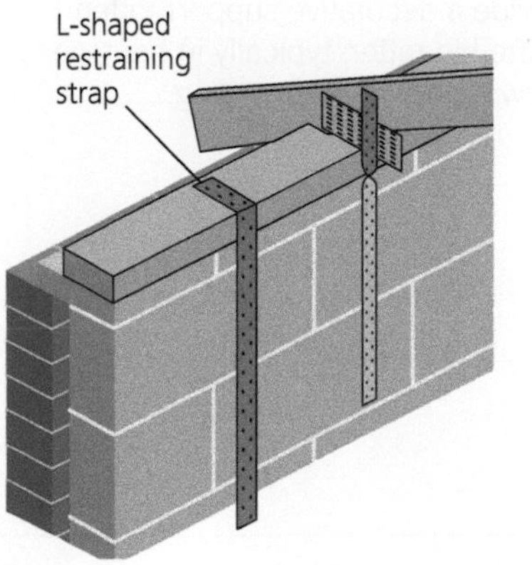	Required by Building Regulations; used to tie the rafters to the gable wall to prevent lateral movement of rafters and to fix down the wall plate. Restraining straps should not be more than 2 m apart.

▼ Table 4.1 The main components that make up a cut roof (continued)

Component	Description
Binders 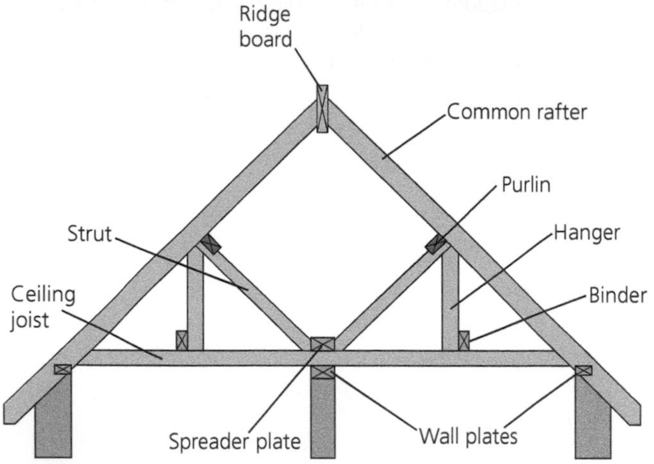	Timbers running at a right angle to the ceiling joists, used to help stop the ceiling joists flexing and twisting.
Struts 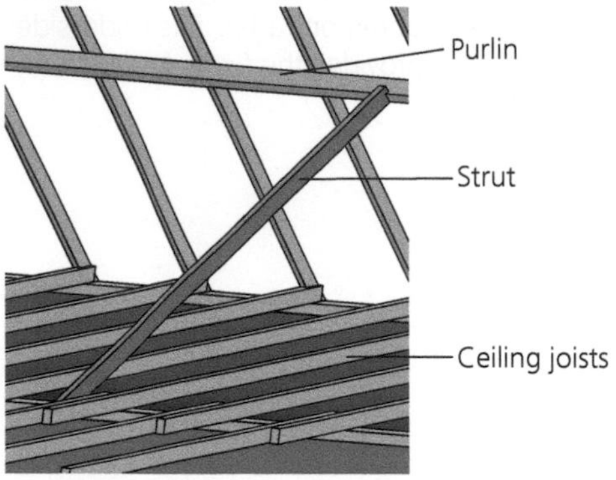	Used to help prevent the purlin from deflecting, these should not be positioned at an angle greater than 35° to the vertical and should be used in conjunction with anti-slip blocks fitted at the ceiling joists.
Angle tie 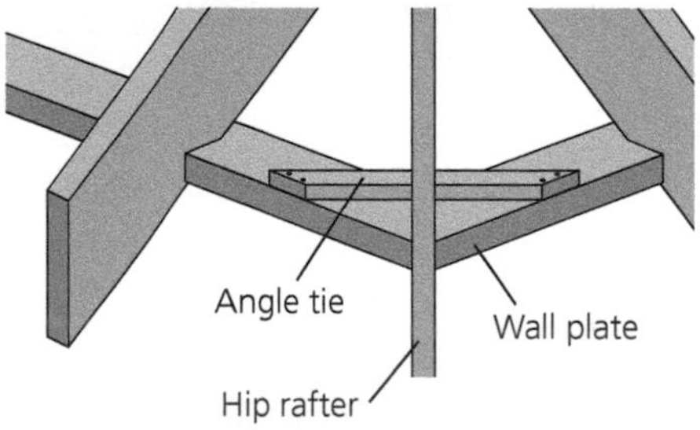	Used with hipped roofs at the corner of the wall plate, providing additional support for the seating area of the hip rafter and helping to resist spreading of the wall plate.
Dragon tie and beam 	Used to provide a decorative support (often carved) for the hip rafter, typically in heritage work and new timber-framed roofing.

▼ Table 4.1 The main components that make up a cut roof (continued)

Component	Description
Trimmers 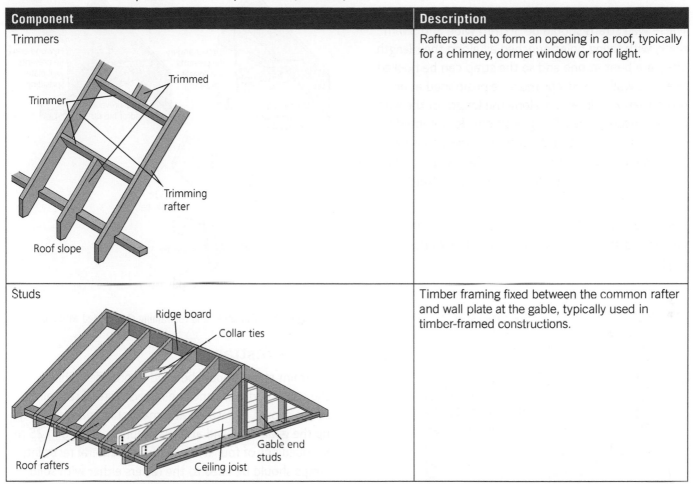	Rafters used to form an opening in a roof, typically for a chimney, dormer window or roof light.
Studs	Timber framing fixed between the common rafter and wall plate at the gable, typically used in timber-framed constructions.

INDUSTRY TIP

Valley jack rafters are nowadays often mistakenly referred to as 'cripple rafters'.

ACTIVITY

Using the components and descriptions in Table 4.1, produce a set of information cards that could be used as information refresher check cards as part of a toolbox talk. Use them to test members of your class or colleagues.

② CONSTRUCTING A TRADITIONAL CUT ROOF

Building Regulations Part A

Part A of the Building Regulations details the requirements for a building's structure. An outline of the guidance given in Part A of the regulations is given below, along with guidance from the NHBC (National House Building Council), in relation to a traditional cut roof.

Wall plates

Wall plates should be a minimum of 3 m in length, bedded parallel and level, extending over at least three rafters. Where joints are required, wall plates should be joined using half-lapped joints including joints at the corners. Where wall plates require joining in their length, a half lap equal to 2.5 times the material width should be used.

Wall plates should be held down to the cavity wall with galvanised mild steel lateral restraint straps. These wall straps should be 30 mm wide and 5 mm thick, with fixing holes punched all along their length. They are bent at one end so the strap can be hooked over the wall. The straps must be positioned at no more than 2 m intervals along the length of the wall plates, ensuring their fixing positions do not interfere with the positioning of the rafters. If the straps fall in the same part of the wall plate as the fixing positions of the rafters, you will need to adjust the straps' position slightly.

Angle ties should be used on hipped roofs, with galvanised steel ties fixing the hip rafter to the angle tie.

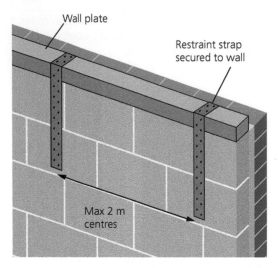

▲ Figure 4.11 Restraining straps used to fix the wall plate to the wall at 2 m maximum centres

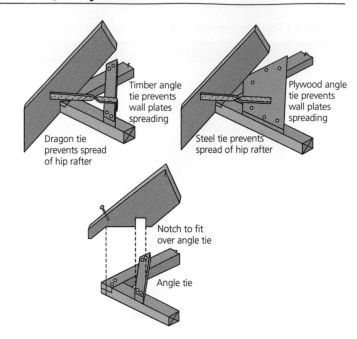

▲ Figure 4.12 Angle ties and restraining straps used with hip rafters

Rafter restraint

Restraint straps should be used to provide stability to the gable-end walls and the rafters. Lateral restraint straps should be located at a maximum spacing of 2 m up the gable wall and at a maximum spacing of 1.25 m for buildings of four storeys or more. Lateral restraint straps should be fixed to the rafters either with solid **noggins** or, alternatively, with 100 mm × 25 mm timber fixed over four rafters.

KEY TERM

Noggins: timbers used as a brace between rafters to help stiffen and provide a fixing point for restraint straps.

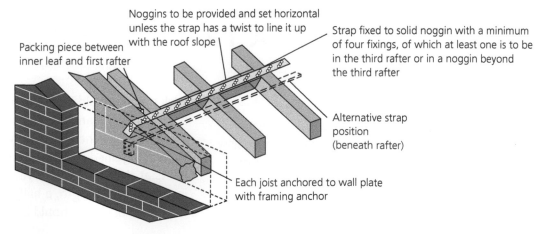

▲ Figure 4.13 Lateral restraint straps up the gable wall

Rafter and purlins

Rafters should be fixed to the wall plate by a birdsmouth joint one-third of the depth of the common rafter, with skew nailing through the rafter to the wall plate. The ceiling joists are fixed to the side of the rafter. Alternatively, truss clips can be used to fix the rafters to the wall plate. Where purlins are used, the purlins should be built into the gable wall. Purlins are usually placed plumb to the roof pitch. The common rafters have a birdsmouth to fit over the purlin; the depth of the birdsmouth should be one-third of the depth of the rafter.

Where purlins require joining in their length, a scarfing joint should be used and supported with a strut. A scarfing joint should also be used when lengthening the ridge board.

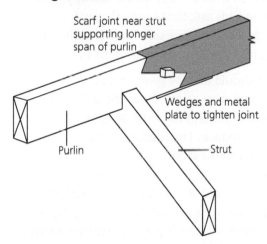

▲ Figure 4.14 Scarf joint in purlin with supporting strut that is birdsmouthed around the purlin and screwed in place

Building Regulations Part L

Part L of the Building Regulations details the requirements for insulation and ventilation in a building's structure. An outline of the guidance given in Part L is provided below, along with guidance from the NSBC in relation to a traditional cut roof.

Insulating pitched roofs

The main purpose of installing suitable insulation is to ensure the dwelling meets the minimum energy performance requirements of the Building Regulations.

Insulating a pitched roof is usually achieved in one of two ways, with the main difference being the positioning of the insulation.

1 A **warm roof** will make the entire structure of the building warm; the insulation is placed on top of the rafters as well as between the rafters. Immediately above the top layer of insulation sits a weather-proof membrane. The warm roof construction has many benefits over a traditional 'cold roof' construction, the main advantage being that the whole dwelling retains the heat, which in turn helps to prevent damp and any associated decay problems. A warm roof is recognised as being the form of roofing most suited to the UK climate, providing both a cost- and thermal-efficient solution.

2 A **cold pitched roof** is where the insulation is placed either between or between and under the rafters or at ceiling joist level.

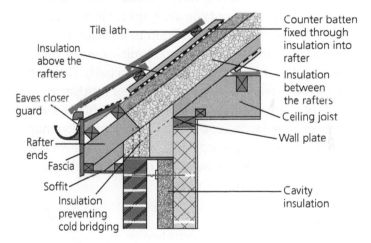

▲ Figure 4.15 Warm roof insulation detail

Ventilating pitched roofs

With a warm roof, only the gap between the underside of the tiles and the vapour control layer requires ventilation. The inner portion of the roof, or the area under the insulation and inside the building, does not require ventilation. Ventilation consists of a 50 mm minimum clear-air path between the insulation and the underlay to ensure a clear airflow from the eaves to the ridge. This can be achieved using counter battens.

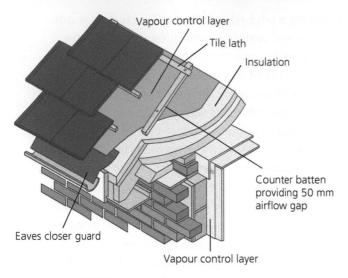

Vapour control layer

Tile lath

Insulation

Counter batten providing 50 mm airflow gap

Eaves closer guard

Vapour control layer

▲ Figure 4.16 Pitched roof with underlay

Ventilation in cold pitched roofs

To avoid condensation in cold pitched roofs that have insulation at ceiling level, the roof void should be ventilated to the outside air. A spacer in the eaves should be used to allow insulation to be installed over and beyond the wall plate, to minimise the cold bridge without blocking the ventilation path. The roof should be fitted with ridge or high-level ventilation. Cold roofs are ventilated at eaves level by fitting soffit, eaves or over-fascia ventilators. Cross-ventilation may also be achieved by using gable-end vents. Where the insulation is placed between the rafters, vents should also be placed along the ridge.

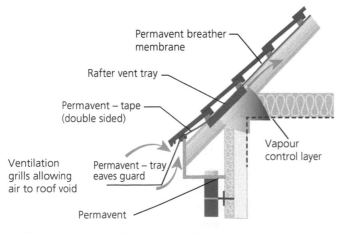

Permavent breather membrane

Rafter vent tray

Permavent – tape (double sided)

Ventilation grills allowing air to roof void

Permavent – tray eaves guard

Vapour control layer

Permavent

▲ Figure 4.17 Cold roof ventilation with insulation placed at ceiling level

Working safely at height

Work on roof structures inevitably requires working at height, regardless of whether the job is new build, refurbishment or alteration. While working at height, particular consideration must be given to safe working practices including the following regulations:

- Health and Safety at Work Act (HASAWA) 1974
- Work at Height Regulations 2005
- Provision and Use of Work Equipment Regulations 1998
- Personal Protective Equipment Regulations 1992
- Manual Handling Operations Regulations 1992.

A detailed risk assessment and method statement will both need to be produced before any work can commence on a roof. These should identify all significant hazards associated with the roof construction, and put in place suitable and sufficient control measures to prevent any risk of injury.

Consideration should also be given to passers-by and how the work may affect them, including noise, plant and equipment, and any dust or debris the roof work may produce. The method statement will identify any control measures and safety equipment required, as well as laying out a safe programme of work. All people on site should be trained, suitably qualified and properly supervised when carrying out construction work; this is particularly important when working at height.

The Work at Height Regulations 2005 place several duties on employers, as follows.

- Working at height should be avoided if possible.
- If working at height cannot be avoided, the work must be properly organised, with risk assessments and method statements completed and adhered to.
- Risk assessments should be updated regularly.
- Those working at height must be trained and competent.

When arranging for personnel to work at height, the following points should be considered.

- How long is the job expected to take?
- What type of work will it be? (It could be anything from fitting a single light bulb to removing a chimney or installing a roof.)
- How is the access platform going to be reached?
- How many people will use the access equipment, and could there be overcrowding?
- Will people be able to get on and off the structure safely?
- What type of access will any scaffold require?
- What are the conditions like? (Extreme weather, unstable buildings and poor ground conditions must be considered.)
- What are the risks to passers-by? Could debris or dust blow off and injure anyone on the ground below?

Fall arrest equipment

It is also necessary to consider whether operatives require fall arrest systems and equipment. If safe working platforms cannot be provided, falls may be prevented if operatives wear safety harnesses with work restraint systems. If such systems cannot be used, effective fall arrest measures must be in place – ideally collective measures, such as safety nets or other soft landing systems. This is particularly relevant when working on cherry pickers and aerial access platforms, and systems should be in place to prevent people falling through open floor joists or roof structures.

> **HEALTH AND SAFETY**
>
> Personal fall arrest systems such as safety harnesses are a last resort after other safety systems have been considered. They should not be the first and only form of protection considered – the ideal situation is not to work at height at all, but this is not always practicable when the job involves a roof.

▲ Figure 4.19 Workers wearing fall arrest safety harnesses on an aerial access platform

Types of access equipment

Access equipment can vary from a simple hop-up to an independent scaffold system, and each type of equipment has its role in working at height. The text that follows outlines different types of access equipment and their typical uses.

Scaffold systems

Independent, or putlog, scaffold systems are the most comprehensive type of access equipment in common use when working at height for prolonged periods. This type of access equipment should be erected and maintained by competent scaffolders and inspected regularly by competent supervisors. These are the best types of access equipment to work from, as they have plenty of room to store materials and, on well-run sites, will have loading

▲ Figure 4.18 Fall arrest equipment

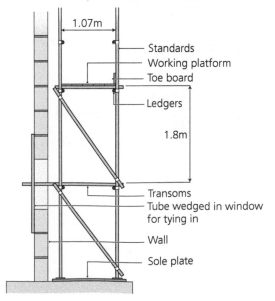

Standards
Working platform
Toe board
Ledgers
1.07m
1.8m
Transoms
Tube wedged in window for tying in
Wall
Sole plate

▲ Figure 4.20 Independent scaffold

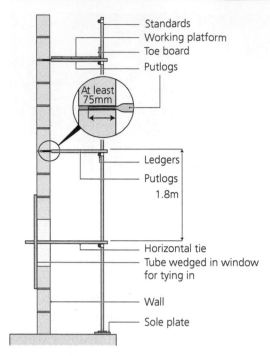

▲ Figure 4.21 Putlog scaffold

bays ready for lifting materials up by crane – or, more commonly, by forklift trucks or all-terrain vehicles with high reach capabilities. These types of access equipment incorporate inward-opening only access gates at access and egress points.

Mobile tower scaffolds are very versatile when working at height, but still need to be properly erected, inspected and maintained.

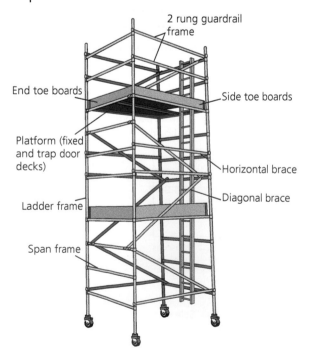

▲ Figure 4.22 Mobile tower scaffold

Ladders

Ladders should be used only for accessing work platforms. It is not acceptable to use a ladder as a workstation when erecting roofs. Ladders should be positioned so they lean at an angle of 75° – one unit out for every four units up. Ladders should be tied off at the top of the scaffolding and should extend at least 1 m above the landing point, to give a good hand-hold position when accessing or egressing the platform. The rung of the ladder should be positioned so that it is level with the working platform.

▲ Figure 4.23 Access ladder extending 1 m above the platform and tied to the scaffold, with the ladder rung level with the working platform

Whatever type of access equipment is used, never adjust or alter it in any way, not even to enable you to work more efficiently. Any adjustments must be carried out only after approval from the installers or other suitably trained personnel and, even then, can be performed only by those who have received suitable training on the erection and adjustment of the access equipment.

ACTIVITY

You have been asked to help select suitable access equipment for use on a two-storey building. You are required to replace the existing roofing members and have been given a time frame of three weeks. Identify the type of access equipment required, stating your reasons, and outline any variable considerations that could affect the length of time necessary to complete the work.

Determining lengths and angles of cut roof components

Information requirements

To correctly calculate the true lengths and angles of roofing components, the following four pieces of information are usually needed.

1 Span: the distance measured from the outside of one wall plate to the outside of the other wall plate.
2 Run: the distance from the outside of one wall plate to the centre of the ridge board, measured along a horizontal plane and equal to half the span.
3 Pitch: the angle at which the roof rises.
4 Rise: the distance from the wall plate to the apex of the roof, measured vertically.

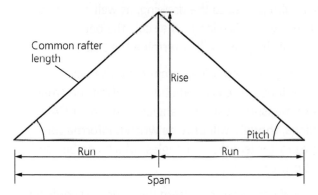

▲ Figure 4.24 Information required to calculate rafter lengths and angles

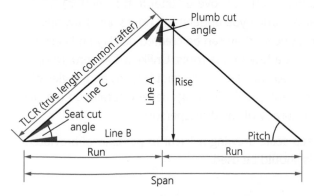

▲ Figure 4.25 Producing true lengths and angles for common rafters

Methods used to determine rafter lengths and bevels

To calculate the correct angles and lengths of roof components, you could use any of the following six methods:

1 full-size setting out
2 scale drawings
3 roofing square
4 rise and run
5 ready reckoner
6 roofing apps and calculations.

Full-size setting out

This can be a useful practical method of establishing rafter lengths and bevels when setting out on the wall plate of the roof. The wall plate needs to contain a right angle, such as with hipped roofing. If the wall plate is not at a right angle on the corner, it will not only give incorrect angles but will also give incorrect true lengths. When the roof does not contain a hipped corner, the setting out can be done on a sheet of plywood or similar material. By joining several sheets together, this method can be used for larger roofs.

Here's how to do full-size setting out:

1 On a sheet of plywood, mark out a right angle. Label the vertical line 'A' and the horizontal line 'B'.
2 Along line A, mark the required rise.
3 Along line B, mark the required run.
4 Join the ends of lines A and B to create line C – this is the true length of the common rafter (TLCR).
5 The angle that is formed by lines B and C is the pitch. It gives the required angle for the seat cut, which will sit on the wall plate.
6 The angle that is formed by lines A and C is the plumb cut, which will sit up against the ridge board.
7 Set up sliding bevels for each of these angles.

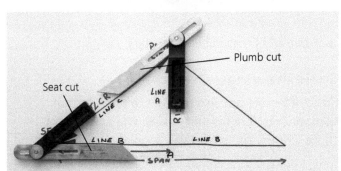

▲ Figure 4.26 Setting and positioning sliding bevels

ACTIVITY

Set out a roof that has a rise of 1000 mm and a run of 1600 mm. Measure the length of the hypotenuse. Is this the same as the length you find if you use Pythagoras' theorem (see Improve your maths, below)? Set up sliding bevels for both the plumb cut and the seat cut.

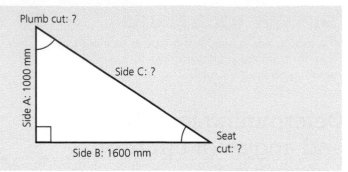

IMPROVE YOUR MATHS

To check you have a right-angled triangle, use the 3–4–5 rule, Pythagoras' theorem:

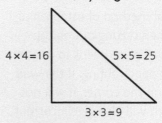

4 × 4 = 16
5 × 5 = 25
3 × 3 = 9

If a triangle has sides measuring 3, 4 and 5 (these can be any unit of measurement as long as the same units are used for each side), it must be a right-angled triangle with a 90° angle between the shorter sides. This method is based on Pythagoras' theorem from geometry, which is as follows:

A^2 (3 × 3 = 9) + B^2 (4 × 4 = 16) = C^2 (5 × 5 = 25)
9 + 16 = 25 for a right-angled triangle

C is the longest side (**hypotenuse**), and A and B are the two shorter sides.

If the distance is less than 5 units, your corner is less than 90°. If the distance is more than 5 units, your corner has an angle of more than 90°. All the angles in a right-angled triangle add up to 180°.

KEY TERM

Hypotenuse: the side opposite the right angle; the longest side of a right-angled triangle.

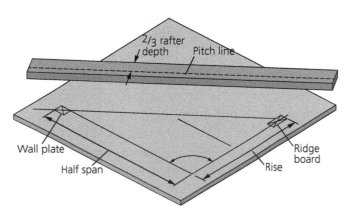

▲ Figure 4.27 Example of full-size setting out

Scale drawings

Scale drawings can also be used to produce roof components' lengths and angles. The accuracy of these lengths and angles will depend on the skill of the person who produced the drawing, as well as the scale of the drawings. The larger the scale the better, but the size of the paper will determine the scale.

Drawings produced using computer-aided design (CAD) will be more accurate but are often too small to take measurements and angles from. Most CAD drawings, though, will provide written information such as rafter lengths and angles, which can be used to set out the rafters.

Using scale drawings can seem daunting and confusing compared with using a roofing square or a ready reckoner (methods covered later in this section). However, once you understand how scale drawings are produced, they can be created quickly and easily with just a few simple tools. Unlike a CAD program, which also calculates the exact lengths and angles of components, scale drawings depend totally on the skill and accuracy of the draftsperson.

To keep scale drawing as clear as possible, abbreviated labels should be used.

▼ Table 4.2 Abbreviated labels for roofing components

Abbreviation	Definition
CR	Common rafter
VR	Valley rafter
TL	True length
PC	Plumb cut
SC	Seat cut
EC	Edge cut
TLCR	True length common rafter
TLHR	True length hip rafter
TLVR	True length valley rafter
TLJR	True length jack rafter
TLCrR	True length cripple rafter
SCCR	Seat cut common rafter
SCHR	Seat cut hip rafter
ECP	Edge cut purlin
ECVJ	Edge cut valley jack

Abbreviation	Definition
HR	Hip rafter
JR	Jack rafter
SCVR	Seat cut valley rafter
SCJR	Seat cut jack rafter
SCCrR	Seat cut cripple rafter
PCCR	Plumb cut common rafter
PCHR	Plumb cut hip rafter
PCVR	Plumb cut valley rafter
PCJR	Plumb cut jack rafter
PCVR	Plumb cut valley rafter
ECHR	Edge cut hip rafter
ECVR	Edge cut valley rafter
ECCrR	Edge cut cripple rafter
SCP	Side cut purlin
SCVJ	Seat cut valley jack

To produce scale drawings, you will need the following:

- scale rule
- tee square
- set square
- protractor
- good-quality compass
- 2H pencil
- paper
- drawing board
- tape to attach the paper to the board.

ACTIVITY

Using the equipment list above and following the steps below, develop the true lengths and angles of roof components.

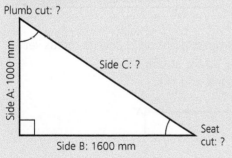

The following steps show the procedure for producing a scaled drawing for a common rafter, as shown in Figure 4.28. You can use your own sizes and scales to suit your paper size.

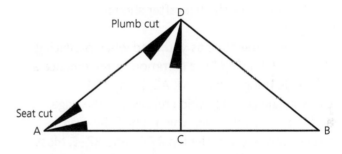

▲ Figure 4.28 Common rafter true length, seat and plumb cuts

1 Choose a suitable scale to enable the developed roof to fit on the paper.
2 Draw line A–B to the required scaled span of the roof.
3 Mark the run of the roof C, which is equal to half the span.
4 From point C, mark the rise of the roof up to point D.
5 Lines A–D and B–D are the true lengths of the common rafter.
6 Angles CAD and CBD are the seat cut angles.
7 Angles ADC and CDB are the plumb cut angles.
8 Mark and label the angles, set sliding bevels to the angles and transfer to a pitch board for use in the workshop.

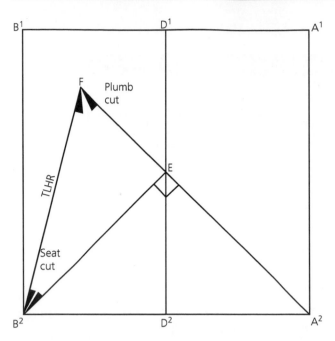

▲ Figure 4.29 Hip rafter true length, seat and plumb cuts

The following steps show the procedure for producing a scaled drawing for the hip rafter shown in Figure 4.29.

1 Using the same scale as you used when producing the scaled drawing for a common rafter, produce a rectangle using corners A1, A2, B2 and B1.

2 Draw in line D1–D2, which represents the ridge and is equal to the run, which is half the span.

3 Draw a plan view of the hip A2–E and B2–E. Most hips are set at 90° so point E is equal to an angle of 45° from points A2 and B2.

4 Extend line A2–E from point E a distance equal to the rise point F.

5 Draw line B2–F. This is the true length of the hip rafter (TLHR). The plumb cut is at point F, being the angle between lines F–E and F–B2. The seat cut is at point B2, being the angle between lines B2–E and B2–F.

6 Mark and label the angles, set sliding bevels to the angles and transfer to a pitch board for use in the workshop.

The dihedral angle or backing bevel to the hip rafter

The **dihedral angle** is the angle at which the two sloping roof surfaces meet against the hip rafter. This angle provides a level surface for the tile battens to be fixed to the hip rafter. The dihedral angle has dropped

out of fashion and is seldom used nowadays, mainly due to increased pressure to produce roof structures quickly.

The following steps show the procedure for producing a scaled drawing for the dihedral angle on the hip rafter shown in Figure 4.30.

1 Using the scale drawing produced for the hip rafter, mark point G along line B2–F. It can be at any distance along this line, but if it is too close to B2 then the angles will be so small that it will be difficult to see; too close to point F and it will become large and over-complicated. About halfway along the line will give an acceptable size to set the sliding bevel.

2 From point G, mark a line at a right angle to line B2–F until it strikes line B2–E at point H.

3 At point H, mark a line at a right angle to line B2–E until it strikes the wall plate at points I and J:

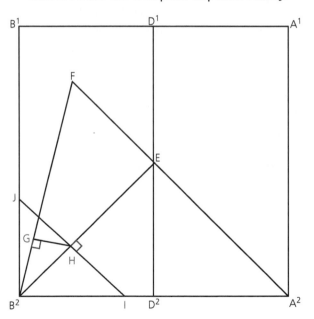

4 With your compass point on H and its radius set at G–H, mark an arc at point K.

5 Draw in lines I–K and J–K:

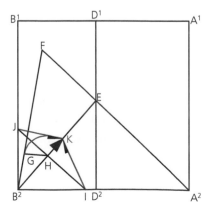

6 The dihedral angle for the hip rafter is the angle between B2–K and I, and between B2–K and J. Mark and label the angles.

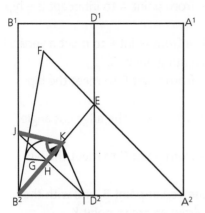

▲ Figure 4.30 Completed dihedral angle

Edge cut for the hip

The following steps show the procedure for producing a scaled drawing for the edge cut to the hip rafter shown in Figure 4.31.

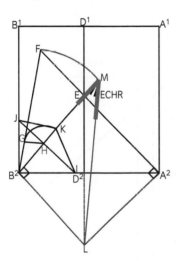

▲ Figure 4.31 Edge cut to hip rafter

1 Either use the scaled drawing produced for the hip rafter or repeat steps 1 to 6 for producing a scaled drawing for a hip rafter to create a new drawing.

2 Extend line D1–D2 down until it meets the lines taken at right angles from line B2–E and A2–E to create point L.

3 With your compass at point B2 and the radius set at B2–F, mark an arc to meet extended line B2–E at point M.

4 Mark line M–L, which is the edge cut to the hip rafter at point M.

Lengths and angles of the jack rafters

The following steps show the procedure for producing a scaled drawing for the jack rafter true lengths and angles shown in Figure 4.32.

1 Repeat the steps outlined for the drawings above to produce a scaled drawing for the common rafters and hip rafters.

2 With your compass centre at A and the radius set as A–D, draw an arc to point N.

3 From point N, draw down a vertical line to point P (which is horizontally across from point E, the position of the first common rafter).

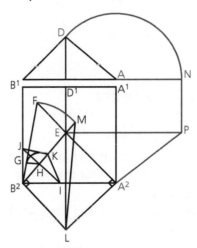

4 Draw on jack rafters at the correct **centres**:

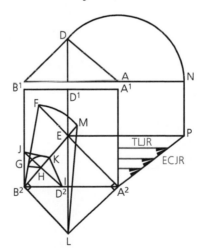

KEY TERMS

Dihedral angle: the angle given to the top edge of the hip rafter running up its length from the eaves to the ridge.

Centres: the distance from the centre of one rafter to the centre of the next.

5   Mark on the edge cut angle and true length:

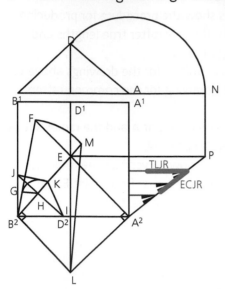

▲ Figure 4.32 Jack rafter edge cuts and lengths

Edge and side cuts for purlins

The following steps show the procedure for producing a scale drawing for the edge and side cuts for the purlins shown in Figures 4.33 and 4.34.

1   Produce a common rafter at the required pitch (40°) and a hip rafter plan view (45°).
2   Draw the purlin on common rafter points ABCD, as shown here:

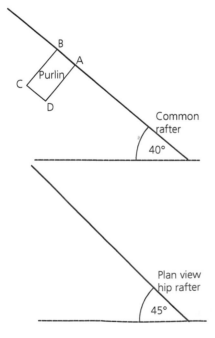

3   Draw a horizontal line through point B.
4   Using a compass centred at B, and with the radius set as A–B, mark an arc to point E.
5   Drop a vertical line from point A to intercept the hip rafter at point F.
6   Take a horizontal line from point F to meet a vertical line from point E, creating point G.
7   Drop a vertical line from point B to meet the hip rafter at point H.
8   Join points G and H to achieve the edge cut angle for the purlin.
9   Drop a vertical line from point C to meet the hip rafter at point J.
10 Using a compass centred at point B, and with the radius set as B–C, draw an arc to point K.
11 Draw a horizontal line from point J to intercept a vertical line from K, creating point L.
12 Join points H and L to achieve the side cut for the purlin.

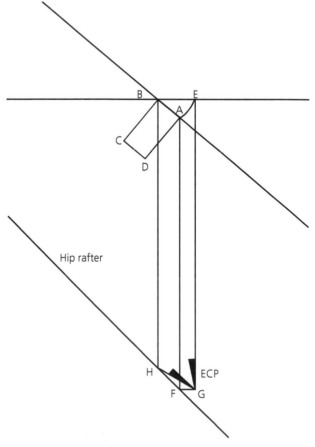

▲ Figure 4.33 Edge cut to purlin

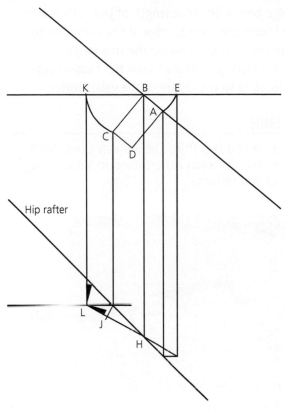

▲ Figure 4.34 Side cut to purlin

Valley rafter and valley jack lengths and angles

For the development of valley rafters, valley jack rafter lengths and angles, follow the same principles as those used for hip rafters and jack rafters.

1 Produce a plan view of the roof, showing the valley.

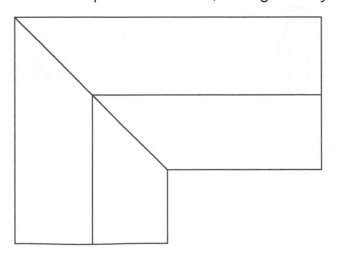

2 Mark the rise at point A, drawing a line 90° from the plan view of the valley rafter position, to give point B.
3 Mark a line from point B to the wall plate in the valley to give point C.
4 Mark the plumb cut angle between lines A–B and B–C.
5 Mark the seat cut angle between A–C and B–C.

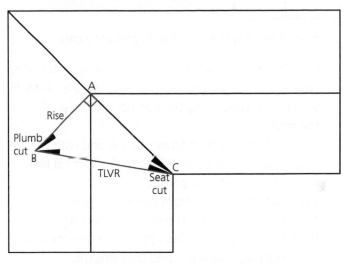

▲ Figure 4.35 Valley rafter true length, seat and plumb cut

The edge cut to the valley rafter can be produced using the following method.

1 Use the drawing produced for the valley rafter. At point C, mark a line at 90° to line A–C to point D, the ridge line.
2 Using a compass centred at point C, and with the radius set to C–B, mark an arc to point E.
3 Draw line D–E, and mark the edge cut angle between lines C–E and D–E.

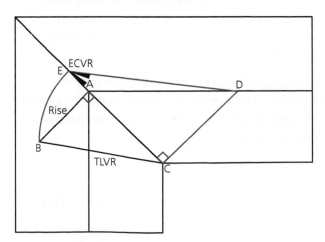

▲ Figure 4.36 Edge cut to valley rafter

The valley jack and cripple rafter lengths and angles can be produced using the following method.

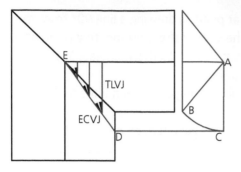

▲ Figure 4.37 Valley jack rafter true lengths and angles

1   Use the drawing produced for the valley rafter edge cut or produce a new drawing. Add the elevation of the roof containing the correct pitch/rise of the roof.
2   Using a compass centred at point A and with the radius set to A–B, mark an arc to the vertical line taken from point A to point C.
3   Take a horizontal line from point C to point D.
4   Take a line from point D to point E, the ridge line.
5   Mark the valley jack rafters at correct centres, and mark the edge cut angle and true lengths.

Roofing square

The roofing square is a valuable addition to the carpenter's toolbag, as well as an aid to setting out right angles on a large scale, such as when working on stud walling. Many other activities fall within its scope, including stairs and roofing.

The roofing square has a wide part called the blade and a narrow part called the tongue. Both the tongue and the blade have millimetres marked along their edges. They can also have pitch degrees marked and a chart giving rafter lengths per metre run for standard pitches.

By calculating the appropriate metre run for the required pitch, the true length of rafters can be worked out. The plumb cut is marked off along the blade and the seat cut along the tongue.

The true length of a rafter is always measured along the **pitch line**. The pitch line for a common rafter runs from the outside edge of the wall plate to the centre

of the ridge board. The true length of jack rafters is measured from the outside edge of the wall plate to the centre line of the hip rafter; the true length of valley jack rafters is measured from the outside edge of the wall plate to the centre of the valley rafter.

KEY TERM

Pitch line: a line two-thirds of the way down from the top of the common rafter, used to set out the lengths of the rafters.

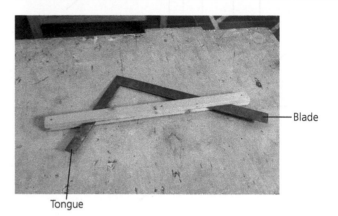

▲ Figure 4.38 Roofing square

Producing the common rafter

The steps in Table 4.3 demonstrate the use of a roofing square to produce common rafters and hip rafter cuts. In the example described, the specification for the roof is:

- span of 3.6 m
- 40° pitch
- eaves overhang of 300 mm
- ridge board thickness of 25 mm.

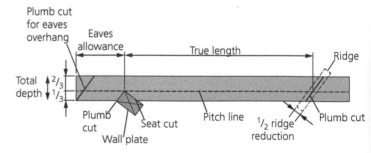

▲ Figure 4.39 Example of common rafter that is set out

▼ Table 4.3 Using a roofing square to produce common rafters and hip rafter cuts: step by step

Step 1	Mark a line along the common rafter at two-thirds of its depth. This is the pitch line.	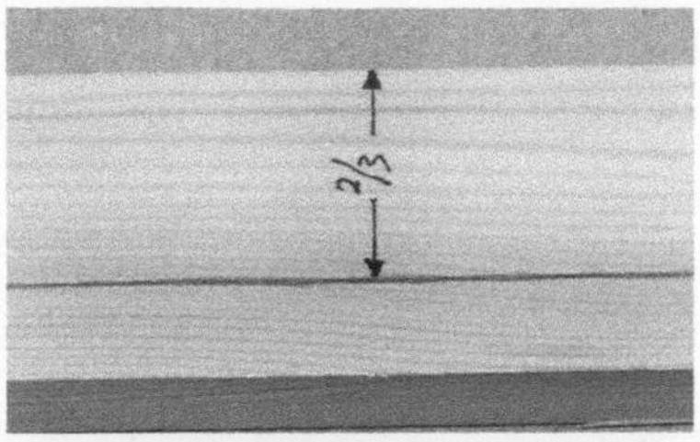
Step 2	Position a roofing square on the common rafter with the required pitch on the pitch line, while the common rafter marking is also positioned on the pitch line.	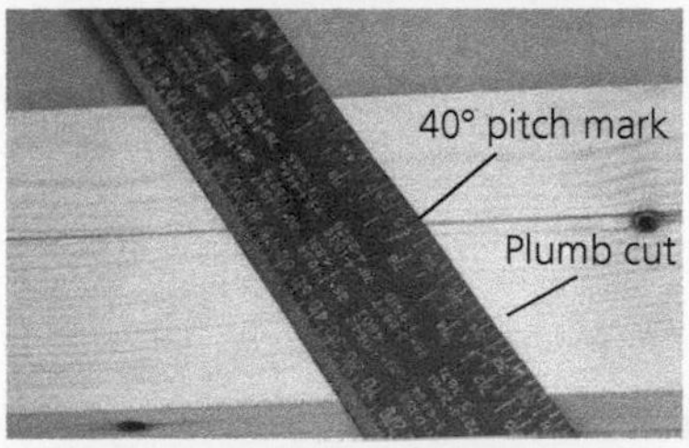 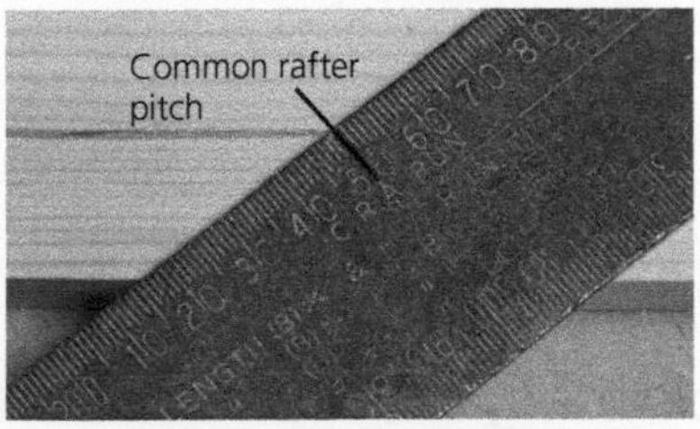
Step 3	Set sliding bevels to the seat cut and plumb cut angles.	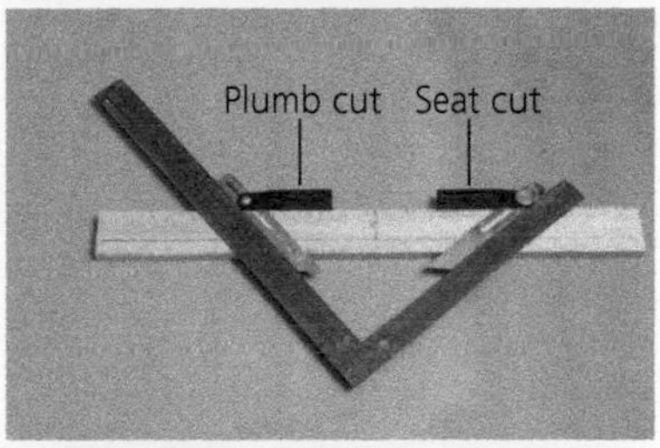
Step 4	Mark the plumb cut for the eaves finish of the fascia board and mark out the birdsmouth for the wall plate, allowing the correct eaves overhang. This should be measured at a right angle from the plumb cut.	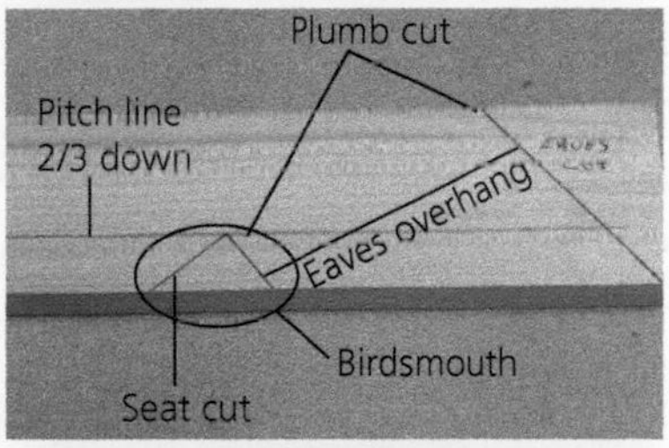
Step 5	Calculate the true length of the common rafter. In this example for 40° pitch, it is 1.3054 mm × 1800 mm (half the span) = 2.3497 m. The bottom number (1.6444 mm) is used to calculate the hip rafter length.	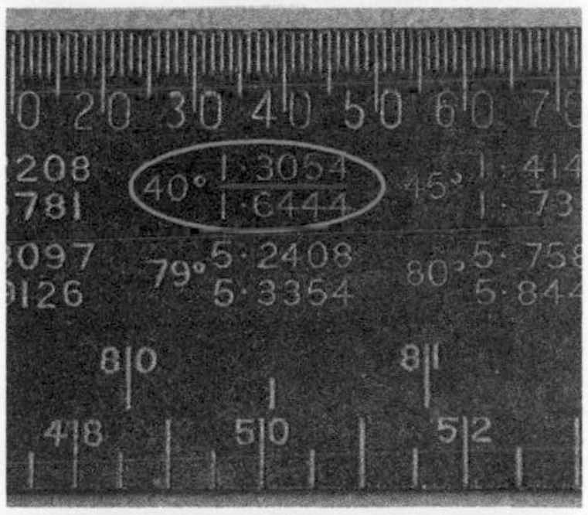
Step 6	Round up 2.3497 mm to 2.350 mm and measure along the pitch line. This is the measurement to the centre of the ridge board. Measure back half the thickness of the ridge board to allow the plumb cut of the common rafter to sit against the ridge board.	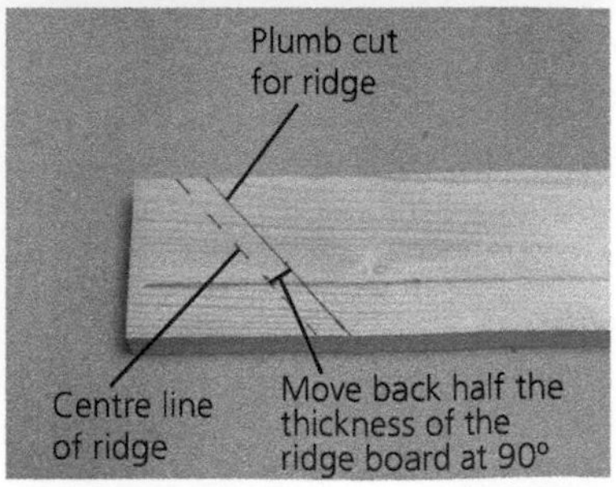

ACTIVITY

Set out a common rafter using a roofing square using the specification given above.

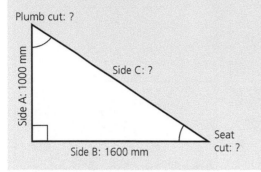

Plumb cut: ?

Side A: 1000 mm

Side C: ?

Side B: 1600 mm

Seat cut: ?

INDUSTRY TIP

The birdsmouth on a rafter incorporates the seat cut and plumb cut. Combined, they always form a 90° angle that sits on the wall plate.

Once the setting out has been completed, the first rafter can be cut (apart from the eaves overhang cut, which is usually done after fitting). The rafter can then be used as a pattern to mark out the remaining rafters.

Producing the hip rafter without the dihedral angle

▼ Table 4.4 Producing the hip rafter without the dihedral angle: step by step

Step 1	Mark the pitch line from the top surface of the rafter **equal** to that of the common rafter pitch line and **not** two-thirds down the thickness of the hip rafter.	Same depth as 2/3 the common rafter and *not* 2/3 the depth of the hip rafter
Step 2	Position the roofing square on the pitch line to the required pitch (40°) and mark point B on square, which is the required positioning of the square for hip rafters.	
Step 3	Set sliding bevels to the correct plumb and seat cuts; label each one to reduce the chances of mixing them up.	
Step 4	Mark the plumb cut for the end of the eaves overhang, and measure up the pitch line the required amount to the start of the birdsmouth. Then mark out the birdsmouth.	

▼ Table 4.4 Producing the hip rafter without the dihedral angle: step by step (continued)

Step 5	Calculate the length of the hip rafter: 1644 mm × 1800 mm (half the span) = 2959.2 m (this can be reduced to 2.959 m). This is to the centre of the ridge board along the pitch line and will require a reduction. The reduction is equal to the diagonal distance from the face of the saddle board to the centre of the ridge board, as if you were to continue the pitch line of the hip rafter. This measurement works only if all other factors with the roof are correct; for example, if the hip is exactly 90°, the pitch is exactly 40°, the run is exactly 1.8 m and the crown rafter is exactly 40°. Considering all the variables, it is better to take a working measurement from the roof, as outlined in the following steps.	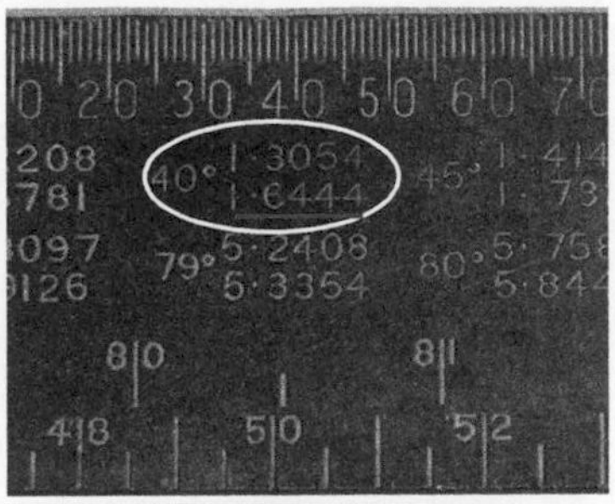 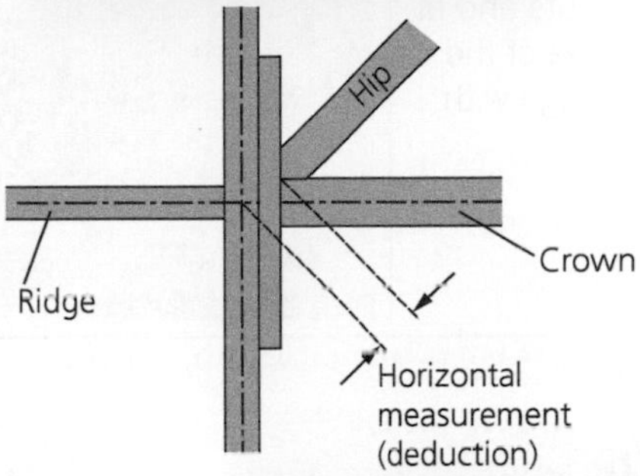
Step 6	Measure the width of the hip rafter across the gap produced at the top of the crown rafter, saddle board and first common rafter; this should be done at 45° to the crown and common rafters. This mark is the point from which to start the measurement.	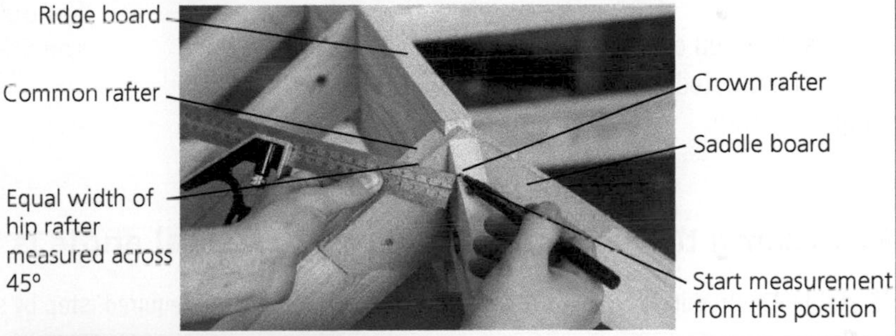
Step 7	Cut back the corner of the wall plate at 45° equal to the width of the hip rafter. The overall measurement can now be taken from this cut-back edge of the wall plate to the mark at the top of the crown rafter marked in step 6 (equal to the width).	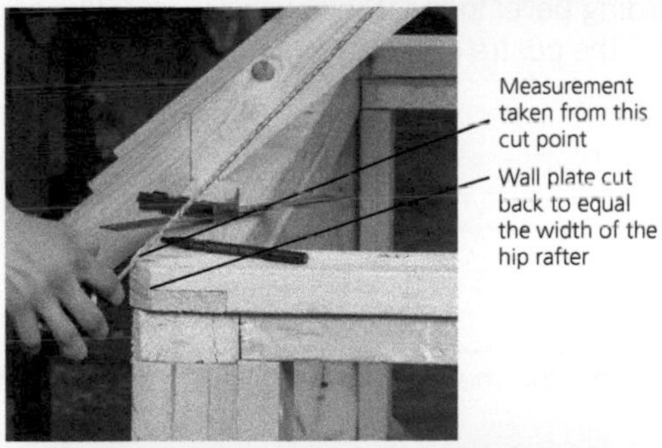 

→

▼ Table 4.4 Producing the hip rafter without the dihedral angle: step by step (continued)

Step 8	Mark this measurement from the pitch line at the birdsmouth to the top of the hip rafter. Mark on the edge cut, creating a compound cut.	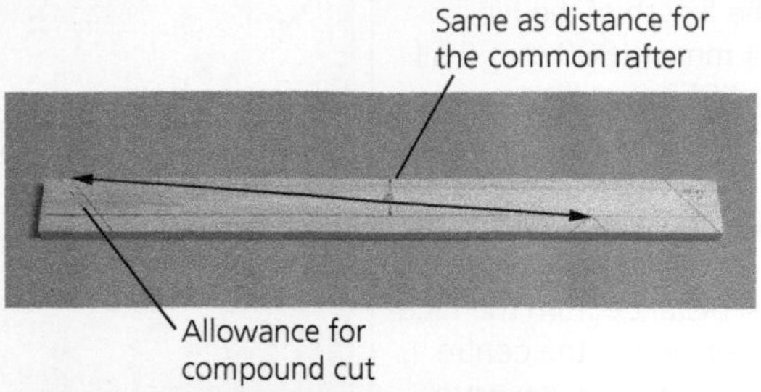 Same as distance for the common rafter Allowance for compound cut
Step 9	Cut the seat and plumb cuts and fit the hip rafter. The top edge of the hip rafter should line through with the common rafters.	

INDUSTRY TIP

In practice, you can take a physical measurement from the part-constructed roof, using a length of tile lath or a long tape measure.

ACTIVITY

Set out a hip rafter using a roofing square from the specification given above.

Producing the hip rafter when the dihedral angle is required

▼ Table 4.5 Producing the hip rafter when the dihedral angle is required: step by step

Step 1	Having set the sliding bevel to the dihedral angle, mark this angle on the hip rafter from the centre of the rafter.	
Step 2	Take the vertical pitch line measurement from the common rafter.	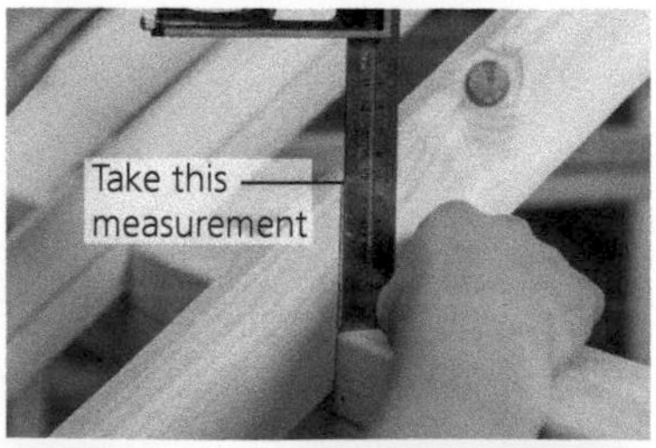 Take this measurement
Step 3	Mark the plumb cut position on the hip rafter, ensuring you allow enough material for the eaves overhang.	

▼ Table 4.5 Producing the hip rafter when the dihedral angle is required: step by step (continued)

Step 4	Measure down the plumb line using the measurement taken in step 2. The measurement must start from the dihedral line and will end at the starting point for the seat cut.	
Step 5	Mark off the seat cut.	
Step 6	Measure across the gap produced at the top of the crown rafter, saddle board and first common rafter. This mark is the point from which to start the measurement up the hip rafter from the seat cut.	
Step 7	After cutting back the wall plate equal to the width of the hip rafter, the overall measurement can be taken.	
Step 8	Measure from the seat cut to the dihedral angle line, applying the measurement taken in step 7.	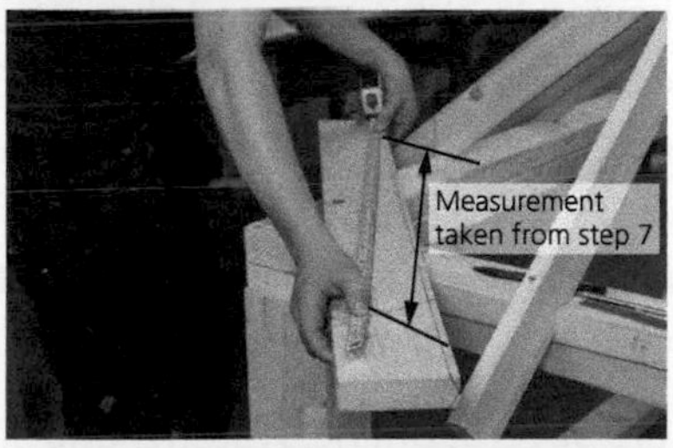
Step 9	Mark the edge cut to the top of the hip rafter, forming a compound cut. Plane the dihedral angle down the hip rafter. Cut both the seat cut and compound plumb cut. Fit the hip rafter, ensuring that the dihedral angle lines through with the top of the common rafters.	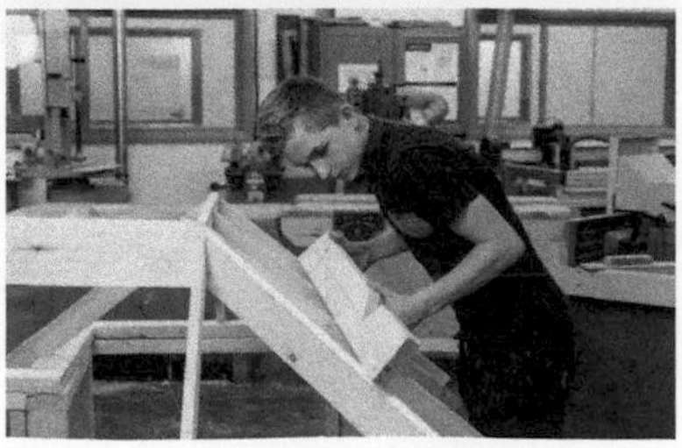

▲ Figure 4.40 Dihedral angle on hip rafter lining through with the top of the common rafters, showing setting out is correct

Valley rafters are produced in a similar manner to hip rafters but the valley rafter's birdsmouth has a compound cut to fit into the corner of the wall plate, ensuring the best possible solid and secure fit.

Rise and run

The rise and run method is a way of finding the pitch on an existing roof. The horizontal distance of the roof is measured as a fraction of the vertical distance and is traditionally represented as a ratio of feet and inches, e.g. 8 : 12. This means that, for every 12 inches of run, the roof rises 8 inches. As it's a ratio, it can also be applied to metric measurements – for example, we could use 4 : 10 (400 × 1000) or 3 : 10 (300 × 1000).

Typically, the roofing square is used at 'one-tenth' scale, i.e. if the run is 3000 mm and the rise is 2500 mm, this would be applied to the square as 300 mm and 250 mm. This is then stepped along the rafter 10 times to achieve the full length. If 300 mm is used for the run and the rise equals 300 mm, then the pitch of the roof will be 45°.

There are several online systems and smartphone apps available to convert the rise into degrees of pitch.

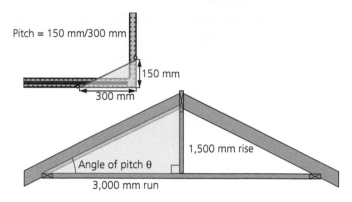

▲ Figure 4.41 Finding the pitch

IMPROVE YOUR MATHS

Use an online app to work out the pitch of a roof having a rise of 1500 mm and a run of 3000 mm or a ratio of 6 : 12.

Ready reckoners

Ready reckoner books consist of a series of tables that provide the required angles for the seat and plumb cuts for common rafters, hip rafters, valley rafters, jack rafters and the purlin. Also provided are the diminish lengths for the jack rafters. The 'diminish' is the amount that each jack rafter is shorter than the last as it steps down the hip.

To allow the true lengths of the common and hip/valley rafters to be calculated, the ready reckoner books also include a table stating the required length of the rafter per metre run or part of a metre – look at Figure 4.43 to see an example of a chart for a roof with a 40° pitch. Using the example shown, we can work out the true length of the common rafter for the roof.

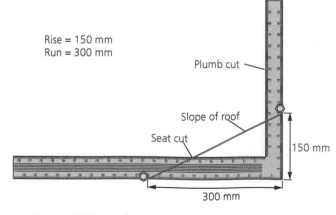

▲ Figure 4.42 Rise and run method for producing seat and plumb cuts

RISE OF COMMON RAFTER 0.839m PER METRE OF RUN   PITCH **40°**

BEVELS: COMMON RAFTER   – SEAT   40
               "   "   – RIDGE   50
           HIP OR VALLEY   – SEAT   30.5
               "   "   "   – RIDGE   59.5
           JACK RAFTER   – EDGE   37.5
           PURLIN   – EDGE   52.5
               "   – SIDE   57.5

JACK RAFTERS 333mm CENTRES DECREASE 435   (in mm to 999 and
              400   "   "   "522   thereafter in m)
              500   "   "   "653
              600   "   "   "783

Run of rafter	0.1	0.2	0.3	**0.4**	0.5	0.6	0.7	0.8	0.9	**1.0**
Length of rafter	0.131	0.261	0.392	**0.522**	0.653	0.783	0.914	1.044	1.175	**1.305**
Length of hip	0.164	0.329	0.493	0.658	0.822	0.987	1.151	1.316	1.48	1.644

▲ Figure 4.43 Extract from ready reckoner/rafter table for 40° pitch (measurements in bold are used in the following example)

Example

A roof has a span of 4.8 m, an eaves overhang of 400 mm and a pitch of 40°.

Step 1 – Divide the span by 2:

$$4.8 \text{ m} \div 2 = 2.4 \text{ m}$$

Step 2 – From the example chart, we can see that the rafter length is 1.305 m for every 1 m of run. Multiply this by 2:

$$2 \times 1.305 \text{ m} = 2.61 \text{ m}$$

Step 3 – The chart shows that for 0.4 m of run, the rafter length is 0.522 m. Add this to the previous calculation:

$$2.61 \text{ m} + 0.522 \text{ m} = 3.132 \text{ m}$$

Step 4 – For the eaves overhang, the chart shows that for 0.4 m run, the rafter length is 0.522 m. Add this to the previous calculation:

$$3.132 \text{ m} + 0.522 \text{ m} = 3.654 \text{ m}$$

The true length of the common rafter in the example is 3.654 m. This is the length of the rafter along the pitch from the edge of the fascia to the centre of the ridge board. Half the thickness of the ridge board must be deducted from this length to arrive at the required cutting length.

Alternatively, deduct half the thickness of the ridge board from the run before calculating the lengths; this will then give a measurement to the face of the ridge board. Using the above information, the rafter can now be set out and cut.

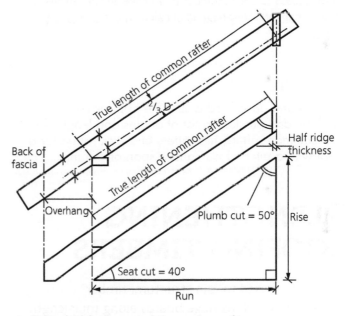

▲ Figure 4.44 Setting out of the common rafter

IMPROVE YOUR MATHS

Using both the ready reckoner book and an online application, follow the above specification but substitute the pitch with a 50° pitch. Work out the true length of a common rafter. Give a short presentation to your tutor or supervisor on which of the two methods you would rather use and why.

Roofing apps and calculators

With modern smartphones and tablets etc. mobile apps and calculators for roofing can be used. There are several available that provide all the information required to set out roof rafters, including angles and lengths of the rafters. On the down side, though, not all work locations allow easy connection to a mobile internet signal and there is always a risk of damaging your phone, tablet or laptop.

INDUSTRY TIP

You can download apps or use websites that have electronic versions of ready reckoner books. However, you won't always have access to the internet or your phone on site, so this traditional hard-copy method of calculation remains very useful.

ACTIVITY

Source and download a suitable app for your mobile, and produce a list of angles and lengths for the common rafters, hip rafters and jack rafters for a roof with a pitch of 40° and a span of 4.5 m.

3 LENGTHENING ROOFING TIMBERS

The method used to join structural timber components such as ridge boards along their length is known as 'scarfing'. This technique produces a joint with a smooth or flush appearance on all its faces. In the case of scarfing joints used in roofing (usually found in the purlins or ridge boards), the length of the scarfing joint should be between two and a half and four times the depth of the component being jointed. These joints incorporate hardwood wedges that are driven into the joint to force together and lock all the surfaces, before the joints are glued with a suitable adhesive such as polyvinyl acetate or polyurethane-type adhesive.

Most scarfing joints are 'fished jointed'. A fished jointed beam also incorporates steel or timber plates fastened to either side of the scarfing joint and then bolted together.

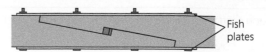

▲ Figure 4.45 Structural scarfing joint that is fished jointed

Fixing and positioning rafters

Before fixing and positioning rafters, ensure that the wall plate has been levelled and is correctly positioned to ensure the span is correct and consistent along its length. The wall plate should be fixed down using restraining straps fixed at centres no more than 2 m apart. The wall plate and ridge board can now be marked out.

HEALTH AND SAFETY

Before starting work on any roof, always double check it has suitable working platforms that are correctly guarded and properly installed.

ACTIVITY

Research different methods of preventing falling from height while working on a roof, then produce a set of toolbox talk safety cards detailing safety measures that should be used with each method.

Gable-ended roof

Below is a method for marking out the wall plate and fixing rafters for a gable-ended roof.

1 Mark out the first set of rafter positions, 50 mm from each gable wall.
2 Following the roof specification, mark out the remaining rafter positions. These will usually be at 400 mm, 450 mm or 600 mm centres.
3 Mark out the ridge board from the wall plate, ensuring enough length is left to allow the ridge to carry on through the gable end to form part of the gable ladder, if required.
4 Temporarily fit the ceiling joists, and plank out to provide a working surface (Figure 4.46).
5 Fit the two outer pairs of rafters onto the wall plate. This is usually done by skew nailing through the birdsmouth on each side of the rafter into the wall plate and/or using metal fixing plates attached to the wall plate and the rafter. Temporarily brace each pair of rafters as you work (Figure 4.47).

6 Temporarily attach a length of tie lath to the top edge of the ridge board to provide a stop at which the common rafters will sit. This tile lath stop also helps to prevent the ridge board from falling during fixing.

7 Push the ridge board up through the gap at the top of the rafters, allowing the plumb cuts of the rafters to sit up against the ridge board and tie lath. It usually takes several operatives to achieve this safely and efficiently.

8 Correctly position the ridge board and fix the rafters by skew nailing.

9 If a purlin is required, it is usually positioned now to aid the fixing and positioning of the remaining rafters, and to help overcome potential problems with sagging rafters. However, this can only be done if the gable wall has been constructed. The purlin is either built into the gable wall or the wall is corbelled out to support the purlin. The purlin can be installed with its depth in a vertical position or at an inclined angle to match the pitch of the roof.

10 Position and fix the remaining rafters.

11 Correctly position and fix the ceiling joists against the sides of the rafters.

12 Allow for the gable wall to be finished, if not already done. Fit noggins for the gable ladder.

13 Fit purlins, if not previously done.

14 Fit diagonal struts for the purlins, to prevent any deflection.

15 Finish the roof at the verge and eaves with the gable ladder, bargeboard, fascia and soffit, as required.

INDUSTRY TIP

Steel or glulam beams are now commonly used as purlins because suitably sized solid timber is becoming more and more difficult to obtain and, as with all structural roof components, the material used for a purlin needs to be suitably stress graded and to have dimensions specified by a structural engineer.

INDUSTRY TIP

Large span cut roofs incorporating a purlin sometimes also have hangers and binders fitted to help stiffen the ceiling joists and prevent them sagging.

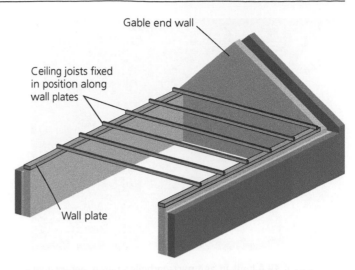

▲ Figure 4.46 Gable roof erection, using ceiling joists, which can be decked out to use as a temporary work platform

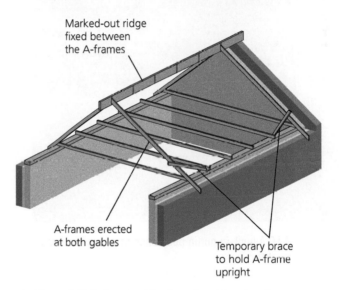

▲ Figure 4.47 Fixing the ridge board and bracing the common rafters

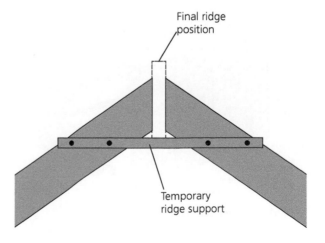

▲ Figure 4.48 Close-up of temporary ridge support fixed to common rafters

▲ Figure 4.49 A built-in and part-corbelled purlin, set vertically

▲ Figure 4.50 A built-in purlin positioned at an incline to match the pitch of the roof

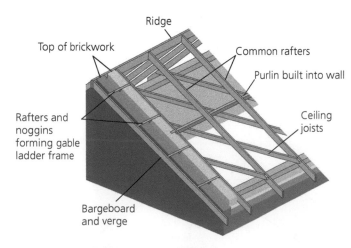

▲ Figure 4.51 Gable end roof detailing

Hipped-ended roof

Outlined below is an example of a method for constructing a hipped-ended roof.

1 On a correctly laid and fixed wall plate, mark the position of the crown rafter. The run of the roof will be the centre of the crown rafter.
2 The same measurement as the run is marked up on both sides of the roof. This is the centre position of the first set of common rafters. Repeat on the other hipped end if required.
3 Mark the positions of the remaining rafters to the correct centres, as specified in the architect's drawing, usually 400 mm, 450 mm or 600 mm.
4 Fit the first set of rafters as described in steps 1 to 8 in the previous section on fixing the rafters in a gable-ended roof.
5 Fit the saddle board to the outer end of the first sets of rafters, those that will form the hip.
6 Fit the crown rafter, remembering to allow for saddle board thickness deduction.
7 Cut the corner of the wall plate at 45° to equal the thickness of the hip rafter.
8 Fit angle ties across the corner of the wall plate as required. These provide an extra supporting surface for the hip rafter and help to reduce the risk of spreading at the corner of the wall plate.
9 Check the length of the hip rafter before cutting. If the roof is even slightly out of square, plumb or pitch, the hip rafter length could alter considerably.
10 Cut and fit any purlins required, along with strutting.
11 Position and fit the remaining common rafters.
12 Fit the jack rafters, allowing for the diminish in lengths.
13 Correctly position and fit the ceiling joists, including the short return joists to the feet of the jack and crown rafters on the hipped end.
14 Fit any collars and hangers, as required.
15 Finish the roof at the eaves overhang, as required.

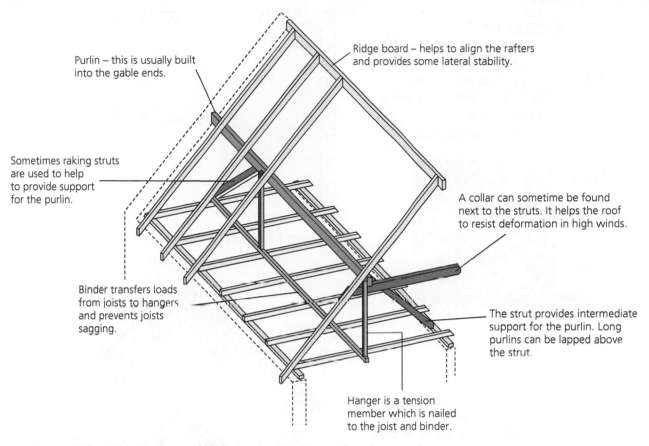

Purlin – this is usually built into the gable ends.

Ridge board – helps to align the rafters and provides some lateral stability.

Sometimes raking struts are used to help to provide support for the purlin.

A collar can sometime be found next to the struts. It helps the roof to resist deformation in high winds.

Binder transfers loads from joists to hangers and prevents joists sagging.

The strut provides intermediate support for the purlin. Long purlins can be lapped above the strut.

Hanger is a tension member which is nailed to the joist and binder.

▲ Figure 4.52 Example of hangers and binder being used with a purlin on larger roof spans

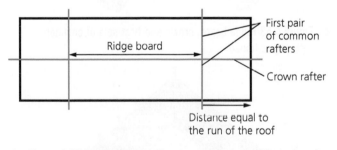

Ridge board

First pair of common rafters

Crown rafter

Distance equal to the run of the roof

▲ Figure 4.53 Setting out for the crown rafters and first sets of common rafters

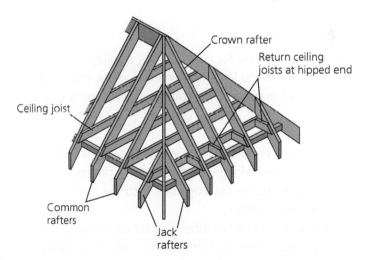

Crown rafter

Return ceiling joists at hipped end

Ceiling joist

Common rafters

Jack rafters

▲ Figure 4.55 Hipped-ended roof detailing

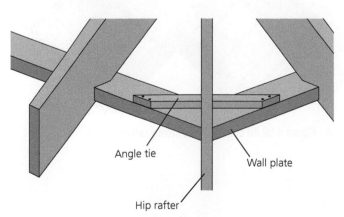

Angle tie

Wall plate

Hip rafter

▲ Figure 4.54 Angle ties used with a hip rafter

INDUSTRY TIP

A 'dragon tie and beam' is a method of providing extra support for the hip rafter at its junction with the wall plate and angle tie. It is often found in traditional building methods and timber-framed construction.

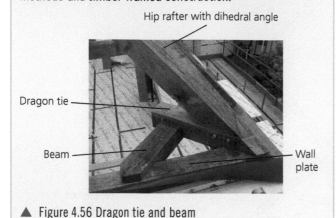

Hip rafter with dihedral angle

Dragon tie

Beam

Wall plate

▲ Figure 4.56 Dragon tie and beam

ACTIVITY

Research traditional timber roofing methods used with hip rafters and dragon tie and beam, and draw suitable jointing methods for the hip rafter and dragon tie and beam to the wall plate.

Valley roof

Below is a method for constructing a valley roof.

1 On a correctly laid and fixed wall plate, mark out the wall plate for the position of the common and crown rafters. The run of the roof will be the centre of the crown rafter.
2 Fix the first pair of common rafters to the wall plate at each end, along with both crown rafters.
3 Position and fix the ridge board to the common and crown rafters.
4 Position and fix the intermediate common rafter.
5 Cut and fix the hip rafter and the hip jack rafters.
6 Cut and fix the valley rafter, which is measured and marked out in a similar way to the hip rafter. The valley jack rafters can now be marked out and cut; valley jacks should always be positioned opposite a common rafter and will have the same plumb cut as the common rafter and the same edge cut as the hip jack rafters.

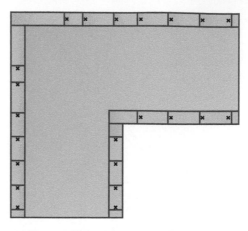

▲ Figure 4.57 Setting out the rafter positions for a valley roof

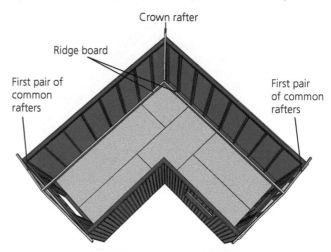

Crown rafter

Ridge board

First pair of common rafters

First pair of common rafters

▲ Figure 4.58 Ridge board, crown and first sets of common rafters positioned

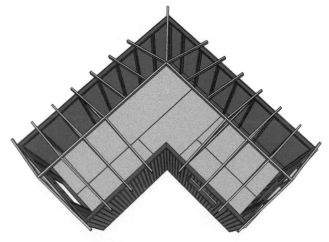

▲ Figure 4.59 All common and crown rafters fitted

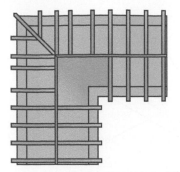

▲ Figure 4.60 Hip rafter and hip jack rafters fitted

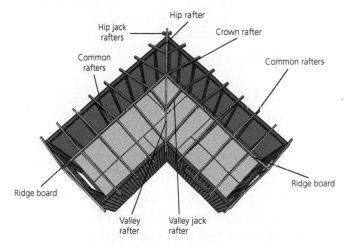

▲ Figure 4.61 Valley roof components

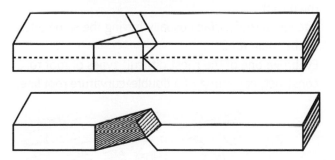

▲ Figure 4.62 Example of seat cut for valley rafter

Forming openings in roof structures

When openings are required in roof structures, for features such as chimneys, roof lights (windows in the roof) or **dormer** windows, the roof structure must be adjusted and strengthened around these features to ensure that the roof is still fully supported and weathered.

To allow for the increased gap between roof members, which usually accrues as a result of accommodating these features, the rafters around the feature will require cutting and strengthening. This process will involve cutting out parts of some rafters and fixing

additional support to carry the weight of the roof over the missing rafter sections. This process is called 'trimming an opening' or 'trimming out'.

> **KEY TERM**
>
> **Dormer:** an opening in the roof surface where a secondary roof projects away from the plane of the existing roof, typically used to give greater head room within the primary roof – for example, in a loft extension.

Trimming an opening

The rafters that are used to surround an opening in a roof have the following names.

- Trimming rafter: the rafter that runs from the ridge board to the wall plate alongside the opening, with a trimmer rafter fixed to its side.
- Trimmer rafter: the rafter that runs between the two trimming rafters. It can have a trimmed rafter fitting up against it.
- Trimmed rafter: the rafter that is shortened at either the plumb cut or the seat cut end, which instead fits up against the trimmer rafter.

As both the trimming and trimmed rafter take increased loads due to the gap created by the opening, each of these rafters will need to be increased in strength. This is a requirement of the Building Regulations Part A and the NHBC standards. This extra strength is achieved either by increasing the rafter section size or by doubling (having another rafter cut and bolted to the side of the first).

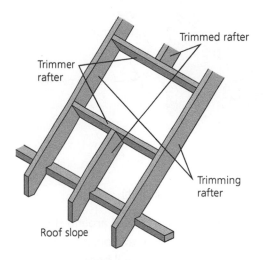

▲ Figure 4.63 Rafters used to trim an opening

To comply with Building Regulations Part A and the NHBC standards, when constructing an opening around a chimney, the rafters need to be placed with a 50 mm clearance gap between the rafters and the brickwork of the chimney. This is to prevent the passage of damp from the chimney wall to the timber. There also needs to be at least 200 mm clearance from the inside face of the chimney flue, to prevent the risk of combustion. Additional rafters may need to be fitted to comply with your specification's maximum spacings.

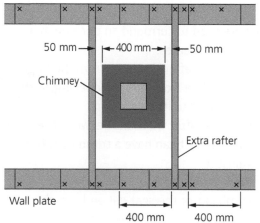

▲ Figure 4.64 Trimming around a chimney (a)

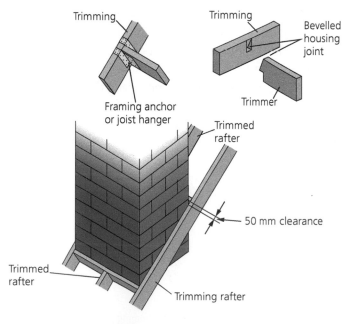

▲ Figure 4.65 Trimming around a chimney (b)

To enable the roof surface to be finished around the chimney, and to prevent water penetration, a back gutter, front apron and side flashing will need to be constructed.

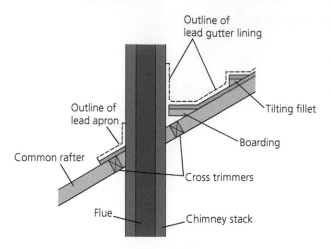

▲ Figure 4.66 Forming a chimney opening

Constructing dormer windows and roof lights

Dormer windows are usually referred to by their roof shape or design, which typically falls into one of four categories.

1 Flat: although these are flat along their horizontal plane, they usually have a fall from the main roof.
2 Segmental or arched: a roof surface that is curved in its horizontal plane running back towards the main roof.
3 Pitched: a roof surface usually having the same pitch as the main roof.
4 Eyebrow: one of the most difficult types of roof to construct, consisting of a double-curvature roof line.

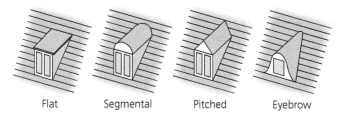

▲ Figure 4.67 Dormer roof styles

Typical framing arrangements for dormer roofing

As with chimney openings, double rafters are required on either side of the opening that is to be used for the dormer window. These will carry the increased loads caused by the opening and its subsequent framework. The pitch used with dormer windows is usually the same as the main roof pitch, and the rafters and eaves overhang are constructed in much the same way as for the main roof components.

The stud framework is used to form the dormer wall cheeks, which sit on top of the trimming rafters of the opening. The wall plate for the dormer roof is fixed to the top of the stud framework, while the rafters of the dormer are constructed and fixed to the dormer wall plate in the same manner as for the main roof.

To allow for the valley jack rafters to be fixed on a dormer roof, a lay board is placed on top of the main roof's rafters at an inclined angle following the dormer roof pitch. The valley jack rafters are fixed at one end to a new ridge board using the same plumb cut as the common rafters. They are fixed to the lay board at the other end using a compound cut comprising the same seat cut as the common rafters, and an edge cut to match the valley rafter edge cut.

Lay board outer edges should line through from the outside edge of the ridge board to the outer edge of the first common rafter

Valley jack rafters have their seat cut resting on the lay board with the top edge of the valley jack rafter lining through with the lay board

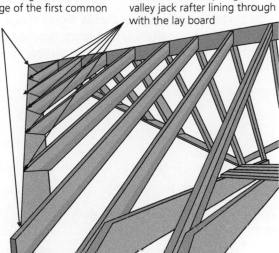

▲ Figure 4.70 Lay board and valley jack rafters

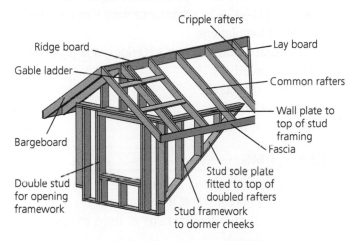

▲ Figure 4.68 Pitched dormer framing arrangement

Cripple rafters
Ridge board
Gable ladder
Lay board
Common rafters
Wall plate to top of stud framing
Fascia
Bargeboard
Double stud for opening framework
Stud sole plate fitted to top of doubled rafters
Stud framework to dormer cheeks

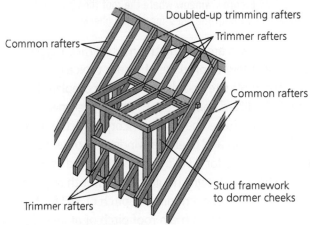

▲ Figure 4.71 Flat roof dormer window

Doubled-up trimming rafters
Trimmer rafters
Common rafters
Common rafters
Stud framework to dormer cheeks
Trimmer rafters

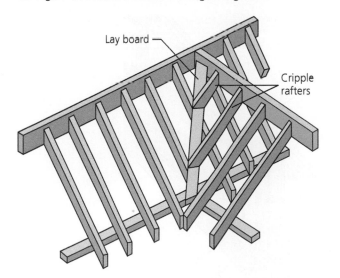

▲ Figure 4.69 Lay board and cripple rafters

Lay board
Cripple rafters

▲ Figure 4.72 Typical framework for segmental window construction

▲ Figure 4.73 Eyebrow window where the roof tiles pass over the window roof without the need for dormer cheeks

Eyebrow windows are among the most difficult type of dormer window to produce; they are typically manufactured off site and delivered as a completed window to be lifted and fixed into place. Eyebrow windows do not have cheeks but incorporate the roof tiles, which run up and over the window, in order to eliminate the need for a valley and a lay board as well as **soakers**. The roof design should be as smooth as possible and should not exceed a minimum pitch of 35° for the eyebrow window, with a main roof pitch of at least 55°.

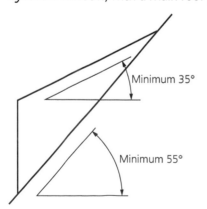

Minimum 35°

Minimum 55°

▲ Figure 4.74 Minimum roof pitches to ensure enough fall to take rainwater away without the use of soakers

Typical construction layers of a dormer cheek

The dormer cheeks will normally be of timber stud construction following the same principles as stud walling and timber-framed buildings. Suitable timber section sizes need to allow enough thickness of insulation to be placed between the studs to achieve good thermal insulation.

The following list describes the components that are labelled in Figure 4.75.

1   Internal plasterboard, 15 mm thick with a foil-back vapour barrier or 2 × 9 mm-thick boards cross jointed.
2   Timber studwork, typically 50 × 100 mm; larger sections may be required to achieve a specific U-value.
3   Insulation of suitable thickness to achieve the required U-value in accordance with Part L of the Building Regulations.
4   Structural **sheathing**, e.g. exterior grade plywood or OSB board.
5   Breather membrane.
6   External cladding or vertical tiles hung on tile battens.
7   Decorative finish, if required.

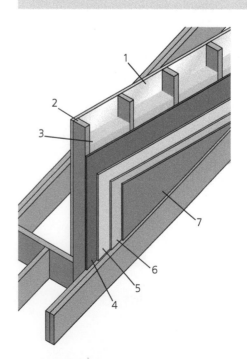

▲ Figure 4.75 Components in a dormer installation

Preventing heat loss: insulation used in dormer roofs

All roof spaces should be insulated to make sure they comply with current Building Regulations Approved Document Part L. Not only does it make sense from a comfort point of view, but with increased fuel prices and rising bills, it makes both economic and environmental sense to insulate homes.

Insulation is incorporated into the building mainly for thermal purposes, but it can also be used for fire and sound protection. Many insulation materials can achieve more than one of these roles, and this is often the factor that determines which is used.

The density and thickness of the material affect its ability to prevent heat loss. A low-density material that traps air within its structure is a good insulator. Typical examples of insulation materials that contain lots of air pockets are as follows.

- Rockwool®, rock fibre or mineral wool/fibre: available as quilted rolls for use in floors, lofts and timber wall construction, as sheets or 'batts' for use in cavity wall insulation, and as loose fill for use in floors, lofts and secondary insulation of cavity walls (by being blown into the cavity of older homes).
- Polystyrene: available in sheet form and loose fill, its uses are the same as above; it is also suitable for use between rafters.
- Polyurethane: available in large rigid sheets and used for floor and wall insulation, often combined with plasterboard for use in insulating older properties with no cavity in the walls.
- Recycled materials: these include plastic bags and sheep's wool, typically available in rolls for loft insulation and timber wall construction.
- Reflective film: typically foil backing to plasterboard, polystyrene or polyurethane sheets, acting as a heat-reflecting vapour barrier, but also used to encase Rockwool quilt rolls.

In addition to insulation, the timber construction must incorporate vapour and moisture barriers.

The effectiveness of insulation in a building is usually measured in two ways:

1 U-value
2 R-value.

The U-value is a measure of heat loss. This could be through a floor, wall or roof, and is measured in watts per metre squared per Kelvin. The higher the U-value, the lower the thermal performance of the building. A low U-value improves the thermal performance of a building, resulting in a warmer building for the same energy usage; this is good for the house owner and the environment.

The R-value is a measure of how well a material insulates at a given thickness and is expressed in m² Kelvin per watt. The higher the R-value, the better the insulation is at retaining the heat.

ACTIVITY

Source different types of insulation commonly used with roof insulation, and establish the U- and R-value for each type. Analyse the differences and suggest a suitable type and thickness for both warm and cold roof installations.

Vapour barriers

Vapour barriers are used on the inside of the dormer cheeks, usually situated between the plasterboard and the insulation. The purpose of a vapour barrier is to help prevent any warm, moist air from passing through the wall from the interior of the building and condensing in the colder areas of the dormer cheek.

Moisture barriers/breather membranes

Moisture barriers or breather membranes are usually a type of lightweight building felt that is both waterproof and breathable. This allows any water vapour to pass through the barrier to the outside. These are fixed on top of any sheathing materials used for the dormer cheeks and act as a secondary line of defence from water penetration.

INDUSTRY TIP

Vapour barriers are placed on the warm side of the wall, while moisture barriers are fixed to the external sheathing on the outer edge.

209

Installing roof windows into existing roofs

Roof windows are often required as an addition to a roof that has already been constructed and finished. The procedure required to fit roof windows in these circumstances is different to the procedure used when they are being built into new roofs.

> **HEALTH AND SAFETY**
>
> Whether the roof windows are being installed in a new roof or an existing one, always ensure that a suitable risk assessment has been completed and all control measures are in place.

The following step-by-step guide takes you through the installation procedure for roof windows in an existing roof.

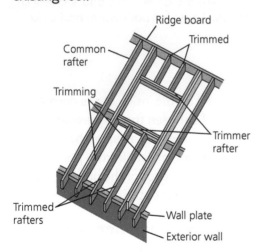

▲ Figure 4.76 Example of a roof with a purlin trimmed out to receive a roof window

1   Always ensure the area around the opening at ground floor level is clear and entry to the area is cordoned off to prevent access by those not working on it. Board out the inside of the roof with suitable flooring around the working area to avoid damaging the ceiling.
2   Working from the inside of the roof, mark out the position of the opening on the common rafters.
3   Starting in the centre of the opening, cut back the roofing felt to expose the underside of the roof tiles and tile laths.

4   Start to strip back the roof covering: remove the first tile or slate by lifting and pushing it up towards the ridge. Repeat this process with the adjoining tiles on either side – this will release the lower row of tiles. Lift the lower tile and release its nib from the tile lath. You can then pull the tile back into the roof between the tile laths.
5   Remove the remaining tiles in a similar fashion and store them safely away from the working area. Ensure that a large enough area is stripped around the proposed site of the window to allow for easy access and to provide installation room for fitting the new roof window.
6   Cut and remove tile battens from the area.
7   Cut and fit an extra set of trimming rafters beside the outer set of rafters not requiring cutting. The new trimming rafters should be securely fixed to the wall plate, ridge board and beside the existing common rafters, forming a doubled-up set of trimming rafters. The new extra rafters forming the doubled-up trimming rafters should be terminated at their seat cut; they do not need to pass through the wall and join with the eaves overhang.
8   Mark out and cut the trimmed rafters to suit the size of the new roof window opening, allowing enough fixing room around the new roof light and for the doubled-up trimmer rafters. The required finished opening size will be given in the manufacturer's fixing instructions supplied with the roof window. Temporary supports may be needed to support the cut ends of the trimmed rafters while they are being cut and trimmed out.
9   Cut and securely install the doubled-up trimmer rafters to the trimming rafters, ensuring a square opening is produced and the trimmer rafters are in full contact with the cut ends of the trimmed rafters. The trimmer rafters can be skew nailed or screwed into the trimming rafters.
10 Securely fix each trimmed rafter to the trimmer rafter.
11 Cut and position any additional framework as required to close the opening width to the required size.

12 Install the new roof window following the manufacturer's instructions and secure it in place.

13 Fit new breather membrane around the new roof opening, ensuring enough overlap with the existing roof underlay.

14 Fix new tile lath where required.

15 Fit the flashing kit for the roof window according to the manufacturer's instructions.

16 Refit the roof tiles.

▲ Figure 4.77 Stripped roof area ready for trimming out

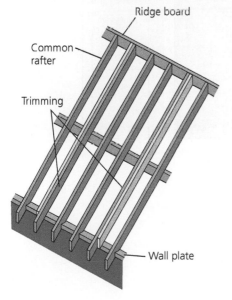

▲ Figure 4.78 Extra trimming rafters fitted

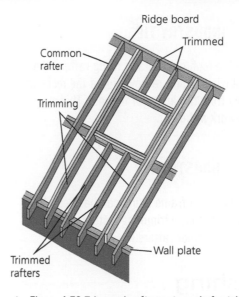

▲ Figure 4.79 Trimmed rafter cut ready for trimmer rafters

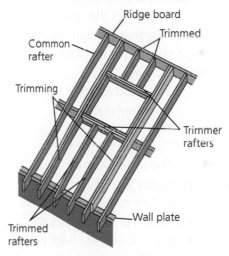

▲ Figure 4.80 Trimmed-out roof

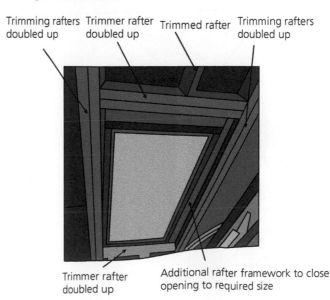

▲ Figure 4.81 Completed trimmed-out framework for roof window

211

Eaves finishing

The projecting foot of the rafters after they join the wall at the birdsmouth is known as the 'eaves overhang'. The rafters are usually finished with fascia board and soffit, providing a closure for the roof. Eaves can be finished in four different ways:

1 flush eaves
2 open eaves
3 closed eaves
4 sprocketed eaves.

Flush eaves

Flush eaves have the ends of the rafters cut with a plumb cut so that the ends project just past the outer walling by about 10 to 15 mm. The rafter end then has a fascia board attached to it, which in turn carries the guttering. The gap between the wall and the back edge of the fascia board creates a ventilation gap for the roof.

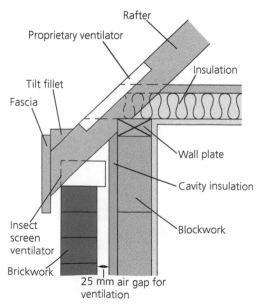

▲ Figure 4.82 Flush eaves

Open eaves

Open eaves have the rafter ends cut to the plumb cut so the ends project past the outer wall the specified amount to provide extra weather protection. The rafter ends are usually boarded on their top edge as far as the brickwork, to provide a decorative finish when viewed from ground level. Fascia board, to which guttering can be attached, is then fixed to the rafter ends.

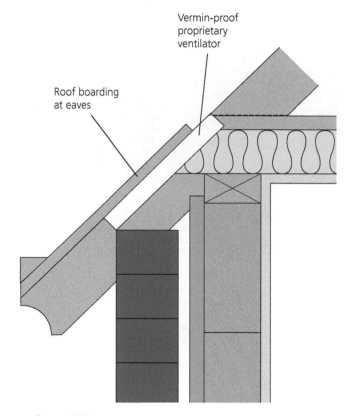

▲ Figure 4.83 Open eaves

Closed eaves

Closed eaves or boxed eaves again have the rafter ends cut to the plumb cut with the rafter ends projecting past the outer wall in the same way as with open eaves. However, instead of boarding being placed to the top edge of the rafters, the rafter ends are enclosed with fascia and soffit. To hold the soffit in place, brackets are fixed to the sides of the rafters and used either to push the soffit down onto the outer wall or to hold it flush against the outer wall. Ventilation needs to be provided in the soffit, either with a continuous ventilation grille fixed into the soffit, or with circular ventilation grills fixed at intervals along the soffit.

Sprocketed eaves

On steeply pitched roofs, rainwater, particularly from heavy downpours, can overshoot the guttering. To eliminate the possibility of overshoot, the roof's pitch or slope is reduced via a sprocket (short rafter) fixed to the sides of the rafters just above eaves level. The sprocket provides the fixings for the fascia and soffit, as with the closed eaves finish.

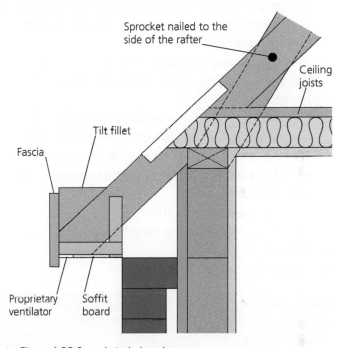

▲ Figure 4.85 Sprocketed closed eaves

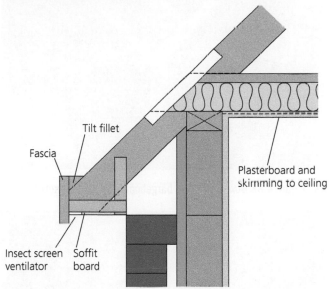

▲ Figure 4.84 Closed eaves

Forming the eaves overhang

The following example outlines one method of forming a closed eaves overhang.

1 Mark the end of the rafter at one end of the roof with the required eaves overhang, as specified in the roof design, and drive a small nail or screw part way in.
2 Repeat at the other end of the roof.
3 Attach a chalk or string line to one of the nails or screws, pull it tight and fix it around the nail or screw at the other end. This line is then used to mark the cutting position of each rafter.
4 Mark each rafter with a plumb cut to the string line and cut off the rafter ends with the required plumb cut.
5 Fix a tilt fillet to the top of the rafter ends, unless a proprietary over-fascia roof ventilation system is being used. These types of system eliminate the need for a tilt fillet and reduce the required height of fascia board.
6 Form hangers for the soffit.
7 Form the box end to the bargeboard.
8 Cut and fix the soffit.
9 Cut and fix the fascia.

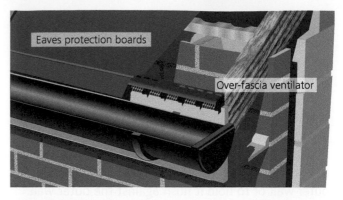

▲ Figure 4.87 Example of an over-fascia ventilator grille and eaves protection boards

▲ Figure 4.88 Box end for the bargeboard, fascia and soffit

▲ Figure 4.86 String line used to mark rafter ends

▲ Figure 4.89 Box end with bargeboard, fascia and soffit fitted

Test your knowledge

1 Which approved document from the Building Regulations specifically covers roofing?

A Part L

B Part K

C Part A

D Part D

2 Which is not a specific requirement for setting out rafter lengths?

A Type of eaves overhang

B Height of wall plate

C Pitch of roof

D Span

3 Which of the following roof designs usually incorporates a purlin?

A Double roof

B Single roof

C Collared roof

D Dormer roof

4 On which type of roof could you find a gablet?

A Gable end roof

B Valley roof

C Hipped end roof

D Dormer roof

5 The pitch line used on common rafters is positioned how far down from the top edge of the rafter?

A ½ down

B ¾ down

C ⅓ down

D ⅔ down

6 On which one of the following roofing components would a dihedral angle be found?

A Hip rafter

B Jack rafter

C Cripple rafter

D Crown rafter

7 With which item would a dragon tie and beam normally be associated?

A Valley rafter

B Jack rafter

C Ridge board

D Hip rafter

8 When forming an opening in a roof, what is the name of the rafters that are cut partway along their length but still fixed to the ridge board?

A Trimmed rafters

B Trimming rafters

C Trimmer rafters

D Common rafters

9 What should the maximum intervals be between restraint straps fixed to wall plates?

A 1 m

B 1.5 m

C 2 m

D 2.5 m

10 Where is the verge?

A Under the eaves

B Along the hip

C Along the valley

D At the gable end

11 What is the standard abbreviation for the hip rafter's true length?

A HRTL

B CRTL

C TLHR

D THRL

12 What is the minimum ventilation airflow gap required between the counter batten and the tile lath in warm roof construction?

A 20 mm

B 30 mm

C 40 mm

D 50 mm

13 What is the name of the type of joint used to lengthen a ridge board?

   A Dovetail joint

   B Scarfing joint

   C Bridle joint

   D Housing joint

14 Which part of the Building Regulations specifically covers building insulating?

   A Part L

   B Part K

   C Part A

   C Part D

15 Where would a lay board be found?

   A Along the top of the hip rafter

   B Along the top of the common rafters for the valley of a dormer window

   C Alongside the restraining straps used to fix the wall plate down

   D Under the underside of the purlin

16 A corbel can be used to support which roofing component?

   A Common rafter

   B Purlin

   C Hip rafter

   D Jack rafter

17 In roofing, an eyebrow is a type of what?

   A Dormer window

   B Valley

   C Hip

   D Lengthening joint

18 Where would you find the vapour barrier in a dormer cheek?

   A On the cold side of the wall

   B On the inside surface of the outer wall and before the insulation

   C On the warm side of the wall behind the plasterboard

   D On the warm side of the wall before the plasterboard

19 Which type of eaves overhang helps to reduce the slope of a steep roof?

   A Closed eaves

   B Flush eaves

   C Open eaves

   D Sprocketed eaves

20 Which of the following is used to represent heat loss in a building?

   A R-value

   B U-value

   C K-value

   D P-value

CHAPTER 5

FITTING DOORS, WINDOWS AND THEIR FURNISHINGS

INTRODUCTION

Working efficiently is the key to success for any project in the construction industry, no matter how large or small. If buildings can be designed with standard-sized components, such as stairs, door frames and windows, the build costs will be considerably cheaper and sourcing these resources will be easier. This approach is normally used by designers and developers for routine new-build housing – however, from time to time, site carpenters may have to make and fit purpose-made, or 'bespoke', frames, and install doors to suit. These types of frames are often needed for double door openings, under staircases (**spandrel frames**), and for wardrobes and cupboards.

Some complex non-standard items, such as stairs, curved headed doors and bay windows, may have to be manufactured by joiners in a workshop, where they will have access to a wide range of specialist machinery. Occasionally, joiners may also fit the work they have constructed, because of the complexity of the job and their understanding of its assembly.

You will have more opportunities to fit custom-made items of joinery in older buildings or on high-end developments, where the customer is usually happy to invest more for one-off items.

As a carpenter and joiner, you must have the knowledge and skills required to install complex non-standard doors, window frames and hatch linings in accordance with current Building Regulations and health and safety legislation.

LEARNING OBJECTIVES

By reading this chapter, you will learn about how to install complex doors, windows and ironmongery, in particular:

1 types of double doors and ironmongery
2 hanging doors
3 types of window
4 fitting bay windows
5 hatch linings.

KEY TERM

Spandrel frame: the panelling used to enclose the space under a staircase. Spandrel frames often have a door built into them to allow access to the void under the stairs, turning it into a usable storage space or a small room where a toilet and sink can be installed.

1 TYPES OF DOUBLE DOORS AND IRONMONGERY

The main function of double door openings is to provide access to and egress from buildings or between rooms. Doors are designed, manufactured and installed to meet performance requirements, depending on the situation and location of the door. When choosing which type of double doors to fit, the following should be considered:

- door design and construction methods
- weather performance (watertightness and wind resistance)
- mechanical performance
- operating forces
- resistance to distortion
- resistance to impact
- performance in use (opening and closing)
- effect of temperature and humidity variation.

▲ Figure 5.1 Bespoke door and lining

Performance

Particularly with external doors, Building Regulations and building standards emphasise:

- weather-proofing
- security
- thermal performance
- sound insulation.

While the door itself may be designed and constructed to perform well against these criteria, the overall performance relies on the correct fitting of the door to the frame, along with its associated ironmongery.

External doors are usually fitted with a 3 mm clearance around all edges to allow moisture movement. Weather seals should be positioned so that, where possible, they are not interrupted or damaged by door locks or hinges. Multi-point locking systems using **espagnolette bolts** help to prevent distortion of the door leaf, and provide enhanced security and weather resistance.

Matchboarded doors are unlikely to meet modern thermal performance requirements for heated buildings. Exposure of end grain should be avoided where possible, and any exposed grain should be fully sealed before the surface finish is applied.

KEY TERMS

Espagnolette bolt: a multi-point locking device that is normally on the locking side of the door.

Matchboarded doors: a series of timber boards jointed together (usually by a tongue and groove arrangement) in their width.

The following factors should be taken into consideration in door design.

- Weather protection: How exposed are the doors to elements such as rain, snow, wind, direct sunlight or extreme temperature change?
- Security: Are the doors intended for internal or external use? Are the materials to be used in the construction of the door, the door's construction method, and its types and quantities of ironmongery all suitable?
- Sound and thermal insulation: External doors need to prevent reasonable levels of cold air and penetrating winds from entering the building. A correctly designed door incorporating double-glazed units, solid door construction and door seals will help with both sound and thermal insulation.

External doors and frames are included when calculating the thermal performance of the external wall, and taken into account in whole-building performance calculations. They do not come under the Building Regulations relating to the passage of sound from outside a building to inside, but planning restrictions may specify particular acoustic performance requirements on exposed sites, for example near flight paths, motorways and railway lines.

Timber choice

Timber is **hygroscopic**, so the door will either shrink or expand, depending on the temperature and humidity of the surrounding air, until it stabilises to **equilibrium moisture content**. It swells as it absorbs moisture and shrinks as it loses it, a response known as 'movement'. Timber species differ in their movement characteristics.

External doors subjected to both internal and external conditions, in addition to seasonal variations, must be designed and fitted to accommodate this movement. Timber known to have a 'small movement' is preferred. Poor timber selection can negate good door design. Low-grade timber is not suitable for external doors.

The timber used for external doors must be of a high quality to stand up to the weather and to offer durability, with good sound and thermal properties. Hardwoods such as oak are extremely durable but also extremely expensive, so other timbers are used by joinery workshops to manufacture doors. Idigbo, utile and European redwood are common alternatives.

Large-scale joinery manufacturers make sure they use timber approved by the **Forest Stewardship Council (FSC®)**, indicating that the forests where the timber is sourced are managed responsibly and are sustainable.

For increased durability, the timber is often impregnated with preservative, often referred to as 'tantalised', 'double vac' or 'vac vac' (short for 'vacuumed').

Privacy

Unglazed doors with one-way door viewers prevent unwanted viewing into rooms or buildings, while still allowing the occupant to see out. Choosing the hanging side of a door carefully can help restrict direct viewing into a room.

▲ Figure 5.2 A typical door viewer in use

▲ Figure 5.3 Escape routes will affect your choice of door

Glazing requirements

- Does the door require full glazing, as might be the case in shops, or partial glazing to allow **borrowed light** from an area?
- Does the glazing need to meet fire-resisting requirements?
- Is there a requirement to use toughened glass in vulnerable situations?

KEY TERM

Borrowed light: light that enters a room or corridor from an adjoining room through a glazed opening, usually above a door.

Fire resistance

Although all doors give a certain level of fire resistance, only approved fire doors can be used at specifically required fire door locations. Provision for an escape route (as specified in the Building Regulations) may govern your choice of:

- door size
- locks
- hardware
- method of operation.

We will look at fire doors in detail later in this chapter.

Access and ease of operation

When designing entrances to buildings or specifying and fitting doors, you should consider those Building Regulations that cover access and ease of operation. These will affect your thinking on issues such as:

- widths of door openings for access by wheelchair users
- types of threshold
- the means by which the door is opened and closed
- how large and heavy the door will be
- who is likely to use the door.

All of these will help to determine the type of ironmongery selected, and whether any door opening and closing assistance will be required.

Appearance and surface finishes

Is the door intended to make a grand statement while providing all the above functions, or is it simply practical and functional?

The surface finishing treatments of external timber doors are not just decorative, but play a vital role in their long-term performance and protection. **High-build exterior stains** and good-quality exterior paints are among the most suitable finishes. If finishing on site, apply at least two coats of protective finish to a door as soon as it is delivered or removed from its wrapping. Leaving a door unprotected for days or weeks will mean it reacts to its environment more rapidly and to a greater extent than a fully sealed door.

> **KEY TERM**
>
> **High-build exterior stain:** a microporous multi-layer timber coating that resists surface mould/algae and protects against UV light, resulting in a humidity-controlling timber finish.

Storage and installation

Poor site storage can lead to damage and/or distortion of the doors. Keep any shrink wrapping or other protection on pre-finished doors in place until the doors are ready for installation, and then keep the wrapping in place as long as possible after installation to protect against wet trades and construction damage.

Store doors flat on a level surface, fully supported along their full length and across their full width, and on at least three bearers. Do not store doors by leaning them against a wall as this will encourage twisting.

If possible, do not install the doors until the building has dried out.

Installation usually involves on-site modification of the doors, including letter plate apertures, glazing, door viewers, and locks and hinges. On-site installation and fitting of doors should not be carried out without considering how this may affect the performance of the doors. Considerations include:
- water leaking into the building
- locks and keeps failing to close the door correctly or coinciding with construction joints in the door framing, such as mortice and tenons from mid-rails; this will reduce the strength and stability of the doors
- the absence, or incorrect design and fitting, of weatherboards.

Health and safety implications

A suitable risk assessment should be carried out before the installation of double doors. All relevant safety considerations should be taken into account when producing the risk assessment. For example:
- How much will each door weigh?
- How far will the doors have to be carried? Will they have to be carried upstairs?
- What power tools are needed? Will these create dust or vibration? How will that be managed?
- Will you need to use access equipment?
- Will you have to glaze the doors? What precautions will you take when handling the glass?
- How will you dispose of the waste?
- Will the building be occupied, and will this cause a hazard to others (noise, trip hazards, etc.)?

These are just some of the questions that should be considered when completing a comprehensive risk assessment. Further information about risk assessments and health and safety information can be found in Chapter 1.

> **ACTIVITY**
>
> Calculate the weight of an FD30 fire door, 1981 mm high × 762 mm wide. Research how many people are needed to hang a door like this safely.

Methods of operation for double doors

▲ Figure 5.4 Bespoke double doors

You should be able to recognise different types of double doors, their associated ironmongery and how to fit them.

Double doors are used for many reasons – for example, as a practical solution to create a wider opening for disabled access, or to create a grand entrance to a room. In some public buildings, such as hotels, hospitals and schools, unequal pairs of doors are often used to improve access when moving trolleys or equipment, or when high volumes of people are anticipated. These usually consist of a full-width door and a half-width door. The full-width door (often called the 'master' or 'live' door) is used for general access, whereas the half-width door (often called the 'slave' or 'dead' door) is usually bolted shut and allows the master door to be shut and secured against it.

Garage doors (external)

▲ Figure 5.5 Side-hung timber garage doors

The majority of garage doors have an 'up and over', roller or sectional action, and are made of steel, timber or **GRP**. The design of these types of doors means they are stored in the loft space of the garage when they are fully open; this is ideal for short driveways with little space in front of them. Alternatively, the doors can have a more traditional design, and be hung as a pair. Double garage doors are usually hung on the outside face, meaning that sufficient space will be needed in front of the garage for the doors to swing open. Heavy-duty ironmongery will also be needed for these types of doors, due to their increased size and weight.

> **KEY TERM**
>
> **GRP:** glass reinforced plastic, a composite material. For example, garage doors are often manufactured with a combination of insulating foam, timber, PVC and GRP, making them resilient to the weather, low-maintenance and well insulated.

Single-action French doors (external)

▲ Figure 5.6 French doors

French doors are traditionally a pair of fully glazed doors that open outwards to allow full access to the width of the opening. They are very popular in domestic housing and are usually found in a rear living room, leading onto the garden. They can also be designed to swing inwards, although this could reduce the space for furniture in the room they open into.

They have an advantage over solid doors in that they flood the room with light and bring the garden into the room. Because they are mainly used at the rear of the property, privacy tends not to be a problem. French doors are also widely used to divide up larger rooms without obscuring the view into the other room.

It's important that there is a weather-tight joint between the meeting styles of the doors. Later in the chapter, we'll look at some methods that can be used to achieve this.

Single-action stable doors (external)

▲ Figure 5.7 Stable doors

This type of door is designed so that the top half can open while the bottom remains closed; the two halves can also be bolted together so they function as a full-height side-hung door. This is a design feature that allows a space to be ventilated with fresh air while still offering some security. The meeting rails between the top and bottom halves of the door are usually rebated together to help protect against the weather. This joint can be protected further from draughts and heat loss, with rubber seals.

Single-action rebated doors (internal or external)

The joint between a pair of double doors may have to be rebated for privacy, security, weather protection or just to provide something solid for one of the doors to close against. Rebating the styles of the doors together is arguably the best way to do this for internal doors, and it's the only way to do this for external doors. Rebated door locks and latches have to be used for these types of doors, to compensate for the shape of the meeting styles.

Rebated doors can be designed to open either from the right-hand or left-hand side first, depending on the rebate position. Normally, one door of the pair will be bolted to the frame and opened only occasionally to allow access; this door is referred to as the 'slave'.

▲ Figure 5.8 Rebated doors

Sliding doors

Sliding doors move parallel to the walls they are fixed to, so they use less space than side-hung doors.

▲ Figure 5.9 Internal sliding doors

▲ Figure 5.10 External sliding doors

Internal sliding doors are normally hung on hangers with wheels fixed to the top rails of the doors. These are then mounted on a track fixed to the face of the wall above the top rails. The sliding door 'kit' (ironmongery) also contains guides that are fixed to the floor to prevent the doors rubbing against each other as they pass, and rubber stoppers to cushion them when they are opened. A **pelmet** may be constructed to cover the track and wheels at the top of the doors, or it may be left exposed as a design feature. Alternatively, the frame can be built to allow the doors to slide within the wall – these are known as 'pocket doors'. A considerable amount of labour is involved in building these types of doors and frames; therefore, they are not commonly used because of the expense.

A similar system can also be used for wardrobe doors, although these tend to have a track fixed to the floor because it is less of a trip hazard.

KEY TERM

Pelmet: boxing-in that is fixed in front of running gear to hide the track of a sliding door and provide a decorative finish.

External sliding doors are known as patio doors. The ironmongery and construction of the door frame will differ from internal linings because it must be designed to withstand the elements and provide increased security.

Side-hung single-action doors (internal or external)

▲ Figure 5.11 Side-hung single-action double doors

These types of doors are the most common double doors in domestic and commercial buildings. Single-action doors swing in one direction only, usually through a full 180°, depending on what kind of hinges are fitted. The detail of the meeting styles will vary,

but will suit the function of the doors (e.g. square or rebated). Either a pair (two) for lightweight internal doors, or a pair and a half (three) butt hinges are commonly used to hang heavier single-action doors.

It is recommended that three hinges are used for:
- fire doors
- external doors
- security
- strength/stability in areas of high moisture, e.g. bathrooms/en-suites and kitchens.

Note: It may be necessary to use four hinges for larger doors.

Types of double doors

▲ Figure 5.12 Master and slave double doorset; the master door opens on the right

Doors are typically divided into eight classifications depending on their method of construction:
1 matchboarded doors
2 flush doors
3 panelled doors
4 glazed or part-glazed doors
5 fire doors
6 double-action doors
7 sliding doors
8 folding doors.

Matchboarded double doors

This group of doors involves a relatively simple form of construction. They are suitable for internal and external use, but are commonly used for sheds, gates and older period properties, as well as industrial and agricultural buildings. These types of double doors are typically side hung but larger industrial doors are often sliding.

Ledged doors

The simplest form is the ledged and matchboarded door. This consists of a set of (usually three) horizontal ledges fixed to the back of the vertical matchboarding, with one fixed at each end and one centrally. Traditionally, boards are nailed onto the ledges through the face using lost head or oval nails, which are then 'clenched' over on the back to prevent the boards from pulling away. Modern practice usually involves screws being used to secure the ledges to the boards instead, which are then counter-bored and filled with timber pellets to conceal the fixings.

▲ Figure 5.13 Timber pellets cut out and fitted

Ledged and braced doors

Ledged and braced doors are constructed in the same way as ledged and matchboarded doors, but have diagonal braces added to prevent the door from

dropping. Bracing must be fitted in the correct direction to support the side of the door opposite the hinges; this means pointing upwards, away from the hinge side. If they are fitted wrongly, the door can drop and fail to function properly. Braces are fitted as close as possible to 45° to the ledges; if they are fitted below this angle, they will not give the door adequate support.

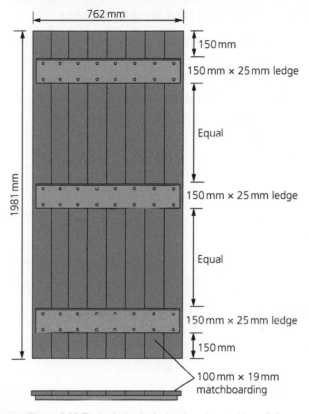

▲ Figure 5.15 Typical single-ledged and matchboard door

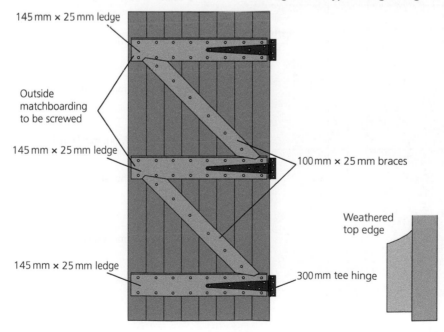

▲ Figure 5.14 Typical single-ledged, braced and matchboarded door

Framed, ledged and braced

Framed, ledged and braced (FLB) doors are an improvement on ledged and braced doors as the matchboarding is fixed into a framed door, increasing the door's strength. These are jointed with mortice and tenon joints. The matchboards are either tongued into a groove that runs around the frame or, more typically, fitted into a rebate that runs around the frame.

To allow the matchboards to pass over the face of the middle and bottom ledges, the ledges are reduced in thickness. This means the jointing arrangement for the middle and bottom ledge requires the use of a **barefaced tenon**.

Non-standard matchboarded doors

Matchboarded doors are manufactured to suit many types of opening size, for internal and external use. Typical locations for matchboarded double doors are:

- domestic garages
- industrial garages
- industrial locations (workshops, factories and warehousing).

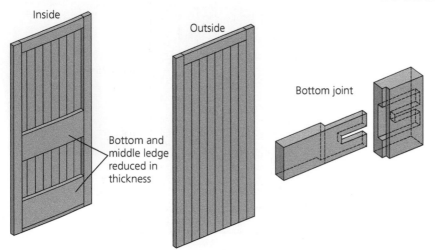

▲ Figure 5.16 Jointing in an FLB door (left); a double barefaced haunched mortice and tenon (right)

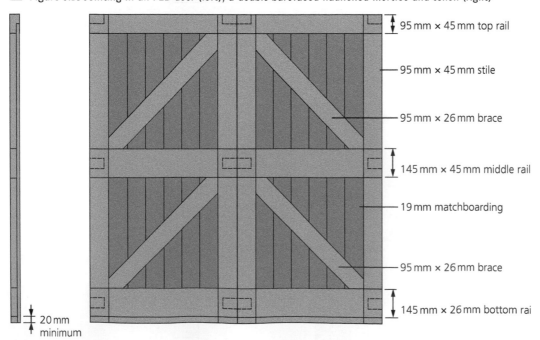

- 95 mm × 45 mm top rail
- 95 mm × 45 mm stile
- 95 mm × 26 mm brace
- 145 mm × 45 mm middle rail
- 19 mm matchboarding
- 95 mm × 26 mm brace
- 145 mm × 26 mm bottom rai

20 mm minimum

▲ Figure 5.17 Framed, ledged and braced doors

These oversized doors typically come as double doorsets that are either sliding or hinged, or a combination of both.

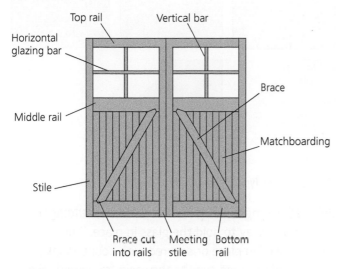

▲ Figure 5.18 Double FLB part-glazed matchboarded doors (inside view)

Double matchboarded doorsets usually have the braces fitted, whereas single doors usually supply the braces loose to be fitted on site (which allows single doors to be used for either hand). Larger doorsets may have four braces, all pointing from the outer corners to the centre of the middle rail. This stiffens up the door and allows more fixing points for the boarding. It also ensures an angle as near to 45° as possible is maintained for the braces, reducing the possibility of the door dropping on the side opposite the hinges.

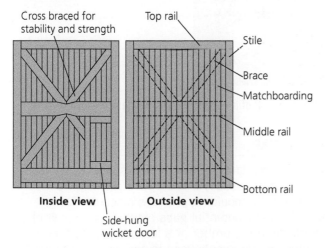

▲ Figure 5.19 A pair of sliding industrial-type matchboarded doors with a wicket door

Where double doors meet in the middle, **weathering** is required. The meeting stiles are rebated over each other to allow the doors to finish flush. It is vital that the rebate is applied to the correct side of the door, so that the correct door is allowed to open first. An alternative method is to fit a planted cover bead to the door, again making sure you fit it to the door that is to open first.

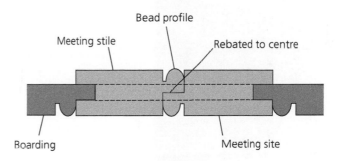

▲ Figure 5.20 Typical design for bead and butt matchboarding

Flush double doors

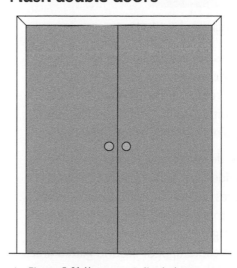

▲ Figure 5.21 Honeycomb flush doors

Flush double doors are constructed with either a hollow or a solid core, faced with sheets of hardboard, plywood or moulded plastic. Joinery workshops rarely manufacture flush doors; they are usually mass produced by larger joinery manufacturers. They are lightweight, cheap and simple to manufacture. The hollow core variety is the most common: a stapled softwood frame in-filled with (usually) cardboard, set in a honeycomb or lattice pattern that is then covered with the facing material.

Because they are hollow, a section of solid timber called a 'lock block' is sandwiched into the core of both sides of these doors to enable a lock or latch and the handles to be securely fixed. The position of the lock blocks is clearly identified by the manufacturer within the instructions that accompany the door or, more commonly, a key symbol or the word 'lock' is printed on the top of the door on the side that contains the lock block.

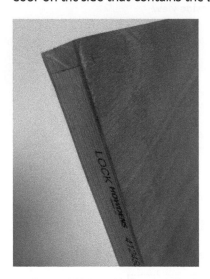

▲ Figure 5.22 Lock block side identified by the manufacturer

Moulded panel doors are a variation on flush doors, using a hollow door construction with the application of facings that are pressed or moulded to give the impression of a traditional panelled door. They often have a textured grain finish to simulate real timber.

Flush doors can be designed to incorporate vision panels. These provide light and a view through to the other side. Apertures cut into hollow doors require additional framing to accommodate the glazing and the glazing beads. Figure 5.23 shows examples of typical profiles that could appear around glazed areas within doors to hold the glass in.

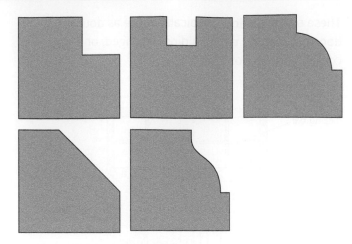

▲ Figure 5.23 Typical door profiles

Rebated glazing beads (called **bolection mouldings**) are the best way to hold the glass in place. They can be nailed in place or secured with countersunk screws and recessed cups to the main framework; the latter method allows easy removal if the glass needs replacing.

These types of door are typically side hung but can be sliding or folding, particularly where rooms are divided up and where there is restricted space for doors to swing.

▲ Figure 5.24 Profile of beading holding glass in place

KEY TERM

Bolection mouldings: small timber sections moulded with a rebate, usually a decorative profile. The rebate on the moulding provides a secure fixing, and hides any potential gaps created as a result of shrinkage in the timber at a later stage. Bolection moulds are commonly used to secure glazing and panels into doors, windows and screens.

Panelled and glazed double doors

Panelled and glazed double doors have a similar construction technique. Both consist of a frame that has a groove or rebate around its inside edges to receive panels or glazing. The framing members for these doors vary in the number and arrangement of the panels. The door construction consists of horizontal members (rails) and vertical members (stiles or muntins). Rails are named according to their position in the door, such as 'top rail', 'intermediate rail', 'middle rail' and 'bottom rail'. When an optional upper intermediate rail (positioned below the top rail) is used, it is often referred to as the 'frieze rail'. The two vertical members on the outside of the door are called stiles, while all intermediate vertical members in panelled doors are called muntins.

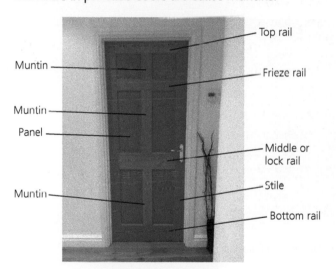

▲ Figure 5.25 Panelled door components

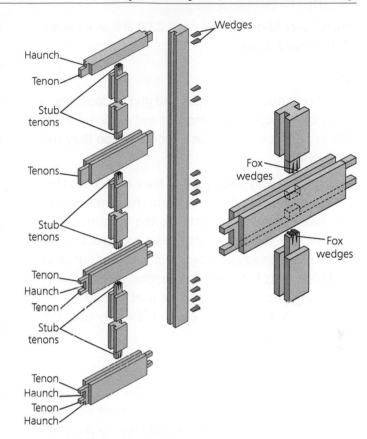

▲ Figure 5.27 Panelled door mortice and tenon joints

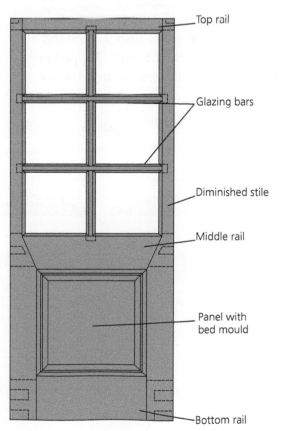

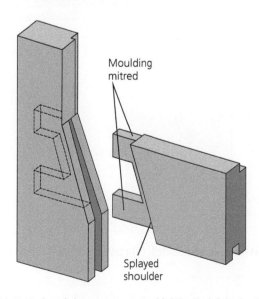

▲ Figure 5.26 Half-glazed door components (right); diminished stile joint detail (left)

French doors (described on page 222) are an example of fully glazed doors.

Joints

In good construction, panelled and glazed doors are usually constructed using mortice and tenon joints, while in poorer, mass-produced construction they are dowel jointed.

Glazed doors may have glazing bars to replace the middle and intermediate rails and the muntins. In half-glazed doors with panels to the bottom of the door, the middle rail should be constructed using a diminished shoulder construction (gunstock shoulders). This form of construction is more time-consuming and difficult than parallel shoulder construction, so this type of joint is rarely used on cheaper, mass-produced doors.

Decorative panels

While panelled doors can have plain flat panels, such as the plywood used in cheaper doors, more decorative and bespoke doors will have decorative panels that give the door a more elegant appearance.

Panels can be divided into three parts. The central part, and usually the largest, is called the field. Around this is the margin, next to the door components. The part of the panel that fits into either the groove or the rebate is called the flat.

The main types of panel are:

- raised panel – having a flat and bevelled section
- raised and fielded panel – having a flat and a margin that is bevelled, leading to a raised field
- raised, sunk and fielded panel – having a flat and a margin that is bevelled but sunk below a raised field
- raised, sunk and raised fielded panel – having a flat and a margin that is bevelled but sunk below a further raised and bevelled field.

We will look at the different types of glass used to glaze doors, and the regulations that control them, later in this chapter.

Side-hung fire doors (internal or external)

▲ Figure 5.29 Side-hung fire doors

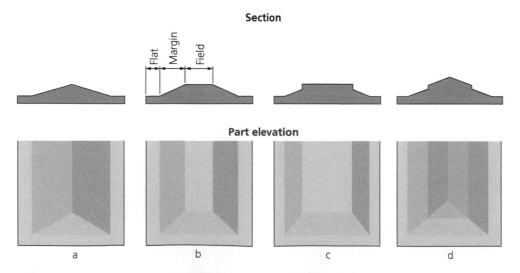

▲ Figure 5.28 Decorative panels: (a) raised; (b) raised and fielded; (c) raised, sunk and fielded; (d) raised, sunk and raised fielded

Fire doors (or fire-resisting doors) are mainly manufactured using the solid core construction method. The core consists of laminated fire-retardant material, faced as a flush door or a moulded panel door. The main function of fire-resisting doors is to delay the spread of flames, smoke and gases into escape routes for a set time, to allow occupants of the building enough time to get out in the event of a fire. Only doors that have been manufactured, evaluated by a UKAS (United Kingdom Accreditation Service) accredited laboratory against British Standards and certified as fire doors should be installed. Certified fire doors have clear labels, or a permanent plug fixed into their edges, to identify the period of fire resistance and their status (e.g. factory-fitted glazing).

Building Regulations provide the exact requirements for doorsets in different locations in a building. The most common specifications for fire-resisting doorsets are shown in Table 5.1. In some buildings, there may be a requirement for increased performance ratings, greater than those shown below; these ratings – FD60, FD90 or FD120 – are rare in domestic dwellings.

▼ Table 5.1 Most common fire-resisting specifications

Performance requirement	Performance rating for doors tested to BS 476-22:1987
20 minutes' fire resistance	FD20
20 minutes' fire resistance with smoke control	FD20S
30 minutes' fire resistance	FD30
30 minutes' fire resistance with smoke control	FD30S

Source: TRADA

Fire doors are available in standard door sizes, and as door blanks to be cut to size and lipped for non-standard door openings. The most common thicknesses are FD30 (44 mm thick) and FD60 (54 mm thick).

To ensure the reliability of the door and frame during a fire, it is essential to use only compatible ironmongery, as referenced in the current Building Regulations, Approved Document B.

Intumescent strips, with smoke seals in some cases, must be fitted to the door frame. If this is not possible then they must be fitted to the sides and across the top edge of the leaves (doors). The seals are in shallow grooves equal to their thickness and width, and held in place with double-sided adhesive tape. When intumescent seals are exposed to heat, they expand to several times their normal size to create a fire stop in the joint between the door and frame.

It is a well-known fact that smoke is the biggest cause of death in fires. To prevent the passage of smoke and dangerous gases through a doorset, brushes or flexible fins can be fitted into the intumescent seals. Alternatively, compression seals can be fitted into the frame rebate or automatic drop-down seals can be fitted at the threshold (bottom).

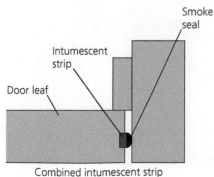

Combined intumescent strip and smoke seal (brush type) fitted in door leaf

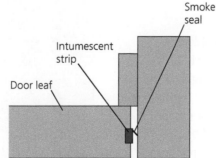

Combined intumescent strip and smoke seal (flexible blade) fitted in door leaf

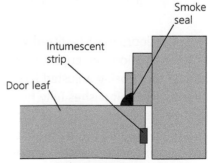

Perimeter smoke seal fitted to door stop; intumescent strip fitted in door leaf

▲ Figure 5.30 Typical methods for fitting intumescent seals

▼ Table 5.2 Types of intumescent seal

Fire seal	Fire and smoke seal (brush)	Fire and smoke seal (single flipper)	Fire, smoke and acoustic seal (double flipper)

Fire doors can have vision panels or be glazed, as long as the glass used is fire rated or wired (refer to the overall specification and the manufacturer's instructions for the exact glazing system). The glass must be bedded in intumescent glazing gaskets and glazing beads, which are manufactured from non-combustible materials or coated in intumescent paint. Beads should again be bolection beads bedded on an intumescent seal.

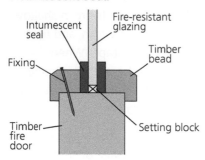

▲ Figure 5.31 A glazing aperture in a fire door

INDUSTRY TIP

Only an approved manufacturer or installer should ever cut an aperture for glass or grilles into a fire-resisting door. Attempting to undertake this type of work on site could be considered 'unauthorised alterations' to the fire door and could invalidate the certification.

Most fire doors must be fitted with a suitable door closer; exceptions are cupboard doors. The closer must be capable of being adjusted to suit the weight of the door to ensure it pushes fully home into the frame. The joint between the frame and door must not exceed 2–4 mm – any bigger and the seal formed by the intumescent strip could be compromised. Fire doors are subject to inspection by Building Control and could fail to comply with Building Regulations if they are incorrectly fitted.

INDUSTRY TIP

Fire doors undergo rigorous testing as a 'doorset'. The term refers to the door, frame, ironmongery, and intumescent seals and smoke sealing devices. The door is not considered a stand-alone fire-rated component. When the doorset is tested and certified as 'fire rated', this means that all future doors must be fitted to the same standards with the same specification frame, ironmongery and seals.

▲ Figure 5.32 Colour-coded plug identifying rating of door

Rebated fire-resisting double doors

Fire-resisting double doors with rebated meeting edges have to twist only half their thickness for fire to pass through; therefore, supporting test evidence must be provided when they are installed. Rebated doors must close in the correct sequence, otherwise the integrity of the doorset will be lost in a fire. Double doorsets with rebated edges are normally fitted with 'selectors and fore-end conversion units', which ensure the doors close fully into the frame in the correct order.

> **INDUSTRY TIP**
>
> There used to be a requirement for door stops to be at least 25 mm thick to meet fire-resistance requirements, but this is no longer the case. Door stops for fire doorsets can now be as thin as 12 mm, the same size as standard door stops.

Double-action doors (internal)

▲ Figure 5.33 Double-action doors

These types of doors swing both inwards and outwards with the use of double-action hinges. They are often used in areas of heavy pedestrian traffic, such as public buildings, schools, colleges and hospitals. You may also see these types of doors in some commercial kitchens; however most restaurants now use two separate door openings, to prevent accidents. The long edges on double-action doors usually have a slight radius on them; this is to maintain parallel gaps between the doors and prevent the leading edges hitting each other as they operate. Various hinges can be used to hang these doors; all of them are sprung, meaning they will return to their closed position once opened. We'll look at these hinges later in the chapter.

Sliding and folding doors

Sliding and folding doors usually contain an equal number of 'leaves' (doors) suspended on a top track mounted on the door lining. The leaves closest to the door jambs are connected to the lining with butt hinges, while the remaining doors are hung on their stiles to each other. As the doors are opened, they fold back onto one another in a concertina pattern. These types of doors are also referred to as 'bi-folding'. Sliding and folding doors are either 'centre folding' or 'end folding'.

Typical arrangement

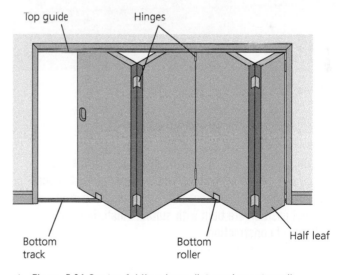

▲ Figure 5.34 Centre-folding doors (internal or external)

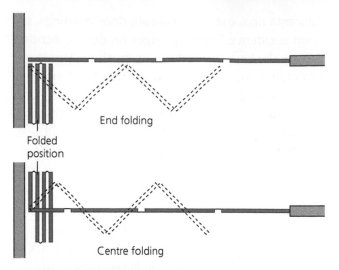

End folding

Folded
position

Centre folding

▲ Figure 5.35 Bi-folding doors

▲ Figure 5.36 End-folding doors (internal or external)

Installing doors (components)

Meeting stiles

The detail on the edge of the meeting stiles on pairs of doors will vary depending on whether they are external or internal, and their operation. The simplest detail is to have 'square' stiles, however these offer no privacy and no protection from the weather or draughts. An easy solution is to plant (fix) a door stop on the face of one of the doors, which will also provide a relatively solid surface for the other door to close against.

A better method is to rebate the meeting stiles, usually 13 mm deep by half the thickness of the door, so they fit flush together. The disadvantage of this method is that rebated locks and latches have to be fitted to compensate for the profiled door edges; this increases the costs of ironmongery and labour. Traditionally, the rebated joint would have had a bead detail profiled along the meeting stiles on both sides of the doors, to make the joint more visually appealing by adding symmetry to the width of the timber.

Another way of forming a joint between a pair of doors is to secure a moulded section of timber to one of the long edges on the styles to form a rebate. These timber sections could have T- or J-shaped profiles. Gluing and pinning timber to the edge of a door is a quick and easy way to form the rebate, and eliminates the need to fit rebated ironmongery.

The long edges on double-action doors, whether single or pairs, should have a slight radius so that a 2–4 mm joint can be maintained when the doors are operated through 180°.

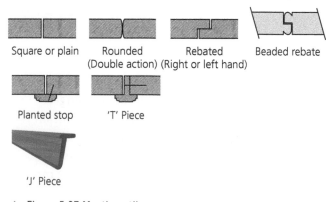

Square or plain   Rounded   Rebated   Beaded rebate
                   (Double action) (Right or left hand)

Planted stop   'T' Piece

'J' Piece

▲ Figure 5.37 Meeting stiles

▲ Figure 5.38 Beaded rebate on a pair of double doors

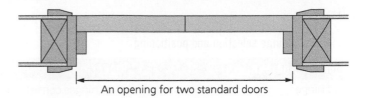

An opening for two standard doors

▲ Figure 5.40 Typical joint to head of double lining

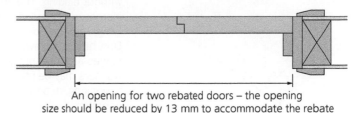

An opening for two rebated doors – the opening
size should be reduced by 13 mm to accommodate the rebate

▲ Figure 5.41 Rebated joint to head of double lining

INDUSTRY TIP

Rebating the meeting stiles offset will allow some standard locks or latches to be fitted, rather than rebated ironmongery kits.

Door ironmongery

Door ironmongery is also known as door hardware or door furniture. As a site carpenter or joiner, you should be able to recognise a range of ironmongery and understand its uses. At your place of work, you may have to advise a client on suitable materials. If the wrong ironmongery is chosen and used, it could be an expensive mistake for both you and the client.

On larger contracts, there will be an ironmongery schedule giving precise details of the ironmongery to be used with each door. These schedules are very useful on site as an accurate check of what is required. The type of door and its intended location will dictate the types of ironmongery used with the doors.

The ironmongery or hardware required for double doors can be divided into categories according to use, as shown in Tables 5.3–5.9.

▲ Figure 5.39 Offset rebated double doors

INDUSTRY TIP

The width of a double door lining should be reduced by 13 mm if the doors have rebated edges.

Hinges

▼ Table 5.3 Hinge selection and positioning

Name of ironmongery	Description
Butt hinge	Butt hinges consist of two leaves joined by a pin that passes through a knuckle formed on the edge of both leaves. They are fitted so that the leaves are recessed equally into the door and frame, with a margin between so that the door operates without binding. The knuckle is usually set to project past the face of the door to give increased clearance. There is normally an odd number of knuckles, and the leaf with the greater number of knuckles is positioned on the door frame ('most to the post'). Butt hinges are available in a range of sizes from 25 mm up to 100 mm. Brass butt hinges are susceptible to wear on the knuckles, so stainless steel or phosphor bronze washers may be fitted between the knuckles to prevent this. The washers also reduce squeaking. In addition, butt hinges are available with ball-bearing joints for a smooth operation.
Ball race butt hinge	High-performance ball-bearing or ball-race butt hinges give a much smoother action and are more durable on heavy doors. They are available in sizes from 75 mm to 150 mm.
Loose pin butt hinge	Loose pin butt hinges enable easy removal of the door by removing the pin from the hinge knuckle. Lift-off butt hinges enable the door to be lifted off when it is in the open position. The hinges are handed and incorporate one long pin hinge and one short pin hinge. The long pin hinge is the lower hinge, to enable easier repositioning of the door.
Rising butt hinge Rise and fall hinge	Rising butt hinges have a spiral-shaped knuckle, allowing the door to rise as it opens. These are particularly useful where a door needs to clear uneven floors, mats and rugs. The shape of the knuckle also gives the door a self-closing action. The top leading edge of the door must be eased to prevent the door fouling the head of the door frame as it opens and closes. These hinges are handed. A similar principle is applied to toilet cubicle 'rise and fall hinges', but here the door is self-opening when the stall is not in use.

▼ Table 5.3 Hinge selection and positioning (continued)

Name of ironmongery	Description
Flush hinge	Flush hinges are suitable only for lightweight doors. They are quick to install as they do not need to be cut into the door or frame.
Tee hinge	Strap hinges are surface-fixed, and screwed or bolted directly to the door and frame. The most common types of strap hinge are tee hinges, which are made from thin-gauge steel and are usually **black japanned** or galvanised. They are used predominantly with ledged and braced doors for sheds and gates.
Cranked, hook and band strap hinge	Heavy-duty hook and band hinges are made from stronger galvanised or stainless steel. They are used for heavier industrial or garage doors, framed, ledged and braced doors, and farm gates.
Double-action hinge (helical)	Double-action spring hinges are designed to make a door self-closing. They have large cylindrical knuckles that can be tensioned to the required closing action. They have three leaves, and open 360°. To correctly install double-action spring hinges, a planted strip of timber should be fitted to the hanging edge of the frame the same thickness as the door, to allow free movement of the cylindrical knuckles. Double-action spring hinges are frequently used in corridors of public buildings such as schools and hospitals, and between kitchen and dining areas in restaurants and public houses.
Hawgood swing door hinge	Hawgood swing door hinges are suitable for heavy-duty and industrial double swing doors. The cylinders enclosing the springs are morticed and recessed into the door frame, and the moving shoes are cut and fit around both sides of the door. This type of hinge design allows the **door leaf** to swing away from the frame due to the design of the shoe that fits around the door; this means there is only a small gap between the door leaf and the frame when the door is closed.
Soss hinge	Concealed hinges are suitable for flush doors with timber, steel and aluminium frames. A common type is Tectus, which is fully adjustable to ±3 mm and is designed for heavy-duty doors. Concealed hinges are commonly used in shop fitting. Soss hinges are used for light-duty applications.

→

▼ Table 5.3 Hinge selection and positioning (continued)

Name of ironmongery	Description
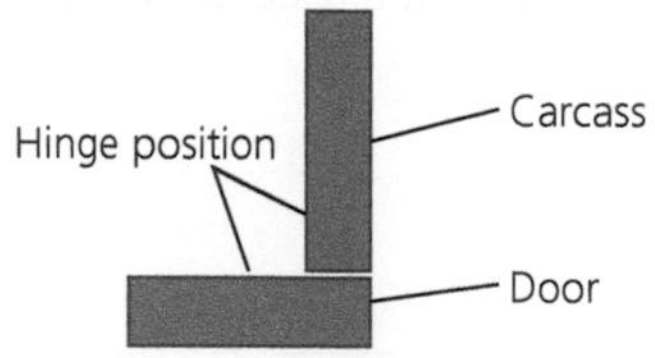 Concealed cupboard hinge  Concealed cupboard hinge fitting options: full overlay hinges; 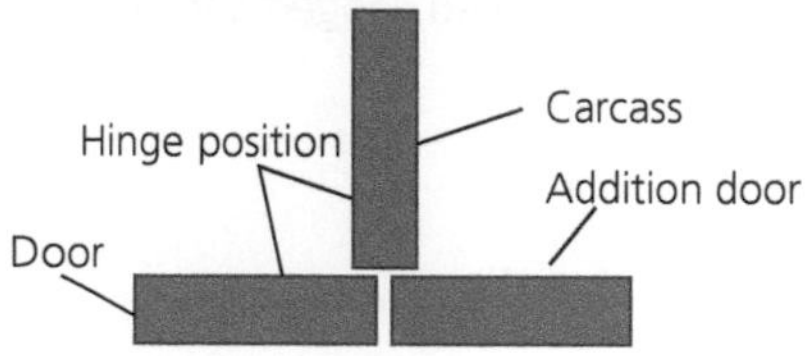 half overlay hinges; 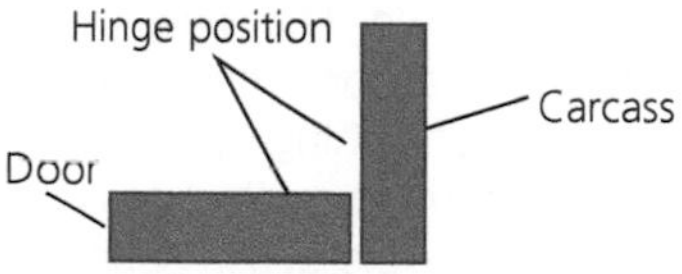 inset hinges	Commonly used to hang kitchen and wardrobe doors because these cabinets are usually made with artificial boards, such as chipboard and MDF, which don't hold screws well in the edges. A 35 mm-diameter 'blind' hole is the standard size to be bored into the back face of the doors to accept the hinge. They are then either secured with 16 mm screws or are simply clipped in place. These types of hinges range from the basic, with limited adjustment, to the soft-close fully adjustable versions.
Flag hinge 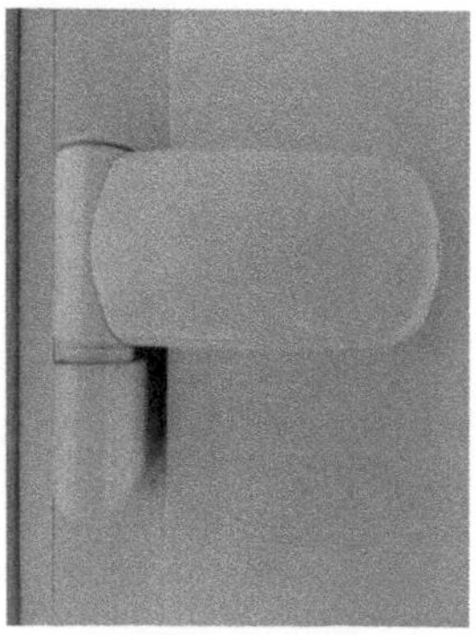	These types of hinges are widely used for uPVC doors but can also be used to hang timber doors. The hinges should be secured with fixing screws to the face of both the door and the frame. One advantage of using flag hinges is that they have a three-way adjustment, meaning the doors can be maintained and adjusted if they 'drop' in the future without having to remove the door from the frame.
Parliament hinge 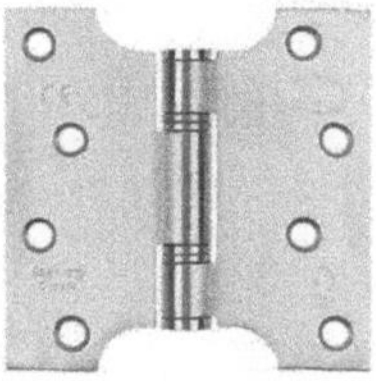	Parliament hinges have knuckles that protrude from the face of the door so that when the door is fully open it stands away from the wall, allowing it to swing a full 180° without binding.

KEY TERMS

Black japanned: finished with a black enamel lacquer originally associated with products from Japan.

Door leaf: one half of a pair of double doors.

Selecting the correct hinges for hanging double doors is essential to ensure the performance of the doors. If the doors do not hang perfectly within the frame, the lock or latch may fail and the doors will not shut correctly.

- Lightweight internal doors (hollow core flush doors) usually require only one pair of 75 mm hinges per door (although bathroom and en-suite doors may be specified to be hung on one and a half pairs of 75 mm hinges, i.e. three hinges).
- For 35 mm-thick timber panelled and glazed internal doors, use one and a half pairs of 75 mm hinges.
- It is best practice to hang all 44 mm-thick doors, whether internal or external, on one and a half pairs of 100 mm hinges.
- Fire doors also need to be hung on one and a half pairs of fire-rated 100 mm hinges.

Hinge positions for doors have regional variations but the standard positions are 150 mm down from the top of the door and 225 mm up from the bottom, while the centre hinge is positioned an equal distance between the top and bottom hinges. On heavier doors, the middle hinge is often moved up to 200 mm below the top hinge.

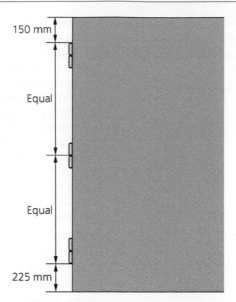

▲ Figure 5.42 Standard positioning for hinges

Locks and latches

Most locks and latches are morticed into the closing edge of the door although a few, like rim locks, sit on the face of the door. There is an increasingly wide variety of locks and latches available, but most fall into five main categories:

1 mortice deadlock
2 mortice locks/latches
3 mortice rebate locks/latches
4 mortice latches
5 rim locks and latches.

▼ Table 5.4 Lock and latch selection and positioning

Name of ironmongery	Description
Mortice deadlock Euro pattern mortice deadlock	Mortice deadlocks are simple key-operated locks that require no handle to operate them. They are commonly used for storerooms or as additional security for external doors. They are available with a claw deadbolt or hook deadbolt for use with sliding doors. They are morticed and recessed into the edge of the door. The keyhole is covered by an escutcheon. The lock is controlled by levers: the more levers the lock has, the more secure it is as it is more difficult to pick. Three-lever and five-lever deadlocks are the most frequently used types. The deadbolt engages into a box striking plate, which is morticed and recessed into the door jamb. European (Euro) pattern cylinder deadlocks, which use a barrel lock to operate the deadbolt, are also available.

▼ Table 5.4 Lock and latch selection and positioning (continued)

Name of ironmongery	Description
Mortice lock/latches Five levers (mortice lock/latch) 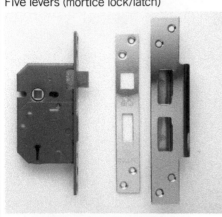	Mortice lock/latches (also known as mortice sash locks) are a combination of a mortice deadlock and mortice latch. They are available as a vertical mortice lock/latch or a horizontal mortice lock/latch. Vertical mortice lock/latches are suitable for fitting in most types of door and come in various sizes, the most common being 75 mm deep with a 57 mm **backset**, and 65 mm deep with a 44 mm backset. Horizontal types are suitable only for fitting in doors with wide stiles, solid flush or set in the middle of the lock rail, as the length of the lock/latch would protrude past the width of a standard door stile. A typical horizontal mortice lock/latch is 150 mm deep, with a spindle backset at 127 mm and keyhole backset at 51 mm. They are typically used with doorknobs and escutcheons. Both types have reversible latches to enable them to be used with left-hand (LH) and right-hand (RH) opening doors. The most commonly used are three-lever and five-lever mortice lock/latches. Three-lever types engage into a striking plate, whereas five-lever types engage into a box striking plate. Mortice lock/latches are available as Euro pattern cylinder lock/latches. Bathroom lock/latches have an additional spindle hole rather than a keyhole.
Privacy lock  Release slot Snib	Privacy locks are commonly used to lock bathroom or toilet doors from the inside, by turning the **snib**. Bathroom mortice sash locks (a type of privacy lock) have to be fitted with bathroom lever handles. This is to enable the door to be unlocked from the outside in the event of an emergency, by turning the release slot on the outside handle with a flat head screwdriver or coin.
Mortice rebate locks 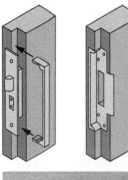 	Rebated mortice lock/latches are available for double doors where the meeting stiles are rebated into each door. Rebate kits convert standard lock/latches to suit rebated doors with a standard rebate of 12 mm.

→

▼ Table 5.4 Lock and latch selection and positioning (continued)

Name of ironmongery	Description
Espagnolette (multi-point lock) 	Espagnolette locks are used extensively with uPVC entrance doors but may also be fitted to timber doors, as they offer high security with multi-point locking.
Mortice latches/tubular latch 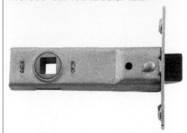	Mortice latches are used mainly for internal doors that do not need to be locked. The most common type is the tubular mortice latch. The latch engages into a striking plate on the lining or frame, which keeps the door shut. It is operated from either side of the door by a pair of lever handles or doorknobs. Mortice latches are reversible, so can be used on LH- or RH-opening doors. They are available in different lengths and backset depths to suit different applications, however the two most common sizes are 63 mm and 76 mm. Rebated tubular mortice latches are used for rebated doors; alternatively rebate kits can be used to convert standard mortice latches for rebated doors.
Rim deadlocks and rim lock/latches Rim deadlock  Cylinder night latch 	Rim deadlocks and rim lock/latches are face-fixed and offer a cheaper alternative to locks that require the additional labour of being morticed in. They are usually operated by a pair of doorknobs. They are often used with ledged and braced doors, as these doors are not thick enough to receive mortice locks. They are screwed to the inside of the door in conjunction with an escutcheon on the outside. Doors fitted with rim deadlocks usually have a pull handle on the outside to operate the door. Rim lock/latches are still used for period properties in conjunction with doorknob sets. Cylinder night latches are mainly used on entrance doors to domestic properties. The cylinder fits through a hole bored through the door and the night latch is fixed to it via a back plate. The door is opened via a key from the outside and by turning a handle from the inside. The latch bolt engages with a staple keep that is fixed to the door jamb. Better-quality night latches have a double-locking facility that improves their security. They are available to fit narrow door stiles.

→

▼ Table 5.4 Lock and latch selection and positioning (continued)

Name of ironmongery	Description
Panic bars 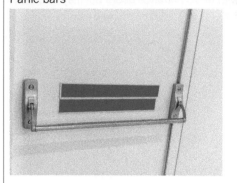 Push pad latch 	Panic bars are a type of espagnolette bolt, sometimes referred to as crash bars or push bars. They are fitted to an exit door, typically a fire door, and are designed to be used by people not familiar with the door's operation. A panic bar allows safe egress through the door by simply pushing on the horizontal bar. They are typically used in shops, schools, hospitals and other public buildings, allowing emergency egress through the doorway. Push pad latches are suitable both for the first opening leaf of a double rebated door and for single timber doors. They are usually fitted on either RH- or LH-opening emergency doors, in areas of low occupancy. They are simply operated by pushing down on the pad to operate the rim latch.

INDUSTRY TIP

Mortice latches generally have a rectangular face plate and striking plate, which will need to be recessed into the door edge and frame/lining. Alternatively, you could install a 'fast latch' with a round face and strike plate. These latches are housed in a round hole in the edge of the door that requires no recessing and only a minimal amount on the frame/lining for the **splinter guard**. For a demonstration of how they are fitted, search for 'UNION FastLatch Installation Video' on YouTube and watch the video tutorial.

▲ Figure 5.43 Fast latch

KEY TERMS

Backset: the distance from the face of the latch to the centre of the spindle.

Snib: a small knob on bathroom lever handles used to throw the bolt on a sash lock.

Splinter guard: a part of the striking plate that extends over the front edge of a door frame or lining.

Door closers

▼ Table 5.5 Door closer selection and positioning

Name of ironmongery	Description
Overhead door closer 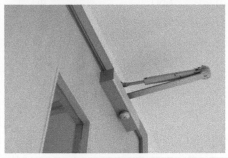	These are used to ensure doors close on their own to prevent the spread of fire, draughts and sound, or to ensure privacy throughout a building. Overhead door closers are fitted to the top of the door and/or the door frame above it. They work by means of either a coiled spring mechanism or a hydraulic system enclosed within the casing, with an arm to either pull or push the door shut. Different strengths of overhead door closer are available to suit the size and weight of the door. Door closers to be used with fire doors must also comply with relevant British Standards. Table 5.6 gives guidance on the correct door closer to use.
Concealed-spring door closer 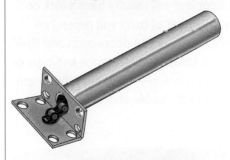 Concealed-spring door closer (fitted) 	Concealed-spring door closers work through a tensioned spring housed in a cylinder. This is morticed into the hanging side of the door and secured with screws. The anchor plate is recessed and screwed into a rebate on the door frame. The closing action is adjusted by using a claw hammer to withdraw the chain and then inserting a metal plate to hold the chain in place. The anchor plate is unscrewed and turned clockwise to increase tension and speed up the closing action, or turned anti-clockwise to slow down the closing action.
Door selector 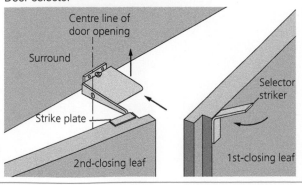	See the section titled 'Installing door selectors', on page 264.

▼ Table 5.6 Guidance on the correct door closer to use

Door closer power size	Recommended door width (mm)	Door weight (kg)
1	750	20
2	850	40
3	950	60
4	1100	80
5	1250	100
6	1400	120
7	1600	160

Bolts and security devices

The slave door leaf will usually be held in place by bolts, while the master door leaf will usually have a lock or latch. The type and number of bolts will depend on several factors, including how the doors open and their construction method, location, appearance and type of finish. Table 5.7 shows some of the common bolts and security devices used with double doors.

INDUSTRY TIP

Building Regulations usually require a door closer to be fitted to fire doors. However, some types of closer can spoil the look of the doorset. One method to overcome this problem is to fit a concealed door closer in the door edge and rebate. Alternatively, you can now install fire-rated butt hinges with a concealed spring in the knuckle that will provide the self-closing action needed to meet current regulations.

▲ Figure 5.44 Stainless steel fire-rated sprung-butt hinge

▼ Table 5.7 Bolts and security device selection and positioning

Name of ironmongery	Description
Inset shoot bolts (flush bolts) 	These bolts are used primarily to secure one half of a pair of French doors, or double doors in domestic or commercial locations. The bolts can either be morticed flush into the face of the door if the edges are rebated, or be morticed into the side for added security if the edges are plain. One bolt should be positioned at the top of the door and another at the bottom. Flush bolts are usually fitted when a superior-quality finish is required.
Barrel bolts Cranked tower bolt (straight)  Straight barrel bolt	Barrel bolts and tower bolts can be used to secure gates and doors from the inside.

▼ Table 5.7 Bolts and security device selection and positioning (continued)

Name of ironmongery	Description
Mortice rack bolt 	Mortice rack bolts are deadbolts operated by a fluted key. They are morticed into the opening edge of external doors, positioned 150 mm from the top and 225 mm from the bottom. They are operated only from the inside.
Hinge bolt 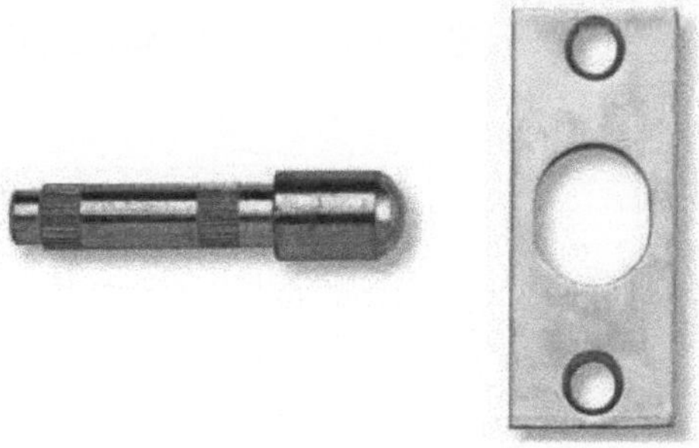	Hinge bolts are morticed into the hinge side of external-opening doors, just below the top hinge and just above the bottom hinge. They are made from hardened steel and improve the security of the door by preventing the door from being levered off its hinges. Security 100 mm butts can also be used; these have interlocking pins to prevent the hinge pin from being attacked.

Door furniture

Door furniture refers to items that are fixed to the door surface, typically to help open or protect the door.

▼ Table 5.8 Door furniture selection and positioning

Name of ironmongery	Description
View hole  	Unglazed doors with one-way door viewers prevent unwanted viewing into rooms or buildings while still allowing the occupant to see out.
Lever furniture  Euro lock brass handle  Round bar handle  Black japanned ornate latch/lock 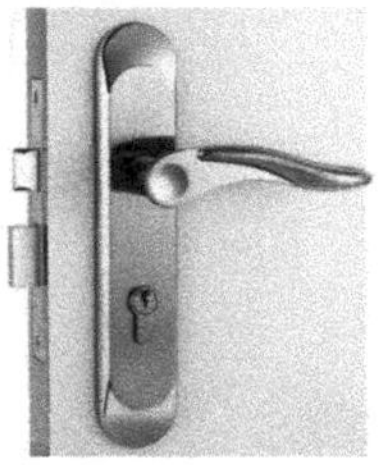 Internal door handle for a mortice latch	Available in a wide variety of patterns and colours, they are used with mortice latches and lock/latches (also referred to as sash locks).
Knob furniture 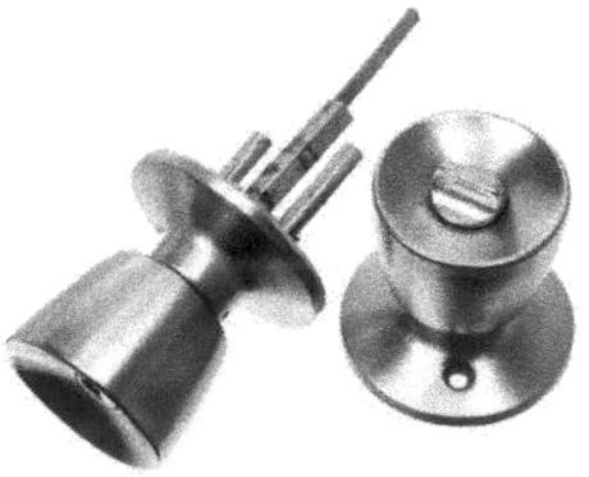 	For use with rim locks and horizontal lock/latches. These should not be used with vertical lock/latches because the knob is very close to the edge of the door and may cause hand injuries when closing the door.
Thumb latch 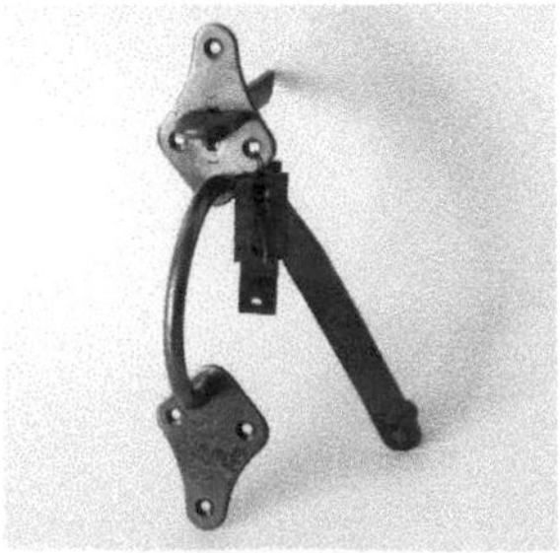 	Usually known as Suffolk or Norfolk latches, thumb latches are used on ledged and braced doors. Traditionally, they were made by a blacksmith from mild steel. A variety of designs are available, usually black japanned.

▼ Table 5.8 Door furniture selection and positioning (continued)

Name of ironmongery	Description
Escutcheon	Used to provide a neat finish to the keyhole on both sides of deadlocks and vertical mortice latch/locks.
Push (finger) and kick plates	To prevent doors from damage, kick plates are positioned along the bottom rail and push plates on the opening or meeting stiles. In factories and hospitals, protection plates can also be fixed in other positions to protect the door from damage caused by trolleys or beds.
Letter plates	Letter plates are usually positioned centrally in the middle rail but can be fitted in the bottom rail, and even smaller vertical letter plates can be fitted into the door stile. Traditionally, letter plates were fitted by drilling a series of holes and cutting out the shape with a pad saw. This is now often done with the aid of a jigsaw, or with a router and a jig to suit the size of the letter plate.
Proprietary threshold	Proprietary thresholds are used to form a weather-proof seal between the bottom of a door and the door frame cill. They are usually made of aluminium with nitrite seals, which can easily be cut to length with a hacksaw to suit the door width. Thresholds are usually supplied with fitting instructions; however, they are normally bedded down on a generous bead of silicone sealant and secured to the cill. Some aluminium thresholds are designed to be fitted with a compatible rain deflector at the foot of the door, to direct water away.
Trickle vent	Trickle vents are used in door and window frames to provide a means to ventilate a building. Most vents have panels on the inside face that can be opened or closed to control the amount of ventilation being supplied through them. In some circumstances, trickle vents are required by the Building Regulations.
Weather seal	There are many different examples of weather seals, each designed to suit the position in which it is to be used, usually in a door or window frame. Weather seals are designed to prevent draughts, water ingress and heat loss. They are available in a basic range of colours to match the natural wood or painted joinery finish. A continuous groove is usually machined around the rebated frame and the weather seal is pushed into this groove so it is held firmly in place.

247

Positioning door locks, latches, door furniture and signs

Positioning of door ironmongery will depend on several factors, typically including the construction method used in the doors, the material from which they are made, the specification and the ironmongery fitting instructions.

Figure 5.45 shows the standard fixing positions for the main ironmongery items in use.

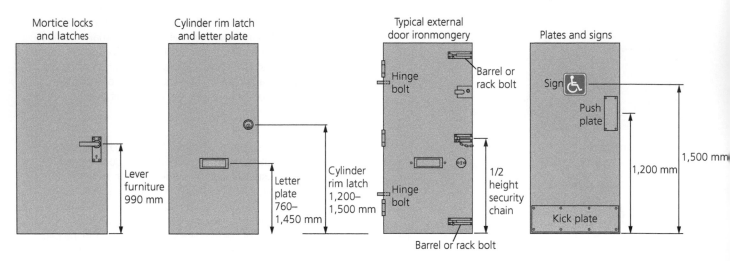

▲ Figure 5.45 Ironmongery heights

Transoms and springs

▼ Table 5.9 Transom and floor spring selection and positioning

Name of ironmongery	Description
Floor spring	Floor springs are used mostly for heavy-duty entrance doors. They can control both single-action and double-action doors. Doors are pivoted via the floor spring at the bottom and a pivot centre at the top. The floor spring is housed in a box that is set into the floor. The bottom of the door is recessed to accept a shoe, which is fitted to the floor spring. The pivot centre is attached to the underside of the head of the door frame and a socket is housed into the top of the door. Floor springs also act as door closers.
Transom spring	The concept of transom springs is similar to that of floor springs, but these are fitted into the head of the frame rather than the floor. Not all floors are suitable for floor springs (e.g. because of underfloor heating or the depth of floor) so transom springs are a better option. They are mainly used to hang heavy aluminium or steel commercial or industrial doors, such as those found in shop fronts, for example. When used in architectural joinery, a bespoke frame must be made to house the transom spring, due to the large hydraulic unit that must be mounted into the head. Transom springs are available with single and double actions.

Materials used for ironmongery

Ironmongery is usually made from either **ferrous** or **non-ferrous** metal.

Ferrous metal

The most common material used for butt hinges is mild steel, which is also the cheapest. The appearance of mild steel can be enhanced by electro-brass or zinc plating. Cast iron is suitable for heavyweight doors but can be brittle. Stainless steel resists staining and corrosion, and is therefore used for hardwood doors and external uses, as well as on fire doors.

The use of iron and steel ironmongery needs careful consideration as these metals rust when they come into contact with moist air or with acidic timbers such as oak. This reaction can cause unsightly staining to the joinery.

Non-ferrous metals

Brass, which is mostly used for ornamental purposes, is extremely resistant to rusting and staining, and is suitable for both internal and external use. It is particularly good for use with oak. However, brass screws are soft and can easily be broken while screwing into hard materials.

> **KEY TERMS**
>
> **Ferrous:** containing iron.
> **Non-ferrous:** not containing iron.

> **INDUSTRY TIP**
>
> Extreme care is needed when using brass screws (usually supplied with the ironmongery). Always pre-drill their holes and add a little petroleum jelly lubricant or wax to the screw thread. This will ease the screw into the material and reduce the risk of snapping.

Metals, finishes and their abbreviations

Table 5.10 illustrates how ironmongery materials and finishes are commonly abbreviated in the construction industry, on drawings, in suppliers' catalogues and on their websites.

▼ Table 5.10 Ironmongery: materials, finishes and common abbreviations

Metal	Finish	Abbreviation
Aluminium	Silver anodised aluminium	SAA
	Anodised aluminium	AA
	Polished and anodised aluminium	PAA
	Etched and anodised aluminium	EA
Brass	Polished brass	PB
	Satin brass	SB
	Polished chrome plated	PCP
	Satin chrome plated	SCP
	Satin nickel plated	SNP
Iron	Armour bright	ABRT
	Berlin black	BB
	Black Japanned	JPD
Steel	Brass plate on steel	EB
	Bright zinc plate	BZP
	Chrome plate on steel	CP
	Galvanised	EG
	Black japanned (painted)	JPD
	Zinc plated	ZP
	Sheradised	SHR
Stainless steel	Satin stainless steel	SSS
	Polished stainless steel	PSS
	Satin copper on stainless steel	SCSS
	Bright copper on stainless steel	BCSS
Zinc alloy	Chrome plated	CP
	Satin nickel plated	SNP
	Satin chrome plated	SCP

Installing a double door frame

When fixing double door frames within openings, it is vital to make sure the frame is not twisted, bowed, out of square or distorted in any way. As with the fixing and positioning of single door frames, poor positioning and fixing techniques will affect the hanging and performance of the doors; the need for care and accuracy is even more vital with double door frames and linings, as any discrepancies will be magnified when the doors are hung and come to meet at the meeting stiles.

The following step-by-step guide should ensure that any problems are kept to a minimum.

1 Check the internal dimensions of the frame against the door sizes.

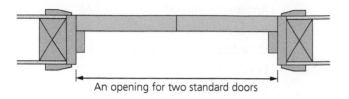

An opening for two standard doors

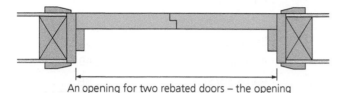

An opening for two rebated doors – the opening size should be reduced by 13 mm to accommodate the rebate

▲ Figure 5.46 Typical joint to head of double lining (top); Rebated joint to head of double lining (bottom)

2 Check the width and height of the frame against the opening in the wall, making sure you have enough clearance for adjusting/levelling.

3 Position the frame within the opening and wedge it in place between the head and wall directly above the jambs. Temporarily pack behind the jambs so there is an even gap on each side between the wall. This will hold the frame in place while further adjustments are made.

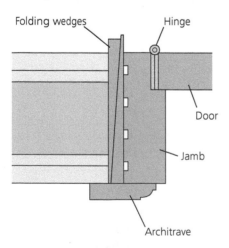

▲ Figure 5.47 Section through an internal frame showing fitting of folding wedges

▲ Figure 5.48 Fixing folding wedges

4 Check the frame is plumb using a 1.8 m spirit level. Square the frame diagonally from corner to corner; the two measurements should be the same. Sight across the frame to ensure it is not twisted. Ensure the frame is not bowed by measuring horizontally across the top, middle and bottom – these measurements should be the same. It is essential the frame is not bowed or distorted as this will affect the correct operation of the doors.

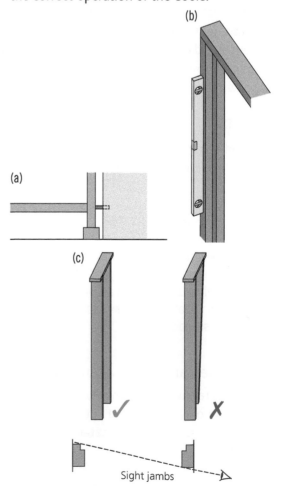

▲ Figure 5.49 (a) Pack off the sub-floor if needed for the FFL; (b) plumb and fix one jamb first; (c) door jambs should be parallel

5 Drill through the frame and into the brickwork or blockwork to the correct depth using an SDS drill, making sure that holes are around 100 mm from each corner with at least another two evenly spaced down the frame and along the head and cill, if there is one. The holes should be lined up so the fixing enters the brick and not the mortar course. This ensures good fixings as some mortar joints are filled with voids behind the surface, allowing the plug to pull free.

6 If you are fixing into studwork or similar timber surrounds, you will need to use a suitable high-speed steel (HSS) drill instead of the SDS drill bit.

7 Insert plastic packers or **folding wedges** (plastic or timber) either side of the fixing hole and insert the fixing into the hole. Screw it home, making sure you do not bow or distort the frame when tightening the screw. The packers should be nipped tight. Check the frame width to ensure it has not been over-tightened, causing the frame to become bowed.

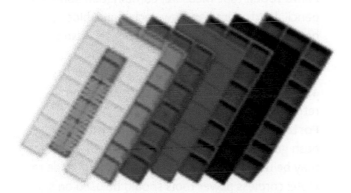

▲ Figure 5.50 Plastic packers used to straighten framework

KEY TERM

Folding wedges: a pair of wedges used 'back to back' to create a pair of parallel faces. By sliding one wedge against the other, the total parallel thickness of the folding wedges can be adjusted to pack out the required gap.

It is sometimes best not to drill and fix the lower fixing on one side of the frame and the cill until the doors have been hung. This will allow for a small amount of movement at the lower end of the frame to compensate for any slight twist within the doors and to ensure the meeting stiles meet flush through their entire height. When the doors are hung and operating correctly, these lower fixings can then be made, again ensuring no distortion takes place.

ACTIVITY

Carpenters will fix items to a variety of materials, including brick, blocks and concrete. In most cases they will use an SDS drill to carry out these tasks. What does the abbreviation 'SDS' mean, and how is it different to standard TCT masonry drill bits?

Produce a chart to show a range of masonry fixings and the type/size of SDS drill bit needed to complete the tasks. You should also provide a brief explanation of why that fixing may be used.

ACTIVITY

Research the amount of vibration and noise that will be produced using an SDS drill in various materials. Compare these results against the same tasks using a standard masonry drill. Analyse the results and summarise your conclusion in several paragraphs.

INDUSTRY TIP

Be careful not to position the frame fixing holes in the same positions as the ironmongery when the doors are fitted.

INDUSTRY TIP

Cut a piece of timber batten exactly the same size as the distance between the jambs, measured at the head. Use this batten to check the distance between the jambs (width) at several points from the top to the bottom of the door frame/lining to double check that it has been fitted accurately.

INDUSTRY TIP

Once the frame has been fitted, the gap between the outside of the frame and the wall can be filled with expanding foam. Once the foam has fully expanded and dried, the excess foam can easily be removed with a saw or sharp knife. The foam provides an excellent additional fixing for the frame, to eliminate any possible movement when the doors are swinging. However, if the frame is poorly secured the foam could distort it as it expands. If expanding foam is used to secure the frame prior to hanging the doors, the frame cannot be adjusted to suit the doors at a later stage.

Further information and tutorials on the use of expanding foam can be found on YouTube.

▲ Figure 5.51 Using a foam gun to fix a door frame

② HANGING DOUBLE DOORS

The process of hanging double doors is similar to that for hanging a single door, so your experience and skills from Level 2 will be useful. However, side-hung double doors require greater care during all stages of fitting the frames and doors, to ensure the doors close correctly at the meeting stiles. Even a small amount of twist or misalignment to either the doors' positioning or the fitting of the frame will have a dramatic effect on where the doors meet, and making it difficult to fit the rebate lock and get the doors to close flush.

Before starting to hang any doors, you must ensure you are fitting the correct door and associated ironmongery in the right position in the building: do this by checking the door schedule against the working drawings. Having the correct tools will determine how quickly and accurately you hang each pair of doors, though of course there are other factors involved, such as competence and experience.

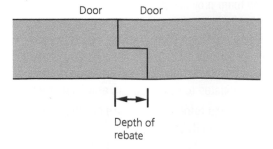

▲ Figure 5.52 A section through the meeting stiles of a door

Tools required for hanging double doors

Hanging doors and fitting and fixing the associated ironmongery can require a comprehensive tool kit and a range of ancillary equipment. Power tools are the most efficient to use, but you will need some hand tools for finishing (e.g. removing the leading edge and arris from the door).

> **INDUSTRY TIP**
>
> An arris is the sharp corner found on the edge of timber. You can remove the arris from a door with a sharp jack or block plane, or a piece of sandpaper. A slightly arrised edge will provide a smoother feel to the joinery, and will also provide a better surface for paint and other finishes to adhere to.

- **Hand tools:** tape measure, combination square, 2H pencil, hard-point handsaw, jack plane, block plane, bench rebate plane (for rebated doors), assorted bevel-edged chisels, mallet, claw hammer, assorted auger or spade (flat) bits, a set of screwdrivers, assorted pilot bits, two marking gauges and a retractable-blade knife.
- **Portable power tools:** plane, router (to include bush guides and router cutters to suit any jigs that may be used), drill/driver, plunge saw and guide rail. Note: cordless power tools should be used on site whenever possible. Mains-powered tools can be used, but they are more hazardous and restricted in their use.
- **Ancillary equipment:** a pair of timber or metal foldable saw horses, hop-up, timber chock and wedge or door stand, door lifter, wedges, air wedge levelling tool, hinge router jig, mortice lock router jig, letter plate router jig, quick clamps, 110V extension lead/spider and transformer, dust sheets, a portable extraction unit (LEV) and PPE.

Figures 5.53 to 5.59 show methods adopted to support a door.

▲ Figure 5.53 Timber chock and wedge

▲ Figure 5.54 Door stand

▲ Figure 5.55 Folding saw horse

▲ Figure 5.56 Door lifter

▲ Figure 5.57 Gravity clamp

▲ Figure 5.58 Gravity clamp in use

Hanging doors isn't difficult as long as you are careful, use sharp tools and follow a process such as the one explained below.

Most new doors will arrive on site packaged in a protective plastic wrap, with folded cardboard over the long edges. They may also have plastic corner protectors and possibly small wooden strips nailed to the top and bottom edges of the doors to protect them while in storage and transit. Care should be taken not only to make sure the packaging is removed and disposed of correctly, but also to ensure all the nails used to secure the protective strips are removed to prevent damage to your tools when fitting the doors. Once the entire packaging has been removed, you should check the doors for damage or manufacturing imperfections and report any issues to your supervisor.

Hanging double doors: step by step

Now you are ready to follow the sequence of hanging the doors in a frame.

1 Use a spirit level to check the head and jambs of the frame are level, plumb and straight. If any discrepancies are identified, you may have to compensate with the margins calculated at steps 2 and 3. If the frame isn't level and plumb within tolerance, you may have to adjust it or remove and refit it.

2 Measure the width of each door plus the margins for the joints between the meeting rails, and the stiles and door jamb. For example, if the width of each door is 726 mm and the gap between them is 3 mm (726 mm × 2 = 1452 mm) plus (3 mm × 3 margins = 9 mm) then the total width of the door lining will be 1452 + 9 = 1461 mm (internal dimension).

3 Check the height of the door frame against the door. You should make an allowance of 2–4 mm (to match the width) between the top of the door (head) and the frame (internal). The margin you allow between the bottom of the door and the floor will depend on the type of flooring used (e.g. carpet, tiles or engineered flooring), and whether it is already laid. It's important that the door just clears the floor as it swings open and closed: if the door is too low, then it could wear or mark the floor covering. You may have to make an allowance for

▲ Figure 5.59 Air wedge lifting tool: deflated (left) and inflated (right)

a threshold or a water bar for external doors (refer to Level 2 Site Carpentry and Architectural Joinery for information on water bars).

Note: An allowance should be made at the bottom for the floor covering when fitting internal doors. External inward-opening doors are usually fitted over a threshold; therefore a suitable clearance is already factored into the design.

4 Check against the plans to find the correct handing of the doors. You may also have to check whether they are left- or right-hand opening first, if they are rebated doors.
5 In turn, place one of the doors onto the two saw horses. Mark on the door with a pencil the hinge side and also the frame (on flush doors or fire doors, the lock side will be marked by the manufacturer on the top of the door, usually with a key symbol). If the doors have horns, these must be squared off and cut flush with the top and bottom of the door. Use either a jack plane or power plane to give a smooth finish to the saw cuts. If the door heights have to be reduced a substantial amount, you could use a plunge saw and guide rail while the doors are laid flat.

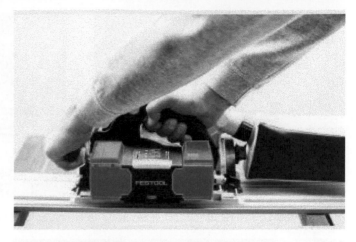

▲ Figure 5.60 Cutting a door to height with a plunge saw and guide rail

ACTIVITY

When reducing the height of a door, always make your saw cuts from the bottom of the door. This will prevent making the top rail of the door smaller than the stiles and potentially weakening the joint (if it is a framed door). Explain the process you would need to go through if the bottom batten in a hollow core door has been removed after cutting the door to size.

6 If the door is to be fitted into a frame with a threshold and water bar, it will need rebating before fitting.
7 Offer one of the doors into the frame and, starting with one of the hanging sides, plane the door to fit against the jamb of the frame perfectly. (If the door frame has been installed correctly, you may not have to plane anything off the edge of the door.)

▲ Figure 5.61 Rebating the bottom of the doors with a router to fit over a water bar

▲ Figure 5.62 Planing a door to size (shooting in)

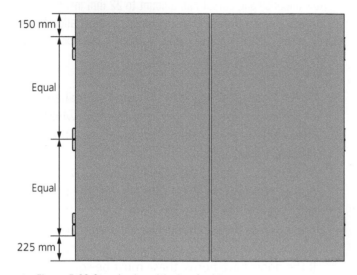

▲ Figure 5.63 Standard positioning for hinges

INDUSTRY TIP

Use a jack plane to remove small amounts of timber from the edges of the door or during final finishing, to give you more control and a better finish free of pitch marks (refer to Chapter 3 for information on pitch marks).

8   Hold the door tightly against one of the jambs of the frame and up against the head. Check that there are no gaps between the door and frame. If the joint between the top of the door and the head of the frame is not parallel, plane that to fit as well.

9   Place the door that has previously been planed on its edge in a door stand; this will hold the door secure while the hinges are marked out and recessed. Mark out the position of the hinges on the hinged side (opposite the rebated side, if they are rebated doors).

Hinge selection and positioning: step by step

The correct selection of hinges for hanging doors is essential for the performance of the door. If the door does not hang perfectly within the frame, the lock or latch may fail to operate correctly and the door will not shut. Lightweight internal doors (hollow-core flush doors) usually require only one pair of 75 mm hinges (although, bathroom and en-suite doors may be specified to be hung on one and a half pairs of 75 mm hinges). For 35 mm-thick timber panelled and glazed internal doors, use one and a half pairs of 75 mm hinges. It is best practice to hang all 44 mm-thick doors, whether internal or external, on one and a half pairs of 100 mm hinges. Fire doors also need to be hung on one and a half pairs of fire-rated 100 mm hinges. (Further details on hinges and hinge positions can be found in *Level 2 Site Carpentry and Architectural*

Joinery, Chapter 5, 'Non-structural carpentry following plastering'.)

Mark the depth and width of the butt hinges on the door edge using the following steps.

1 Adjust a marking gauge to the width of the butt hinge leaf. Transfer this to the edge of the door and mark the width of the hinge recesses:

2 Adjust a second marking gauge to the thickness of one butt hinge leaf and transfer this to the face of the door to set the depth of the hinge recesses:

3 Using a bevel-edged chisel and mallet or hammer, accurately chop out the hinge recesses on the door. Trial fit each recess with a hinge to ensure it finishes flush. Note: Some butt hinges may need to be recessed slightly lower into the edge of the door and frame:

INDUSTRY TIP

You can use a 'rod' to mark out the hinge positions on doors and frames more accurately. To make a rod, first mark out the hinge positions on a length of batten or spare door stop. Align the top of the rod with the top of the door, before transferring the hinge positions to the door edge. The same rod can be used to mark out the hinge positions on the frame by repeating the process, this time with a packer between the rod and head equal to the margin allowed around the door edges; usually 2–4 mm. This method will improve your speed and accuracy if you are fitting more than one door.

INDUSTRY TIP

Some carpenters prefer to use a sharp utility knife to mark out the length and width of the hinges on the door edge. This is done by turning the hinge over so that the knuckle can be placed tight against the face of the door and to the desired height; the knife is then used to scribe a line around the outside edges of the hinge. The process is repeated for the remaining hinges on the door and frame.

HEALTH AND SAFETY

Extra care should be taken when using a sharp utility knife, to avoid cutting your hands. One way to prevent accidents is to keep your idle hand behind the blade at all times. You could also use a utility knife with a retractable blade for safe storage when it's not in use.

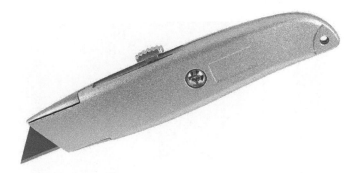

▲ Figure 5.64 Utility knife

4   Reposition the door in the frame and wedge it up to the head of the frame, using either a wedge or an air wedge lifting tool, and allowing a 2–4 mm gap (use a plastic packer in between the top of the door and the head of the frame). The door should also be wedged tightly against the hinge side. The positions of the hinges can now be transferred from the door onto the face of the frame. It is usual to position the top hinge 150 mm from the top of the door and the bottom hinge 225 mm from the bottom of the door, with the middle hinge being positioned centrally between the other two. Repeat the process of recessing the hinges into the frame.

▲ Figure 5.65 Transferring the hinge positions from the door to the frame

INDUSTRY TIP

Always mark both sides of the hinges when marking out. This will reduce the likelihood of recessing the hinges the wrong side of the lines at the next stage.

▲ Figure 5.66 Using a marking gauge to mark out the door frame

▲ Figure 5.67 Recessing the hinge position on the door frame

5   Screw each hinge into the hinge recesses on the door, ensuring that the leaf fixed to the door is the leaf with fewer knuckles. Use a cordless drill to provide pilot holes for the screws.

6   Offer the hinged door up to the door frame. Use a door lift or air wedge lifting tool to lift the door so that the hinges line up with the recesses, and – starting with the top hinge – pilot a hole and secure with one screw to each hinge. Use a hinge drill bit for this stage, if possible, to make sure the pilot holes are drilled centrally in the countersunk holes in the hinges. Check that the door opens and closes correctly in the frame and does not bind on the hinges, and that the clearance gap is correct. Put the remaining screws in the hinge recesses on the frame, and re-check the fit.

▲ Figure 5.68 Hinge drill bit

Drilling only one pilot hole in each hinge at this stage will provide you with an opportunity to adjust the hinge position, if necessary, at a later stage. If all the holes are drilled initially and the hinge position needs to be moved, it can be difficult to add new fixing holes without the screws finding the original positions, which would result in them pulling the hinge away from the back of the recess to reveal a gap.

▲ Figure 5.70 The planed leading edge of the door

9 Once a satisfactory margin has been achieved between the two doors, the door can be laid on its edge in a door stand and the hinge positions marked out on the hanging side. Repeat the process of recessing the hinges in both the door and frame, and secure the hinges with suitable screws.

10 Check that the door closes into the frame correctly, without binding and still maintaining an equal margin of between 2 and 4 mm.

11 Remove the sharp edges (the arrises) from the doors to ensure that paint will stick properly to the corners and to improve the finish. This is best done with a jack or block plane when the doors are off, or with abrasive paper.

Recessing a hinge with a router and jig

Hinges can be recessed into a door and frame using a cordless or mains-powered router and a jig. This method is popular as it can be very quick and accurate after the initial setting up. There are many different types of jig on the market, from a basic single hinge – either 75 mm (or 76 mm) or 100 mm (or 102 mm), to more advanced jigs with multiple positions along their length. Alternatively, you can make your own jig from a piece of plywood or MDF and a batten.

▲ Figure 5.71 Purpose-made hinge jig

▲ Figure 5.69 The line shows the position of a pilot hole initially drilled

7 Offer the second door into the opening between the swinging door, in closed position, and the frame. If necessary, plane the hinge side, top and bottom edges of the door until a satisfactory margin is achieved. Do not plane the meeting stiles if the doors are rebated, as this will cause difficulties at a later stage when fitting the lock/latch.

8 Plane a slight bevel on the edge of the meeting stile of the door that leads into the other door. There shouldn't be any need to remove the leading edge from the doors if they are rebated, because of their reduced thickness along this edge.

The steps shown in Table 5.11 demonstrate the stages of marking out a hinge and setting up a cordless router to form a recess with the use of a hinge jig.

▼ Table 5.11 Recessing a hinge with a router and jig: step by step

Step 1	Use a pencil to mark the height of the hinge on the door edge. Turn the hinge over, so that the knuckle of the hinge can be placed against the face of the door before drawing around it with a sharp pencil or knife.	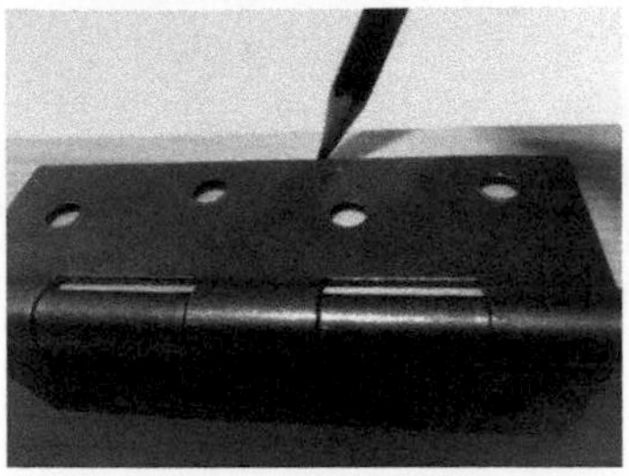
Step 2	Use a marking gauge to mark the thickness of one leaf onto the face of the door.	 
Step 3	You should always refer to the manufacturer's instructions for setting up details of the hinge jig. With the jig illustrated, you simply have to insert the two plastic pins into the location holes indicated, to suit the length of the hinge.	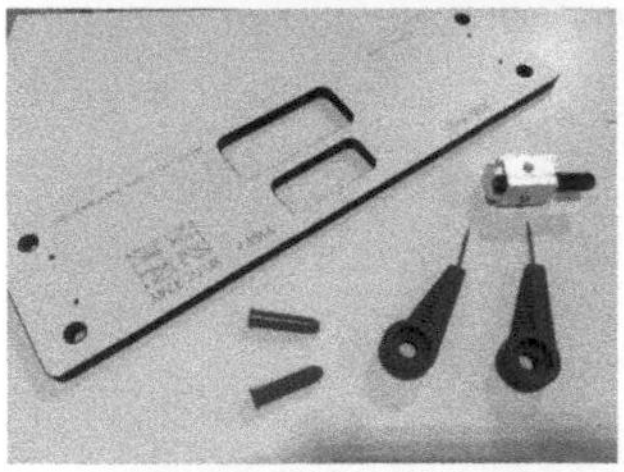 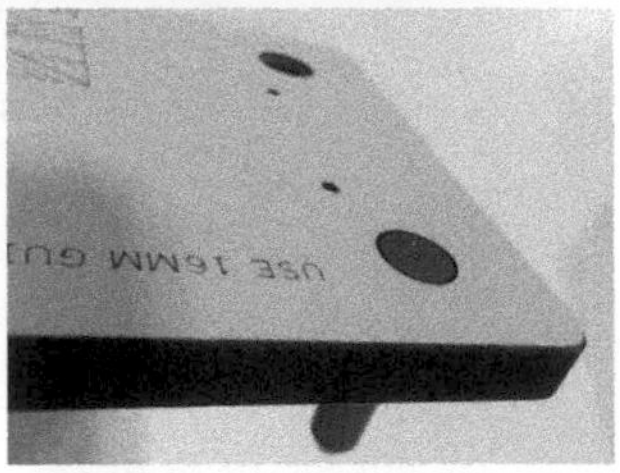
Step 4	Place the jig on the door edge, with the pegs tightly against the face of the door. Align the jig aperture centrally over the marking out on the door edge.	 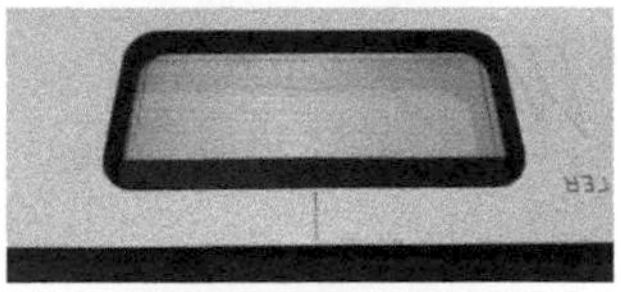
Step 5	Secure the jig to the door with the bradawls provided. These holes will have to be filled once all the recesses have been completed. This could be seen as a disadvantage of using this type of hinge jig.	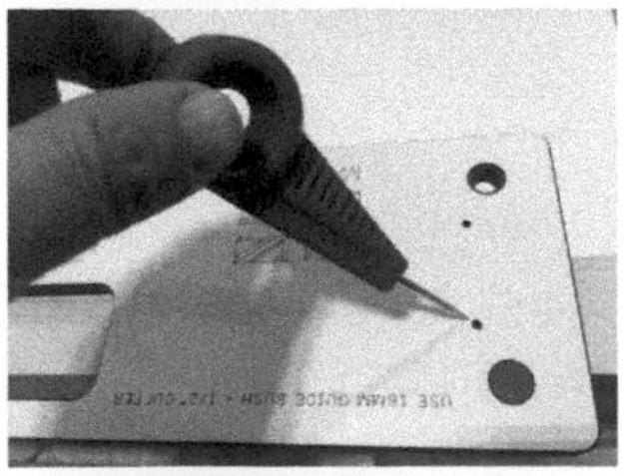 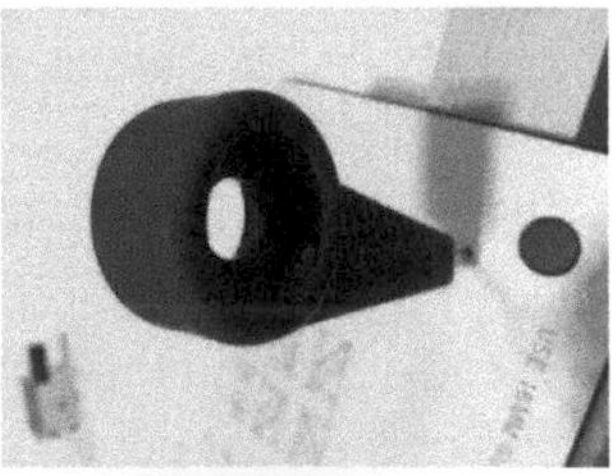
Step 6	Attach a suitable bush guide to the base of the router to follow the jig. Set the router cutter to the depth of the hinge recess while resting on the jig.	  Bush guide

▼ Table 5.11 Recessing a hinge with a router and jig: step by step (continued)

Step 7	Use the router to recess out the area marked.	 
Step 8	Square out the rounded corners with a sprung-loaded corner chisel or a bevel-edged chisel.	 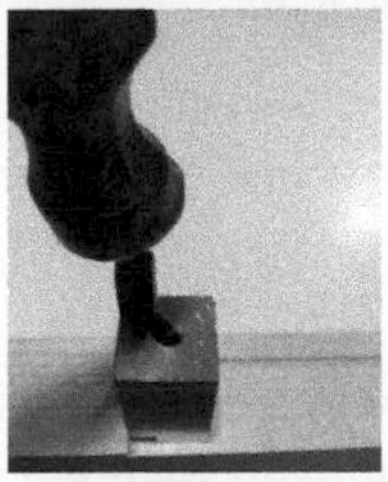  
Step 9	Check the hinge fits in the recess at the correct depth. Drill pilot holes and secure the hinges. You'll need to repeat the process of transferring the hinge positions onto the door frame and recessing the hinges as illustrated above to complete the task.	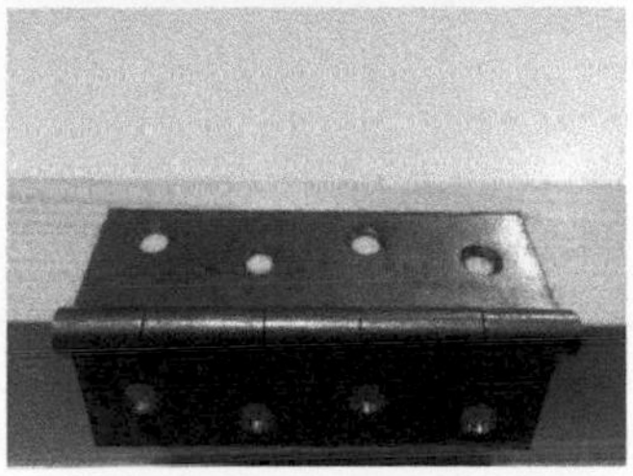

HEALTH AND SAFETY

Ensure the power source (the battery in this case) is disconnected when setting up a router or any other power tool to prevent accidental start-up.

Remember to connect the router to an LEV and to wear suitable PPE while completing this task.

ACTIVITY

How many decibels does your router produce when in use? Using a sound meter, measure the noise created (your college, training centre or employer will probably have one that you can use). Suggest reasonable practical measures that could be taken to prevent the risk of hearing damage.

Binding doors

A common problem occurs when hinges are recessed too deeply into the frame of a door, causing the door to bind on the hinges as it shuts. With rebated door frames and linings, this can take quite a bit of work to put right. One solution is to pack out the hinges by inserting into the hinge recess a piece of card (or, commonly, abrasive paper) that has been cut to the same size as the hinge leaf. (There are also plastic hinge shims available from manufacturers to suit hinges of most sizes.) After the screws have been re-fixed, the door should be checked again for fit.

Another problem that can be more difficult to correct is when the door binds against the rebate on the frame. This usually occurs when not enough back clearance has been allowed. To rectify this, you can remove the screw and adjust the position of the hinge in the recess on the frame away from the rebate, and then (using a different hole) fix it again with a single screw in each hinge; unfortunately, the result is an unsightly gap between the hinge and the edge of the recess. This means either filling or cutting in a thin piece of timber to fill the gap. As the piece to fill the gap is usually very thin, it is difficult to fix, and looks untidy and unprofessional (although, in some cases, the gap may be hidden by the weather-proof and draught-proof seal).

A more time-consuming solution is to remove the door from the frame and then remove the hinges from the door, and increase the width of the hinge recess by the width of the gap. This might mean the existing screw holes will need to be filled with timber fillets so the hinges can be fixed securely to the new recess. When the door is re-fixed in the frame, re-fix the hinges tightly to the hinge recess. This adjustment ensures no unsightly gaps and no binding on the frame, and is the preferred method of correcting this problem.

With plain door linings and frames, the planted stops used to create the rebate can be repositioned if binding occurs.

Fitting door ironmongery (rebated mortice lock/latch)

Once the double doors have been hung in the frame or lining, the door ironmongery can be installed. It's important to refer to the manufacturer's fitting instructions at this stage, to prevent mistakes.

The method shown in Table 5.12 is one way of fitting a vertical mortice lock/latch (sash lock).

▼ Table 5.12 Fitting door ironmongery (rebated mortice lock/latch): step by step

Step 1	Determine the height of the spindle – refer to the specification or, where this is not available, the standard height is 990 mm to the spindle from the bottom of the door. With framed doors, the lock's mortice hole must not interfere with any of the tenons from the middle rail and/or glazing bars to the door stiles.	
Step 2	Wedge the door leaf open to a convenient position; gently wedge it from both sides to prevent it from swinging. Fit temporary packers to level out the rebate. This makes it easier to drill the holes for the lock in the centre of the door.	
Step 3	Holding the lock against the door's meeting stile, lightly mark the upper and lower edges of the lock case. An additional allowance of 5 mm will give the lock a little clearance and aid the lock's fitting and removal. With a combination square, mark the centre of the door leaf and the amount of backset.	
Step 4	Measure down the backset the distance for the keyhole. Using an auger bit or flat bit (also known as a spade bit), bore a 16 mm spindle hole, and a 10 mm and 6 mm keyhole. Ensure you work from both sides of the door. Using a pad or keyhole saw, leave the removal of the keyhole waste until the lock case mortice hole has been cut out. For speed, a larger hole may be drilled for the keyhole, eliminating the need to shape it. However, this process causes increased problems with locating the key smoothly in the lock.	

▼ Table 5.12 Fitting door ironmongery (rebated mortice lock/latch): step by step (continued)

Step 5	Using an auger bit or a flat bit of a diameter slightly larger than the width of the lock case thickness, bore a series of holes to a depth equal to the full depth of the lock body plus 3–5 mm.	
Step 6	Set a mortice gauge to the edge of the drilled holes and mark down the sides of the mortice lock. This will give a neat position for the chisel to locate in and will help to produce a good, clean edge to the lock mortice.	
Step 7	Chop and clean out the mortice hole using a mallet and chisel. Make sure you maintain clean, accurate sides to the mortice, which will permit the lock to be fitted easily and allow for its correct function.	
Step 8	Position the lock in the hole. Mark around its **forend face plate**, using a marking gauge to carefully score along these lines; this will help to prevent the edges from splitting when the forend housing is chopped out. Chop out a recess until the forend fits flush. Remove the temporary batten, fit the rebate lock mortice kit and screw the lock into the door. Fit the spindle and fix the handles to the door, ensuring they are parallel to the edge of the door. Check that the latch operates and that the key activates the dead bolt.	
Step 9	Close the door and mark the position of the latch and lock onto the other meeting stile. Gauge the distance from the back of the door to the front edge of the lock and mark on the door stile, ensuring you allow for the additional thickness of the rebate kit. Form the keep by chopping out the latch and bolt mortice hole in the closing edge of the stile. Recess the striking plate into the door frame until it is flush with the front lip of the door; it may require a deeper bevelled housing to help the latch close smoothly when the door is pushed shut. Note the use of a cramped block to the edge of the door; this will give additional support to the door and help prevent thin, weak edges from breaking away.	
Step 10	The finished doors should finish flush with each other, with a smooth closing operation. The handles should operate smoothly, without any stiffness. The key should enter the lock smoothly from both sides, turning cleanly and operating the lock smoothly.	

KEY TERM

Forend face plate: the part of the lock that is seen when the lock is housed in the door.

Installing overhead door closers

Fitting overhead door closers depends on where the door is situated, the size of the door frame and the amount of room available above the door frame or within the head of the door reveal. They can be fitted on either side of the door, to suit the situation. Overhead hydraulic door closers can be adjusted to speed up the closing action, and have a hydraulic check called a dampening action, which prevents the door slamming shut. They come complete with instructions and paper templates to make installation easier.

Installing door selectors

Where rebated double doors are used, they must be closed in the correct sequence. To aid this, a door selector is used. Door selectors have an arm that drops down onto a strike plate fitted to the top of the leaf that should close second; this prevents the incorrect door leaf from closing fully. The leaf intended to close first has the selector striker fitted to its top; this lifts up the door selector arm as it closes, allowing the second leaf to close behind the first door leaf.

Installing barrel bolts

Two bolts are normally used, one at the top of the door and one at the bottom. They are available as cranked (or necked) for doors and gates that open outwards, and straight for those opening inwards. For heavier-duty industrial applications and taller doors, there are tower bolts with bow and monkey tail handles, which have longer handles to allow the bolt to be operated without having to over-reach. The bolt is secured into a mortice in the head of the door frame, finished with a recessed striking plate. A bolt socket secures the bolt in the floor.

Installing double-action hinges (helical)

Double-action hinges are designed to make doors self-closing. They are typically used in corridors in public buildings such as schools and hospitals, and between kitchen and dining areas in restaurants and public houses.

To allow the door to close on its own, a helical spring is fitted in the large knuckle of the hinge. The spring's tension is adjusted using a hardened steel bar (supplied with the hinge) that is placed in one of the holes in the top of the hinge and turned to apply tension, moving the ring until the required tension is achieved. Turning the cog increases the tension on the spring in the hinge. The tension is needed on the hinge to create the self closing action. Pins are then inserted into the hole to maintain the tension.

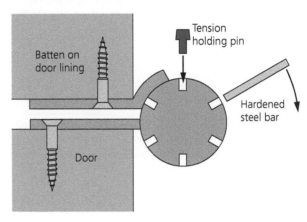

▲ Figure 5.72 Adjustment for double-action door hinges

HEALTH AND SAFETY

When using spring hinges, take great care with the hardened steel bar and pins during adjustment, as the tension built up can be considerable and any sudden slip could result in an injury. Wear impact-resistant glasses or goggles for this stage to reduce the risk.

After tensioning, swing the door open and allow it to close by the hinges. Adjust the tension so that the door over-swings by approximately 200–300 mm – the door should have an easy and comfortable resistance when pushed open.

Double-action hinges are usually fitted onto a batten that is the same width as the door's thickness. This batten is fixed centrally to the door lining. One half of the hinge is fixed directly to the batten while the other half is attached directly to the edge of the door. Although the hinge can be recessed into the batten and door (as with traditional butt hinges), it is difficult to make a neat fitting around the hinge and knuckles of the hinge.

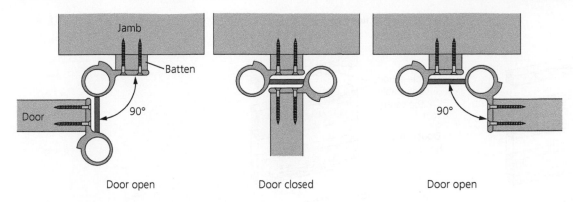

Jamb

Batten

Door

90°

90°

Door open   Door closed   Door open

▲ Figure 5.73 Double-action hinge operation

The meeting stiles of the door will need to be fitted with a radius known as 'heeling'. This allows close fitting along the meeting stiles while still allowing either door to swing in either direction without one door leaf contacting and sticking on the other. Typically, this type of doorset involves the use of a brush seal on the meeting stiles.

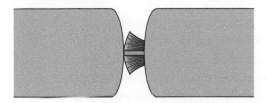

▲ Figure 5.74 Meeting stiles for double doors using double-action hinges

ACTIVITY

Hawgood double-action hinges can be used as an alternative to helical butt hinges. Use YouTube to discover how they are installed. Produce a step-by-step instruction sheet to explain how these hinges should be installed and the benefits of using them.

▲ Figure 5.75 Hawgood double-action hinge

Pivoted floor and transom springs

Double-action floor spring hinges allow the door to swing in both directions and are common in shops and banks. The weight of the door is passed through to the floor instead of the door frame or lining. There are two main types of floor spring hinge:

1 a spring mechanism in the bottom of the door with a pivot point in the floor, used with timber doors
2 a spring mechanism in the floor with the pivot point fixed to the bottom of the door, typically used with glass doors.

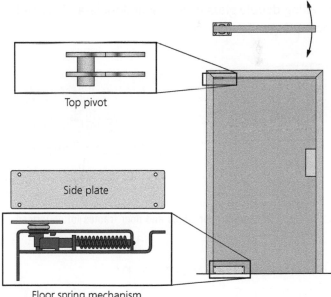

Top pivot

Side plate

Floor spring mechanism

▲ Figure 5.76 Double-action floor spring

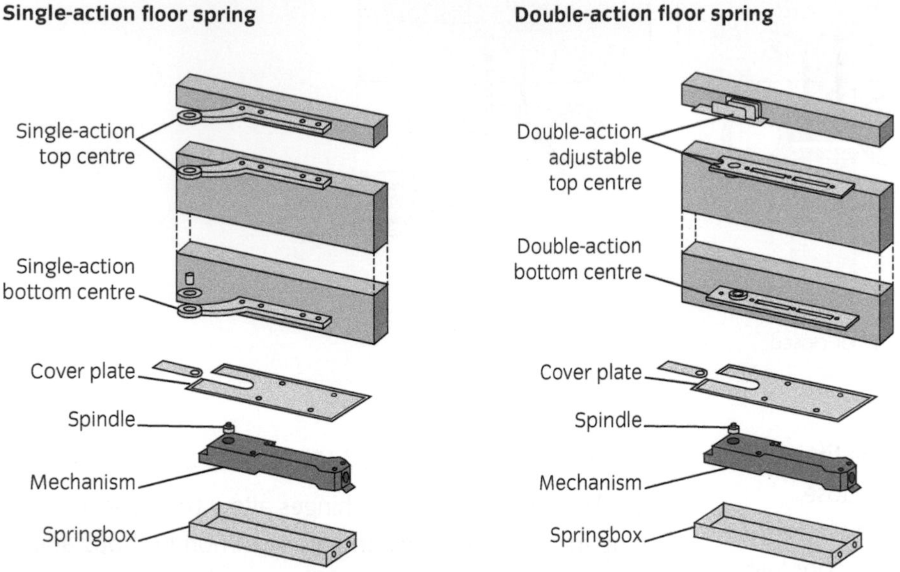

Single-action floor spring

Single-action top centre

Single-action bottom centre

Cover plate

Spindle

Mechanism

Springbox

Double-action floor spring

Double-action adjustable top centre

Double-action bottom centre

Cover plate

Spindle

Mechanism

Springbox

▲ Figure 5.77 Single and double-action floor springs concealed in the floor

Timber doors using this type of pivoting mechanism need both edges of the door shaped, as well as the door frame, with the bottom of the door chopped out to take the spring mechanism. The fitting instructions for floor spring hinges will come with the hinges but follow similar stages as those shown below for installing double glass doors using double-action pivot floor springs.

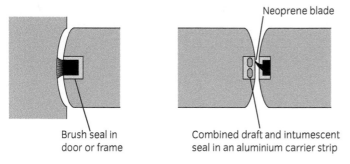

Neoprene blade

Brush seal in door or frame

Combined draft and intumescent seal in an aluminium carrier strip

▲ Figure 5.78 Shaped door edges and door frames (often termed 'heeling')

Sliding door mechanisms (also referred to as sliding door 'gear')

Sliding a door across an opening provides a useful alternative to hinging when there is limited or restricted floor space. The disadvantage is the restricted access to wall space. There is a considerable range of proprietary sliding door systems available, all of which come with the appropriate fitting and installation instructions.

Most systems work along the same principle, with an overhead track fixed to the wall or ceiling from which the door is suspended. The door runs along the track on rollers of nylon (in the case of cheaper lightweight systems) or heavyweight roller bearings (for large industrial doorways). The whole of the top of the sliding gear can be covered with a pelmet to provide a suitable decorative finish.

The bottom of the door can be guided in three ways:
1 a groove, allowing the bottom guide to run along it – the bottom guide is fixed to the floor and positioned so that it runs centrally to the door's thickness
2 a channel tray cut into the floor with a guide pin fixed to the bottom of the door – this method is typically used with industrial doors but can be problematic as the tray tends to fill with debris, etc.
3 guide blocks screwed to the floor either side of the outer faces of the door to prevent the bottom from pulling away, or rubbing on the face of the wall – typically used in domestic situations.

The same principles are applied to pairs of doors where they are allowed to pass over the face of the wall. If the doors are to sit within an opening then a double track would need to be used, allowing the doors to pass one another.

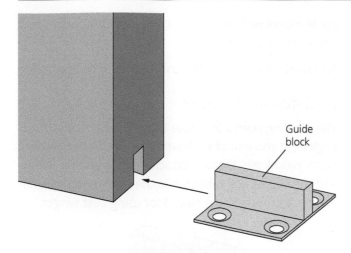

▲ Figure 5.79 Grooved bottom door

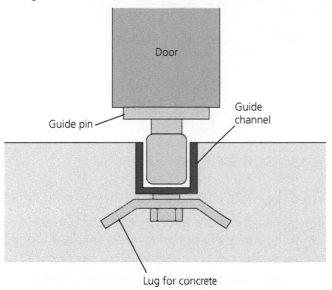

▲ Figure 5.80 Channel tray cut into the floor

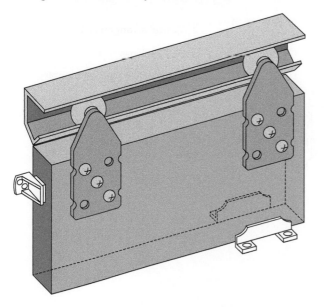

▲ Figure 5.81 Guide block screwed to the floor

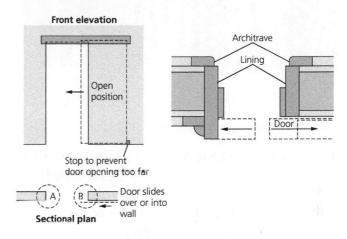

Front elevation

Open position

Stop to prevent door opening too far

Architrave

Lining

Door

A B Door slides over or into wall

Sectional plan

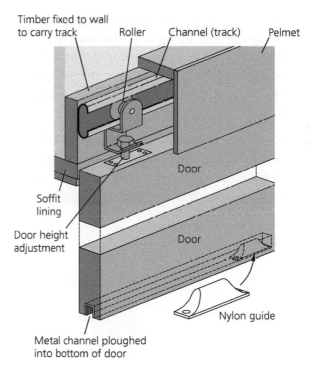

Timber fixed to wall to carry track

Roller

Channel (track)

Pelmet

Door

Soffit lining

Door height adjustment

Door

Metal channel ploughed into bottom of door

Nylon guide

▲ Figure 5.82 Sliding door fitting up to a wall or stop

Folding doors

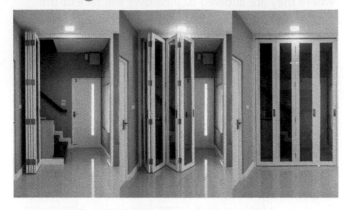

▲ Figure 5.83 Folding and sliding doors (bi-folding)

Like sliding doors, folding doors are a useful alternative to traditional side-hung doors. They are particularly useful for dividing up larger rooms using temporary partition screens, with fitted wardrobes or as fully glazed rear access doors to a garden.

These types of doors have the advantage of closing within the opening or up against the wall, and as a result take up little wall space compared with sliding doors. They also do not have the problem that large doors have of needing an area in which to swing.

Folding doors typically run along a suspended roller system that is fixed to the head of the door lining. The door can be further supported by a roller at its base that runs in a track sunk into the floor, or by using guide rollers with the bottom hinges attached and also run in a sunken trackway.

Folding doors can be either end-folding or centre-folding.

End-folding doors

The pivoting points for these types of door are at the beginning and end of each pair of doors. Each pair of doors has hinges that are attached to both door leaves and the supporting roller, while the other edge of each door is attached to the next door using butt hinges.

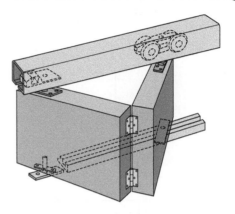

▲ Figure 5.84 End-folding and sliding door

Centre-folding doors

The pivoting position for this type of folding sequence needs to be in the centre of the door. The pivoting sequence starts with a half-leaf door hinged to the framework, typically with butt hinges, as shown in the following illustrations.

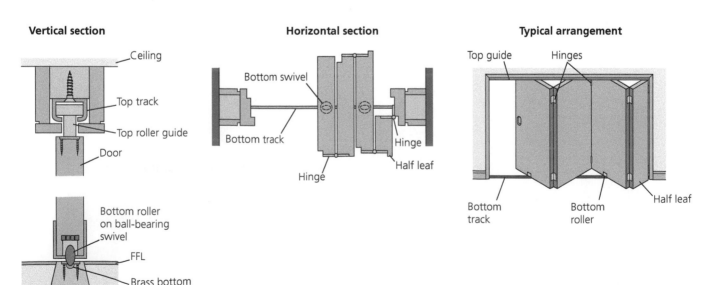

▲ Figure 5.85 Centre-folding door sections

Glazing

Doors may be supplied either pre-glazed by the manufacturer, or unglazed. There are benefits to both. If the glass is fitted in a factory, it can increase the speed of the installation process and reduce the risk of handling the glazing. However, it will increase the weight of the doors and the risk of damaging the glass while hanging them.

Glazing in doors and low-level frames can be extremely hazardous, so it is subject to Building Regulations. Since 6 April 2013, the functional requirements and technical guidance have been covered in Approved Document K 'Protection from falling, collision and impact'.

As a carpenter, you may have to install glass into doors and **combination frames**, so it's important that you understand the regulations that control this. For example, a fully glazed door would require safety glazing, as would a door with glazing at a height below 800 mm. The regulations state that the glazing in these areas must:

- break safely, in a way that does not produce pointed shards with razor-sharp edges
- be robust and of adequate thickness to reduce the likelihood of breakage
- be permanently protected by screens or similar devices.

To meet these requirements, **toughened** or **laminated glass** is used in larger areas. For smaller glass panes, standard **annealed glass** can be used as long as it:

- is as least 6 mm thick
- is not larger than 0.5 m²
- has a maximum width of 250 mm.

Figure 5.87 demonstrates the areas in which safety glass must be used to comply with Building Regulations Part K.

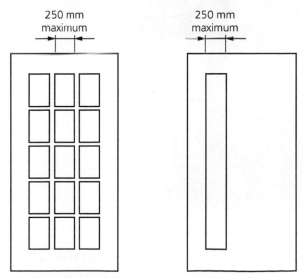

Maximum area of single pane not to exceed 0.5 m², small panes of annealed glass should not be less than 6 mm thick

▲ Figure 5.86 Dimensions and areas of small panes

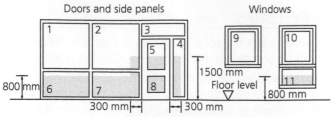

Shaded areas show critical locations to which requirement K4 applies (i.e. glazing in areas numbered 2, 4, 5, 6, 7, 8, 11)

▲ Figure 5.87 Critical glazing locations in internal and external walls

③ TYPES OF WINDOW

The main functions of windows are to provide daylight, ventilation and a view of the outside. They can also be used as a means of escape in an emergency. They must be designed to withstand the weather and conserve heat, as well as to enhance the appearance of the building they are planned for.

Building regulations specify the amount of light and ventilation that a building will require, and planning permission will stipulate the allowed size and position of windows.

▲ Figure 5.88 Purpose-made timber window

The components for windows are discussed in detail in Chapter 7 of *Level 2 Site Carpentry and Architectural Joinery*.

A wide range of window designs is available, the most common types being:

- traditional casement windows
- storm-proof casement windows
- sliding sashes
- bay and bow windows
- pivot-hung windows
- tilt and turn windows
- roof lights.

Section profiles can vary considerably depending on whether the construction is for new or restoration work. This applies particularly in window construction, where **high-performance** windows are required to conform to Part L of the Building Regulations (covering conservation of fuel and power).

KEY TERM

High-performance: modified to give superior performance in terms of thermal and sound insulation.

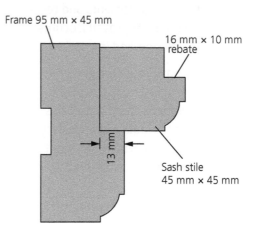

▲ Figure 5.89 Traditional casement sections

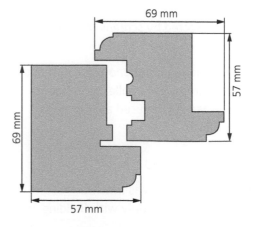

▲ Figure 5.90 High-performance storm-proof casement sections

Installing sliding sash windows (box sash)

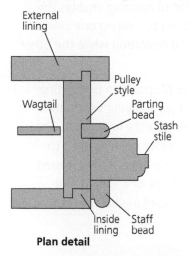

Plan detail

▲ Figure 5.91 Traditional sections for a box frame window

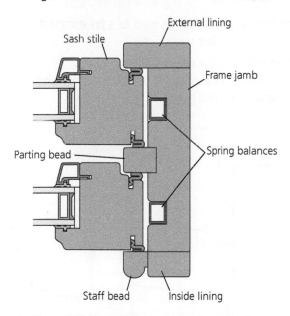

▲ Figure 5.92 High-performance box frame sections

It is usual to supply box sash windows with the sashes in the frame ready to accept sash cords and sash weights when the frame gets to the site. Boxed-frame sash windows are usually fixed into a rebated brick reveal and secured with carefully positioned folding wedges. Traditionally, a solid fix was also ensured by driving an angled nail through the edge of the boxed frame through the wedges into timber pads in the brick reveal. The nailed fixing was then hidden by a quadrant mould or architrave.

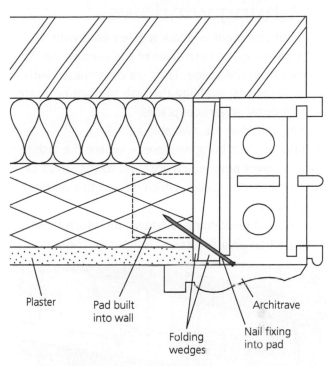

▲ Figure 5.93 Fixing a sash window using timber pads and a nail

Modern installations use fixing foam as an efficient way of securing boxed-frame sash windows. This method has the advantage of securing the whole of the window to the brickwork, and creates a solid packing, allowing for flexibility in positioning mechanical fixings. Galvanised steel brackets are a modern way of fixing the boxed-frame sashes to the brickwork. The brackets are concealed by the plasterwork.

ACTIVITY

Traditionally made sliding sash windows weren't very energy efficient, but nowadays the frame profile has been developed to conform to current Building Regulations to reduce heat loss. Use the internet to find high-performance sliding sash windows, and list the differences between them and the original detailing. Can you think of a situation where the traditional window may still be used?

ACTIVITY

List the tools, equipment and materials required to fix a boxed-frame sash window. If the frame was being fixed using fixing foam, what precautions would you take to ensure the window was fixed safely and accurately?

Pivot-hung windows

This type of casement window is often used with high-rise buildings, as both sides of the window can be cleaned from the inside. They are constructed with centre-jamb pivots, enabling the sash to pivot or rotate in the horizontal or vertical planes.

Pivot hung windows are available as:

- traditional pivot-hung windows – where the sash is hung on pivot pins, which are fixed to the jambs of the frame, and a socket, which is fixed to the stiles of the sash
- storm-proof pivot-hung windows, for which face-fixed friction pivots are used to operate the sash.

▲ Figure 5.94 Horizontal pivot windows

Installing pivot-hung windows

These windows can be built in or fixed in, as with other casement window types, using the same fixing methods. Sash fasteners and stays are used to secure traditional pivot windows; with storm-proof windows, espagnolette bolts are used with locking handles. Vertically hung pivot windows incorporate similar details to the horizontal type.

Tilt and turn windows

This type of specialist window is ideally suited to high-rise buildings, as the method of opening enables the sash to hinge in two directions by bolting one side of the mechanism in the closed operation while the other operation is activated.

It usually operates with the tilt position permitting the sash to open in at the top, allowing ventilation while preventing children or objects falling out. The turn position allows the inwardly opening sash to open fully, enabling the outside glass to be cleaned safely. Window components are of a similar size to the storm-proof sections used in storm-proof casement windows.

> **INDUSTRY TIP**
>
> Tilt and turn windows can also be used as a fire escape if the window is of sufficient size.
>
> Tilt and turn windows may need a temporary brace/stretcher to prevent them going out of square during fixing, especially when using fixing foam.

▲ Figure 5.95 Tilt and turn windows: closed, tilted and turned

Installing tilt and turn windows

Tilt and turn windows can be installed using the same methods as pivot-hung windows and other casement windows.

Roof lights

▲ Figure 5.96 A top-hung roof light in a loft conversion

Roof lights are manufactured in a variety of materials, including timber, uPVC, aluminium and galvanised steel. They consist of a surrounding frame and removable sash. Sophisticated and reliable designs utilise horizontal pivot operation and tilt and turn variations, with durable weather seals and multi-point locking systems available with some types. They are fitted with factory-sealed double-glazed units, and ventilators are also incorporated into the designs.

There are roof lights suitable for most roof pitches, and they come with integral flashing kits that ensure the roof light remains watertight. They are suitable for loft conversions, attic situations and conservation areas.

Installing roof lights

Roof lights are usually fixed to the tops of the rafters, using galvanised steel L-shaped brackets and screws that come with the window. Full fitting and fixing instructions are supplied. Windows can also be incorporated within dormer roof structures, which are covered in pitched roof construction.

▲ Figure 5.97 A centre pivot roof light fitted with flashing kit

INDUSTRY TIP

A common mistake when installing roof windows is to use the wrong flashing kit. It is important that you seek advice to ensure the kit you are about to use matches the roof covering, e.g. double Roman tiles, slate or plain roofing tiles.

Roof windows must be installed exactly as the manufacturer intended. If the manufacturer's instructions are not followed and corners are cut, this could invalidate any guarantee that may have come with the product.

INDUSTRY TIP

The leading manufacturer of roof windows (Velux) provides instructions with clear pictures (no text) and an easy to follow labelling system to ensure that the right parts are used at each stage. You should never force any parts of the window cladding or flashing together during installation unless the instructions tell you to do so; otherwise you could damage parts and this could prevent the window from being sealed properly.

Materials for windows

Windows can be constructed from a variety of different materials, including softwood, hardwood, laminated softwood and hardwood (smaller sections of timber glued together to form larger sections), uPVC, aluminium and steel. Modern production methods use up-to-date technology to design window components to withstand the rigours of the weather. Windows have to offer good security while still being easy to operate, and should provide good thermal and acoustic insulation. The improved quality of high-performance timber windows and their increased length of life in use have been made possible by weather-proof adhesives and the preservation treatment given to timber.

▲ Figure 5.98 Aluminium window frame

Preparing to fix window frames

Before fixing windows, it is good practice to check the plans and specification to ensure they are the right ones for the job. A window schedule may be available for the contract; this will be invaluable for quickly checking the window sizes for each opening.

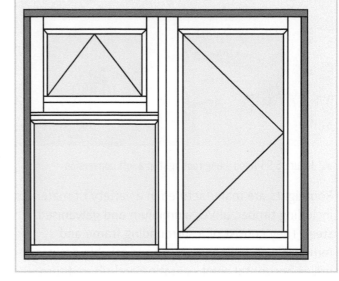

Installing window frames

Installing built-in casement window frames

Casement window frames can be built in the same way as door frames. The bricklayer will position the window onto a bed of mortar, with a DPC (damp-proof course) sandwiched between the bed and the window, and then plumb and level the window with temporary struts to keep the frame upright. The struts are temporarily nailed to the head, and weighted with bricks or blocks at the base to prevent the window moving. The bricklayer must ensure the correct datum is maintained for the height of the window by checking it with a gauge rod. The window can be secured using frame cramps, screw ties or holdfasts.

Windows may have separate cills. These can be made of stone or pre-cast concrete, or formed from a **brick-on-edge** projecting cill. It is common practice to leave the bedding and building of these cills until a later stage to prevent damage. The built-in window would in this instance be packed up to the correct position with spare timber, leaving space for the separate cill to be inserted.

▲ Figure 5.99 Brick-on-edge cill

Installing fixed-in casement window frames

The majority of windows will be fixed in after the openings have been formed. To ensure openings are the correct sizes, the carpenter will make dummy frames to match the sizes of the windows that are to be fixed in later. The frames will be made 5 mm wider than the actual window to allow for the use of packers under the dummy frame; this will ensure the correct height datum can be achieved and allow for easy removal of the dummy frame after the brickwork has set. The plumbing and levelling of the dummy frame will be treated in the same way as for built-in windows.

▲ Figure 5.100 Dummy window frame

After the mortar has set, the dummy frame is removed and the correct window is put in its place. The frames are then levelled, plumbed and fixed using masonry screws through the rebates in the jambs to conceal the fixings behind the glazing and casements. An alternative method of securing the window frame is to use galvanised or stainless-steel metal straps. These are fixed at intervals to the back of the jamb, before

the frame is offered into the opening. The straps are then fixed to the blockwork either with screws and plugs or with masonry screws. Galvanised steel straps are usually 5 mm thick. They may have to be housed into the back of the frame to ensure that, when they are offered into the opening, the overall width of the frame plus the thickness of the straps is not too tight. Care must be taken not to house the straps all the way through the frame, otherwise the housing will be seen on the front face of the window frame.

▲ Figure 5.101 Galvanised steel strap

INDUSTRY TIP

Fixing in window frames after the walls have been built will reduce the likelihood of damage from the elements (e.g. water marks and general dirt) and careless operatives (e.g. gouging, dents and splits). It will also avoid the risk of theft on site.

Another way of forming the openings in new-build construction is to use cavity closers to determine the size of the window frames to be fitted. This method will also prevent thermal loss around unprotected cavities and reveals. The cavity closers are usually positioned and levelled by the bricklayers and will remain in place until the window frames are installed. The fitting of cavity closers is required by Building Regulations Approved Document Part L.

Window frames are often fixed in position with fixing foam. The method is the same as for fixing door frames. Sealing around window frames also uses the same materials and methods as for sealing around door frames.

275

▲ Figure 5.102 Cavity closers used to form window openings in a wall

Fixing foam

Polyurethane fixing foam is now commonly used when fixing in all types of frames and linings. It is applied using hand-held tins with a pre-fitted adaptor or, if greater precision is required, an applicator gun. In both cases, there are various types of nozzle available. Expanding foam adhesive is similar in composition to polyurethane (PU) adhesive, and fills and bonds most products to all surfaces and materials. Different grades of foam can give excellent acoustic and thermal insulation properties and fire rating. They are particularly useful in fixing and sealing window frames in conjunction with mechanical fixings.

When replacing frames and linings in existing buildings, you may find that the brickwork is not in good condition, making it difficult to obtain good fixing points. One way to overcome this is to use PU fixing foam. The sequence of operations necessary when using fixing foam is as follows.

1 The frame or lining is wedged accurately in position, plumb, level and parallel (this also involves placing additional stretchers across the assembled lining or frame to prevent it bowing out as the foam cures and expands). Mask the frame and any brickwork or finished surfaces with masking tape and protective dust sheets.

2 Use a hand-held garden spray to apply a fine mist of water between the surface of the wall and the back of the frame. This aids the adhesion and curing process of the foam.

3 Shake the can well, attach the applicator nozzle to the can or gun and select the best side to apply the foam. (When fixing a replacement external door frame, it is often not possible to apply the foam from the inside of the property due to the existing plasterwork.)

4 Starting at the bottom of the frame, insert the nozzle into the gap between wall and cill (if a closed frame), gently squeeze the trigger and apply foam into the gap (hand-held cans should be held upside down during application). Half fill the gap with foam and work along the cill – the foam will expand into the rest of the gap. Release the trigger.

5 Repeat the process on the jambs and finally, if necessary, on the head of the frame.

6 Allow the foam to expand and cure fully. The time required for this varies with temperature and the thickness of the applied foam, but is usually no more than 40 minutes. Carefully cut back excess foam with a broad-knifed scraper, disposing of the residue in the correct skip. Remove any wedges used to secure the frame (the resulting gaps will require filling with additional foam).

7 The completed frame can now be fixed into the masonry with mechanical fixings in the appropriate positions up the jambs. The cured foam acts as a solid packer and enables a secure fixing to be completed. The temporary braces and stretchers can now be removed.

Sealing window frames

After window frames are fixed they require sealing between the wall and the back of the frame to achieve adequate protection from rain, wind and insects. In addition, the sealant provides an acoustic or thermal barrier. This is often a Building Regulations requirement (the specification will give precise details as to which sealant is to be used).

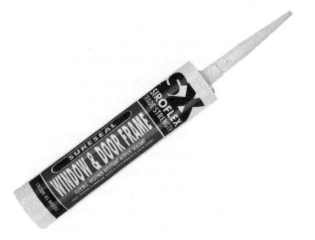

▲ Figure 5.103 Acrylic sealant

Silicone, butyl and acrylic sealants are available, many in different colours to match the frame material and finish. These are supplied in cartridges and applied using a skeleton gun (also referred to as a 'silicone or mastic gun'). The surface of the frame and wall to be sealed should be cleaned off so that no loose particles will hinder the sealing process, and any protruding mortar or fixing foam should be cut back. The bead of sealant should be forced well into the gap, and the sealant should be tooled to create an angled, even finish.

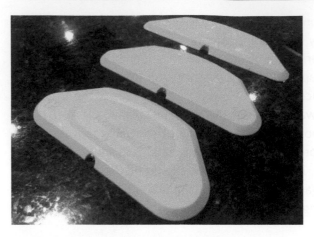

▲ Figure 5.104 Smoothing tools (concave)

4 FITTING BAY AND BOW WINDOWS

These types of window can be constructed from either casement or boxed-frame design windows. They can either be purpose-made or consist of several window frames fixed together with special corner posts that project outwards from the face of the external wall. The shape of the windows can be categorised as:

- bow or segmental
- splayed or cant bay
- square or rectangular bay
- combination bay or square and splayed.

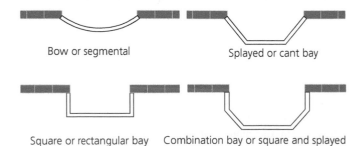

Bow or segmental   Splayed or cant bay

Square or rectangular bay   Combination bay or square and splayed

▲ Figure 5.105 Types of bay and bow window

Work on segmental bay/bow windows is fairly rare due to the expense of building curved brickwork and the fact that most replacement windows are now high-performance uPVC. Occasionally, however, this type of window is specified for new work or as a replacement in a conservation area where an identical window must be installed.

Technically, a true bow window has curved glass, but – due to the expense of making the former to enable the glass company to bend it – flat glass is usually used instead. The glass will sit in straight rebates and form facets around the face of the window. This being the case, the head, cill and any horizontal bars will have the inside and outside faces following the curve, and the rebate produced as a chord to the shape to allow for the flat glass.

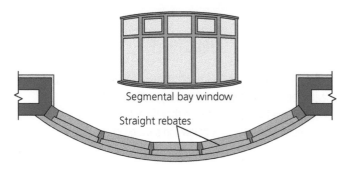

Segmental bay window

Straight rebates

▲ Figure 5.106 Segmental bay window

Segmental bay/bow windows can be constructed in one of two ways.

1 Bay and bow windows can be supplied to site from the joinery workshop fully constructed, ready to be built in by the bricklayers or fixed in by the carpenters. The brickwork to support the window will have been built using a template to match the opening size required. The same methods for securing the windows to the brickwork can be adopted as with casement and boxed-frame window installation.

2 A series of flat frames (one for each facet) can be delivered to site ready to be assembled by the site carpenter. The ends of the cill are to be mitred, glued and handrail-bolted together. The frames are

then connected with screws through the mullions with exterior-grade wood adhesive. It is important to achieve a good joint between the frames to prevent water penetrating the joint through this vulnerable position.

Square bay, splayed and combination bay windows can be supplied in sections that have been dry-fitted in the workshop ready for assembly on site. This allows for easier transportation and storage of the windows on site. The corner posts and mullions can be tongued together with plywood strips, which reinforce the joint. Again, the joints would be glued with a good-quality exterior adhesive and secured with wood screws resistant to corrosion.

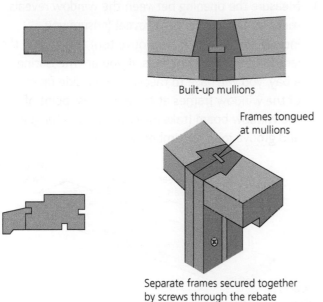

Built-up mullions

Frames tongued at mullions

Separate frames secured together by screws through the rebate

▲ Figure 5.107 Built-up segmental bay construction

▲ Figure 5.108 Handrail bolts

▲ Figure 5.109 Zipbolt™ used as a modern alternative to a handrail bolt

Handrail bolts are principally used to join sections of handrail. To join the cill sections of a bay window, an accurate hole the same diameter as the bolt must be bored through the mitred ends of the cill, allowing for the bolt to be inserted. The nuts are supplied as one round 'castellated' nut, which has slots cut into it to allow for tightening, and one square 'dead' nut, which secures the joint. Access to the bolt ends is via mortices that are cut into the underside of the sections of the cill. The dead nut and a washer are inserted into the mortice, and the bolt is fed through the bolt hole and applied into the dead nut. The joint is then completed by positioning the two sections together and applying the castellated nut and a washer through the remaining mortice, using a screwdriver to tighten the nut. The final tightening is achieved by lightly tapping the screwdriver in the castellated nut slots with a hammer. The accurate positioning of the bolt and the dowels will produce an extremely strong joint.

ACTIVITY

Discuss the advantages of using a handrail bolt to secure the cill sections of bay windows. What are the disadvantages of using handrail bolts? Can you think of an alternative mechanical fixing that would be equally strong?

Bay and bow windows can sometimes be required with no supporting brickwork. They are then supported by special timber corbels or gallows brackets. These are known as 'oriel' windows.

▲ Figure 5.110 Bay window supported on two gallows brackets

Window boards

Window boards provide the finish to the top of the inner wall at cill level, where it abuts the window. After the window has been fixed and glazed, window boards can be fitted. They are usually fixed before plastering to ensure a professional finish. They are usually formed from solid timber, plywood, MDF or uPVC.

Window boards made from solid timber are available in manufactured widths of up to 225 mm. They are usually 25 mm or 32 mm thick and manufactured with a bull-nosed front edge, and are tongued on the rear edge to fit into the groove on the window cill (timber windows only). This hides any potential shrinkage, as windows often have radiators situated below them. Some timber window boards have a groove on the underside to provide a finish for the plaster and to hide shrinkage gaps.

The ends are cut to run past the brick reveals, and are usually finished with the moulding returned to their ends.

- Plywood window boards can be cut to suit wide window reveals. They have a piece of solid timber planted on the front edge to provide a decorative finish.
- MDF window boards come in widths up to 300 mm and are usually pre-finished with primer ready for fitting. They are finished in the same way as timber window boards.
- uPVC window boards have no tongue on the rear edge. They are butted up to the window (usually uPVC as well) and the gap is filled with caulking or covered with a small bead.

The same fixings can be used for window boards as for frames and linings. Modern adhesives provide a range of solutions for fixing window boards in conjunction with, or as an alternative to, traditional methods.

The sequence for fitting and fixing window boards is as follows.

1 Measure the opening between the window reveals, measure the depth of the reveal (ensure you include the depth of the groove too) and add on the length of the returned ends. If you are measuring a bay, you will have to measure the inside faces of the window frames at the narrowest point of the window board (take into account the tongue and groove joint for timber windows).

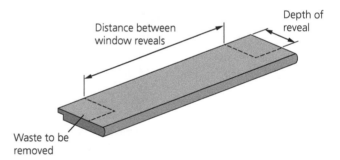

Distance between window reveals

Depth of reveal

Waste to be removed

▲ Figure 5.111 Measure the opening and depth of reveal

2 Determine the angles to be cut at each end of the window boards if fitting into a bay window. To do this, you may have to bisect the angles using the following method.

The rule to follow when setting out mitres is that the line of the mitre is always straight and always bisects the overall angle of the window boards that are to be mitred (provided they are equal widths and straight lengths). If the overall angle is 90° then the mitred angle will be 45°; if the overall angle is 60° then the mitred angle will be 30°; and so on (see Figure 5.112).

KEY TERMS

Obtuse angle: an angle between 90° and 180°.
Acute angle: an angle between 0° and 90°.

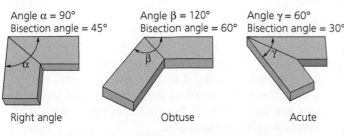

Angle α = 90°
Bisection angle = 45°

Angle β = 120°
Bisection angle = 60°

Angle γ = 60°
Bisection angle = 30°

Right angle   Obtuse   Acute

▲ Figure 5.112 Bisection angles for equal-sized window boards

3   Mark these measurements and angles onto the window board and cut out the required shape with a chop-saw. Return the ends with a block plane and finish with abrasive paper or a router with a round over cutter.

4   Dry-fit the window board into the opening to ensure it fits properly into the groove.

5   Check the window board for level by using a boat level to check for front-to-back level and a longer spirit level to check along the front edge. Pack the window to the right level using flat plastic packers at regular intervals under the window board.

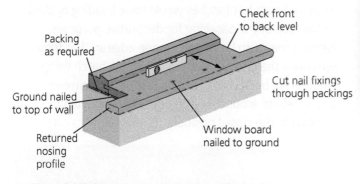

Packing as required

Check front to back level

Ground nailed to top of wall

Cut nail fixings through packings

Returned nosing profile

Window board nailed to ground

▲ Figure 5.113 Checking the window board for level

6   Fix the window board using nails or screws when fixing into timber, and screws and wall plugs when fixing into masonry (pellets should be used to hide screw fixings for hardwood or clear finished window boards). Fixings should go through packers to achieve a flat surface and prevent packers being displaced.

INDUSTRY TIP

uPVC window boards look very appealing when they are first installed, but the surface can develop scratches through regular use, which are very difficult to repair. Many window fitters and carpenters prefer to use solid timber or painted MDF rather than uPVC because these materials can easily be repaired.

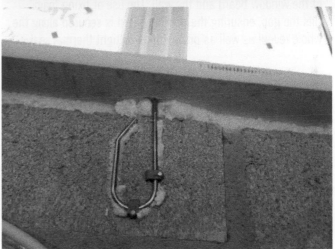

▲ Figure 5.114 Concealed brackets and fixing foam used to fix MDF window boards

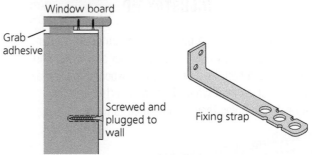

Window board

Grab adhesive

Screwed and plugged to wall

Fixing strap

▲ Figure 5.115 Fixing the window board with adhesive and fixing straps

With manufactured materials commonly being used for window boards, grab adhesives or fixing foam can be used as an alternative to fixing them. The cut window board is bedded on generous beads of adhesive and then held in position with temporary struts off the head of the window reveal. Another fixing method is to use fixing brackets or straps screwed to the underside of the window board, and then fix them to the face of the wall with screws and plugs, or masonry nails. Window boards can be secured to timber frame openings by simply nailing directly through the window board into the material behind it, and using a nail punch to hammer them below the surface. Alternatively, a nail gun can be used to fix the window boards for quickness.

INDUSTRY TIP

All methods may leave a varying gap between the underside of the window board and the wall. The use of fixing foam fills the gap, ensuring the window board is secured along the whole reveal as well as providing an airtight thermal seal.

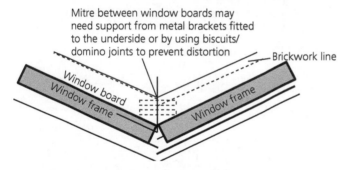

Mitre between window boards may need support from metal brackets fitted to the underside or by using biscuits/domino joints to prevent distortion

Brickwork line

Window board
Window frame
Window frame

▲ Figure 5.116 Additional support is needed for mitres between window boards

INDUSTRY TIP

It's always best to use a plug cutter mounted in a pillar drill, to ensure that wooden plugs are cut accurately and safely. Using a handheld drill to make wooden plugs may result in a slack joint when they are inserted into the counter-bored holes they are intended to fit.

▲ Figure 5.117 A plug cutter

ACTIVITY

Draw to scale a detailed section through a bay window cill in a cavity wall construction, showing two methods of fixing a timber window board to the wall. Label all the parts of the drawing, and ensure you use the correct BS drawing symbols when identifying the components.

⑤ HATCH LININGS

As heat generated in a building rises, it is usually lost through poorly insulated ceilings and loft hatches into the loft space and eventually through the roof. A loft lining isn't very different from a door lining – they both line the space in the wall, or ceiling in this case, and are trimmed with architraves (one side only) to conceal any gaps between the joists and the hatch lining. It was common for a door stop to be nailed around the inside face of the lining to form a rebate for a loft hatch/panel to sit in. Some loft hatches would have beading applied to the face of the panelling for decorative purposes; hatches may also be hinged on one edge so that they open into the loft space. More expensive hatch linings would have been made with heavier section timber, with a solid rebate around the top edges for the hatch to rest in when in the closed position.

▲ Figure 5.118 Traditional loft hatch/lining

Modern loft hatches are well insulated to conform to current Building Regulations. They usually have rigid insulation fixed to the back of the loft panel, draught excluders between the joints and a fixed double extension ladder for safe and easy access to the loft space. Many loft hatches are made from PVC because this is lightweight, and therefore easy to manage while installing. They have the architrave moulded into the lining to eliminate the need for trimming and decorating. However, timber loft hatches are still available and used, and also meet current Building Standards. The disadvantages of timber loft hatches are that they are substantially heavier than PVC, so are usually a two-person job to install, and some mass-produced examples are poorly constructed.

It is important that, whatever material is used for the loft hatch, it is installed following the manufacturer's fitting instructions. It is good practice to measure diagonally from corner to corner on both sides of the frame, making sure the measurements are equal so that the lining is square. Care should also be taken to fix the lining free from twist; both issues could prevent the hatch closing properly and lead to heat loss.

INDUSTRY TIP

Loft hatches with extension ladders fixed to them usually open downwards rather than into the loft space.

ACTIVITY

Research the Building Regulation that controls heat loss though loft hatches. Draw to scale a suitable timber loft hatch that would meet current standards of energy efficiency.

▲ Figure 5.119 A modern energy-efficient loft hatch

Test your knowledge

1 What does the bracing used in framed, ledged and braced doors help to provide?

   A Additional resistance to distortion

   B A means of fixing a lock

   C Additional fixing for hinges

   D A means of fixing a wicket door

2 Rebate locks are normally fitted to which of the following?

   A Hanging stiles

   B Top rails

   C Meeting stiles

   D Bottom rails

3 What is an escutcheon used for?

   A To cover the lock

   B To assist with door closing

   C As a type of closing hinge

   D Finish to a keyhole

4 What type of hinges would need to be used to enable French doors to open a full 180° without binding?

   A Butt hinges

   B Flange hinges

   C Parliament hinges

   D Storm-proof hinges

5 Which of the following is not classed as safety glass?

   A Float

   B Laminated

   C Toughened

   D Georgian wire

6 What is the purpose of an intumescent strip?

   A To seal the floor, preventing water gaining access

   B To prevent water running down and under the door

   C To seal the door when it is exposed to fire

   D To seal around the letter plate

7 Timber can be described as 'hygroscopic'. This means it can:

   A absorb moisture only

   B lose moisture only

   C both absorb and lose moisture

   D neither absorb nor lose moisture.

8 An obtuse angle of 120° has a bisection angle of:

   A 45°

   B 60°

   C 90°

   D 240°.

9 Which one of the following hinges encourages a door to self-close?

   A Butt hinge

   B Expanding hinge

   C Parliament hinge

   D Hawgood hinge

10 What is an 'air wedge' used for?

   A Collecting dust in an LEV

   B Adjusting, holding and levelling

   C Disposing of glass

   D Tensioning double-action hinges

11 Explain the process of bisecting a bay window to establish the angles needed to return a window board, and how the window board could be fixed into position.

12 List the sequence of using a smoothing tool while using a sealant.

13 Explain how you could protect the surrounding area of an occupied property when installing a replacement bay window.

14 Explain the precautions that should be taken while using expanding foam.

15 Sketch five methods of finishing the edge detail on the meeting stiles on a pair of double doors.

CHAPTER 6

MANUFACTURING CURVED JOINERY

INTRODUCTION

Doors and frames vary in style, but the majority are flat and rectangular because these shapes are easier to manufacture and therefore reduce the cost of the items. Curved or shaped joinery is generally produced only for bespoke architect-designed residential, commercial or public buildings. Shaped work of this nature will be considerably more expensive but will provide very pleasing architectural features. We are fortunate that architectural design is currently in favour, even in speculatively built residential accommodation. Architects are using shapes and styles inspired by Edwardian-style properties, for example, but the buildings are constructed to modern and energy-efficient standards.

The principles that you have already learned form the basis of all joinery construction. This chapter will provide the additional knowledge required to take on shaped work successfully. For most operatives, the issue is often not lack of ability but lack of opportunity to practise the additional knowledge required. Often, curved components are bespoke or one-off, or are produced for heritage, restoration or conservation purposes. Clients often have high expectations of quality, and the quickest or cheapest methods are not always appropriate. You may even find that traditional production methods are preferred for some jobs, or that you are required to replicate an unusual shape, so it is important to be aware of a variety of ways of producing curved joinery.

LEARNING OBJECTIVES

By reading this chapter, you will learn how to:
1 prepare and set out curved joinery
2 mark out and manufacture curved joinery.

1 PREPARING AND SETTING OUT CURVED JOINERY

Information sources

Environmental policy

This is the commitment of an organisation to the laws, regulations and policies concerning environmental issues. A policy aims to regulate resource use or reduce pollution to protect natural habitats and the welfare of people nearby. Examples relevant to joiners include having a policy of purchasing timber only from managed timber sources, ensuring minimum waste as far as practically possible, and disposing of wood waste safely.

IMPROVE YOUR ENGLISH

Search for information on the Forest Stewardship Council (FSC®). Write a paragraph about this organisation and how it is important to the joinery manufacture industry.

Safe systems of work

This document is often referred to as a method statement. Whether on site or in a joinery workshop, the hazards involved with specific tasks should be recorded and the safe system of work document made available to all those involved. It should be in a format that is clear, uncomplicated and easy to understand.

Drawings

As with all setting out, you will have to interpret a range of information to create a product that meets the client's needs and wishes, but also meets the requirements of the current Building Regulations. Generally, you will be provided with architect-produced scaled detail drawings, showing the required elevations and sections of the product, and a specification providing information not usually shown on these drawings. (Refer to Chapter 2 for detailed information on drawings, specifications and Building Regulations.)

ACTIVITY

1 Research the profiles listed below and sketch a jamb profile incorporating each moulding type:
   - rebates
   - grooves
   - ovolo, chamfer and ogee mouldings.
2 List the order in which they would be machined.

KEY TERM

Obscured glass: 'frosted' glass that provides a level of privacy by obscuring the view through the glass. Different patterns are available depending on the level of privacy required, graded from 1 (least) to 5 (greatest). This type of glass is commonly used in bathroom windows and front entrances to residential properties.

Specifications of work

A specification for joinery work includes information such as the required species of timber, the finish required (for example, painted, varnished, colour tint, glass type) and any specified ironmongery.

INDUSTRY TIP

Where ironmongery is specified, it is a good idea to use manufacturers' catalogues or websites to check whether any special considerations need to be taken into account when manufacturing the product.

Sometimes, the details of the work will be agreed with the client and the joinery manufacturer. If this is the case it is important that both the scope and detail of the work are signed off by the client before work starts. This will help manage their expectations of the finished work.

Effective communication is key to managing client and supplier expectations. Sufficient information must be given so that a confident decision can be made, and recorded.

For example, consider the information needed to produce a shaped frame with **obscured glass**. Regardless of whether it is for a door or a window, it may also be defined within a door or window schedule. You will remember from Chapter 2 (page 54) that a schedule provides information about a product in table form. It usually gives the item a reference code or number (e.g. D1 to indicate 'door 1' or W1 to indicate 'window 1'). This schedule will be referenced to the site plans showing the final position of the product. It will also state the type of glass required (whether single or double glazed), any pattern type if the glass is to be obscured, and whether the glass is float, laminated or toughened, together with the thickness required.

It will take a little time to interpret this information and take on board what is required, before having a sound understanding of what has to be produced. It is quite common at this stage to have a couple of queries, or even a whole list of them, to discuss with the architect. The architect will have a great deal of knowledge about many aspects of building but will not be a specialist in any of them. With this in mind, they may have drawn something that it is not practical to make. A discussion at this early stage will manage expectations of the completed product. Very often they are only concerned about the final look of the product and will be happy to leave the construction detail to the joinery manufacturer, but it must be agreed at this early stage.

Building Regulations

Generally, it is the architect or architectural technician's responsibility to ensure that what is proposed conforms to the current Building Regulations. However, it is worth having a working knowledge of the main regulations to be able to identify areas of concern and raise them in a timely fashion. The main regulations to are:

- Protection from falling, collision and impact – Approved Document K
- Access to and use of buildings – Approved Document M
- Material and workmanship – Approved Document 7.

You can find out more about the Building Regulations in Chapter 2.

Information from the site survey

Before any setting-out takes place, a site survey is generally required to obtain accurate information.

This could include:

- taking detailed measurements of sizes and shapes required
- creating templets (the traditional name for 'templates') of any door/window opening shapes if required
- noting details of any mouldings or profiles of existing work that have to be matched
- making notes on any access problems that will affect delivery, such as parking restrictions, etc.
- taking photographs or video clips, with voiceover recording particular issues.

Generally, the site survey details for an arch shape can be quite easily ascertained, particularly for new arches, but sometimes the dimensions of the rise and span are not as expected. For example, the drawing might state that it should be a 2080 mm span with a 1040 mm rise, yet the site measurement for the rise is 1000 mm. This should raise concerns and will require you to double check the dimensions. In this example, the dimensions indicate either that the arch centre used was not accurate, or that the centre for the arch was eased too early by the bricklayer and the brickwork has sagged. Alternatively, if it is an old arch, settlement has probably taken place. In this case, a templet of the shape needs to be taken back to the joiners' shop and a frame made to fit it. It is more likely for a low-rise arch to drop or settle, such as those with a segmental or elliptical shape.

INDUSTRY TIP

Often the survey site is a long way from where the product will be manufactured. It is always best to take as many details as possible and take plenty of photographs to refer to. The last thing you want is to have to travel back to the site again because you missed a vital detail. Too much information is better than too little!

After carrying out the site survey, all relevant parties must be informed if there are discrepancies between the information supplied by the architect and the information recorded during the survey. Initially, you would inform your supervisor; depending on the size of the joinery manufacturer this could be the:

- joinery shop supervisor
- joinery works manager
- senior setter out
- owner.

▲ Figure 6.1 A templet being drawn from an arch shape

INDUSTRY TIP

Where an issue arises, such as when dimensions do not tally, there is usually a simple explanation. Double-check your measurements in a logical order to identify where the error has occurred. Do not leave the site until you are happy the measurements are accurate.

Whoever you report to should then liaise with the architect to agree how any discrepancies are to be overcome. This must be confirmed in writing by the architect to prevent any disagreements later. This confirmatory letter, sometimes known as a 'variation order' or 'architect's instruction', will ensure that the agreed changes form part of the contract and the expectations of the finished contract have been managed effectively.

INDUSTRY TIP

To ensure an accurate brick arch shape is produced, the centring of the arch is best manufactured in the joinery shop rather than on site. This means there is a greater chance of a good fit to the finished product and a templet of the arch shape will not be required.

Often, some form of access equipment is required to measure arch openings. The risk assessment that is being produced for the setting-out operation should also cover the survey. (More information about risk assessments can be found in Chapter 1, pages 7–10).

IMPROVE YOUR ENGLISH

Using the five stages of risk assessment, write a risk assessment that covers site survey activities. Using your existing knowledge of joinery, make a note of which parts of this risk assessment might also apply to setting out shaped doors and frames. Then draft a 'safe system of work' for this task.

Arch terms

Most curved joinery will be fitted into a brick arch, although some will be fitted to a concrete or timber-framed structure. Standard terms are used to refer to parts of an arch. While the shape of the arch may alter, the terms remain the same.

For curved joinery, it is also important to know the standard terms relating to circles. These are given in Table 6.1 and Figure 6.3.

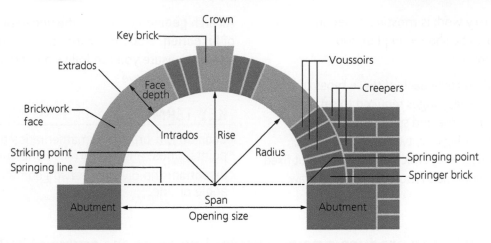

▲ Figure 6.2 The parts of an arch

▼ Table 6.1 Standard terms relating to circles

Term	Description
Circumference	The line bounding the circle, or around its edge
Diameter	A straight line passing through the centre point of the circle and touching the circumference on both sides of the circle
Radius (plural: radii)	A straight line drawn from the centre of the circle to its circumference
Arc	A portion of the circumference
Chord	A straight line (shorter than the diameter) that touches two points on the inside of the circumference of a circle
Tangent	A straight line touching the circumference of a circle
Normal	A straight line drawn through the circumference from the centre of a circle (which will always be at 90° to the tangent at that point)
Quadrant	A quarter of a circle
Sector	A part of a circle contained between two radii and the circumference of the circle
Segment	A portion of a circle contained between an arc and a chord

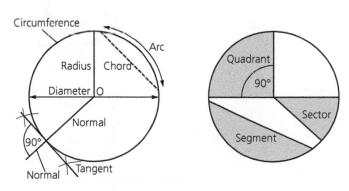

▲ Figure 6.3 Standard terms relating to circles

Arch shapes

Shaped joinery can be created to fit to many different arch shapes. This chapter will include only those shapes shapes listed in the qualification. These are:

- segmental
- gothic
- semi-circular
- elliptical – true and pseudo (not true).

289

While shaped joinery work is mostly shaped in elevation, it can also be shaped in plan, most commonly segmentally.

The following sections cover each of these types of arch, showing the step-by-step stages required to set them out full size using given rise and span dimensions. The examples are drawn using one, two or three centres, except for the elliptical arch, which is drawn without fixed centres and is therefore much more complex to draw.

All arch **geometry** requires the understanding and use of **bisection**. Study the instructions on the following pages to ensure you can follow and reproduce the arch shapes listed above.

KEY TERMS

Geometry: a branch of mathematics that deals with points, lines, angles, surfaces and solids.

Bisection: the division in two geometrically of an angle or line.

Bisecting a level line to produce a perpendicular line along the centre of its length

▼ Table 6.2 Bisecting a level line to produce a perpendicular line along the centre of its length: step by step

Step 1	Draw the required span A–B.	
Step 2	Set the compasses to a distance that is more than half the length of the line A–B.	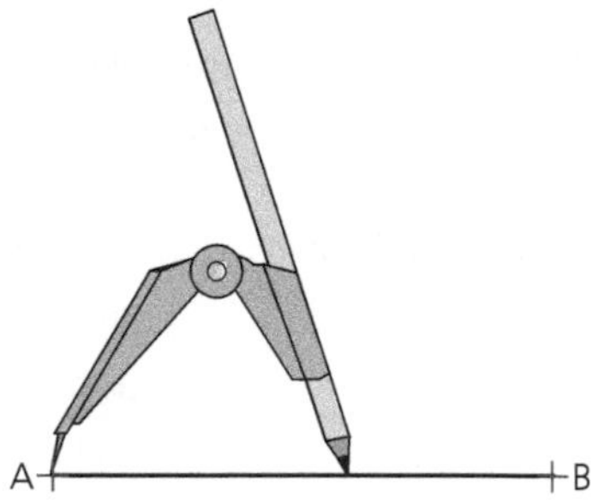
Step 3	Keep the compasses locked and draw arcs through the line from points A and B.	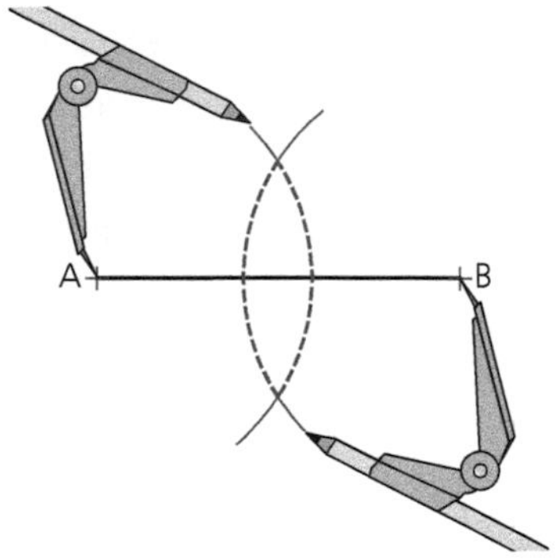
Step 4	Draw a straight line through the points at which the arcs intersect.	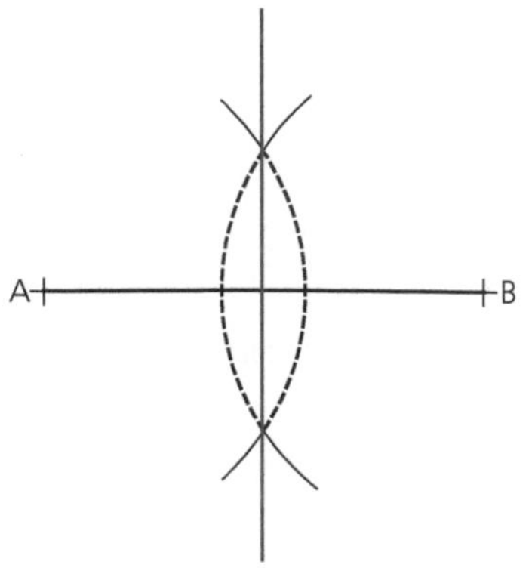

ACTIVITY

1 Draw a 200 mm long line and bisect it to find the centre. Measure each side of the centre line. Are both parts the same length?

2 Describe any other method of dividing this line accurately into two.

Segmentally shaped joinery (in elevation or plan)

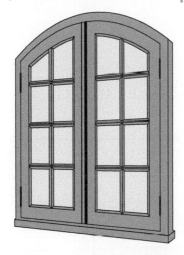

▲ Figure 6.4 Segmental headed window

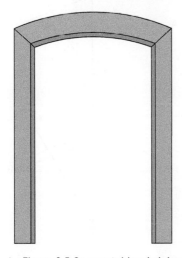

▲ Figure 6.5 Segmental headed door frame

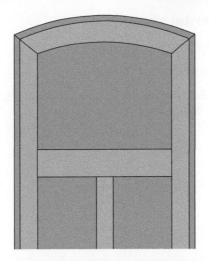

▲ Figure 6.6 Segmental headed panelled door

Segmental arches are still in common use. They give some architectural shape to the **façade** of a building but are not as expensive to produce as semi-circular work because the curved work is limited. A segment is a portion of a circle so, to be able to draw a segment, we must find the centre of the circle that contains the segment.

KEY TERM

Façade: the exterior face of a building; also referred to as the 'front face' of the building.

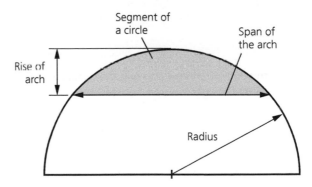

▲ Figure 6.7 Relationship between segment and semi-circle

Drawing a segmental shape

▼ Table 6.3 Drawing a segmental shape: step by step

Step 1	Draw the required span as line A–B.	 A——————B
Step 2	Bisect line A–B to find the centre, point C (see Table 6.2). Extend this line above and below the centre point.	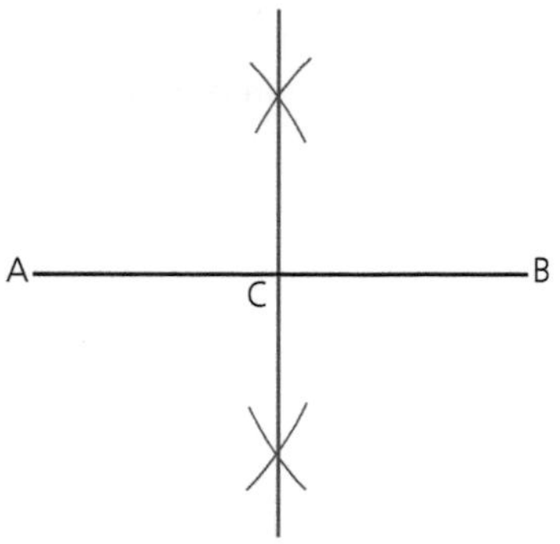
Step 3	Measure the rise D above the centre point on the bisection line.	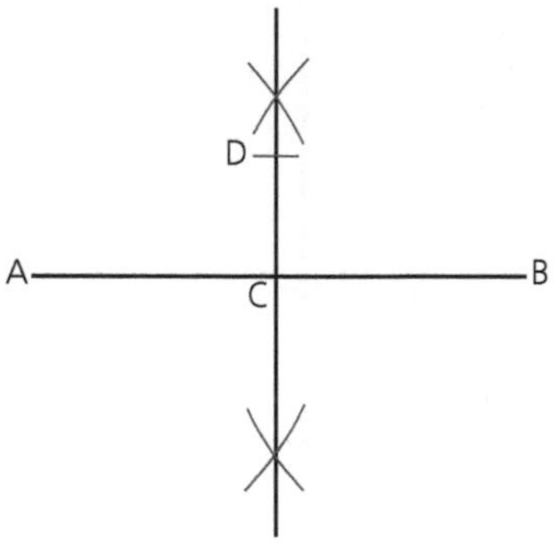
Step 4	Draw chord B–D and then bisect it. Extend the bisection line to cross the extended line D–C at point O.	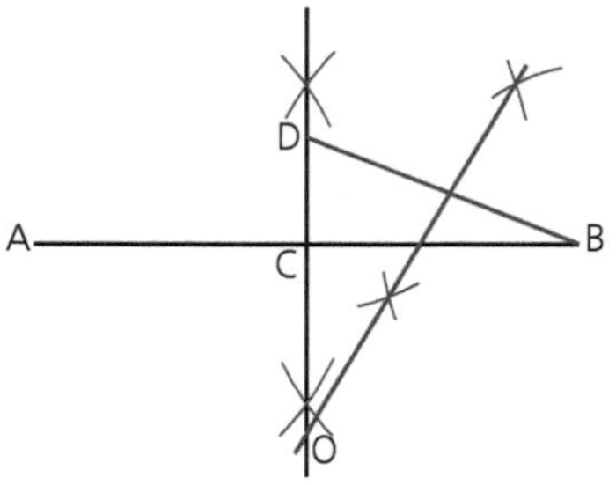
Step 5	With point O as the centre and the distance O–D as the radius, strike the arc A–D–B for the arch.	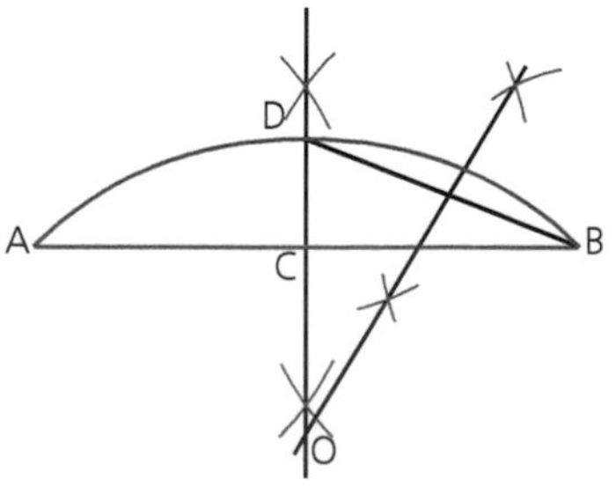

ACTIVITY

Following the method outlined here, draw a segmental arch shape with a span of 150 mm and a rise of 50 mm.

Method of calculating the radius from a given span and rise

You can calculate the radius of an arch if you know the span and rise. The mathematical formula, where R = rise, is:

$$radius = \frac{(½\ span^2 \div R) + R}{2}$$

Example

Using the formula above for reference, let's work out the radius where the span is 900 mm and the rise is 75 mm, working out one step at a time.

Step 1

$$\text{radius} = \frac{(450^2 \div 75) + 75}{2}$$

Step 2

$$\text{radius} = \frac{(202{,}500 \div 75) + 75}{2}$$

Step 3

$$\text{radius} = \frac{2700 + 75}{2}$$

Step 4

$$\text{radius} = \frac{2775}{2}$$

$$\text{radius} = 1387.5 \text{ mm}$$

IMPROVE YOUR MATHS

Calculate the radius of a segmental arch where the span is 1800 mm and the rise is 300 mm.

Where segmental arches form part of an architectural feature, a problem can occur where openings of different widths are required. Rather than the same radius being used, to give a uniform appearance the arch rise is maintained. This is achieved as shown in Figure 6.8.

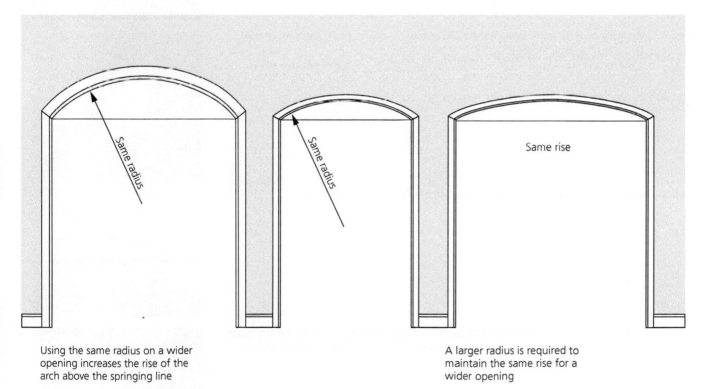

Using the same radius on a wider opening increases the rise of the arch above the springing line

A larger radius is required to maintain the same rise for a wider opening

▲ Figure 6.8 Problematic effect of having differing spans on segmental arches adjacent to one another

There is one other way of producing a segmental curve without a compass or trammel rod. This method uses a segment frame made with three laths of timber or ply. Table 6.4 shows this method step by step.

▼ Table 6.4 Producing a segmental curve without a trammel rod: step by step

Step 1	Follow steps 1–3 from Table 6.3.	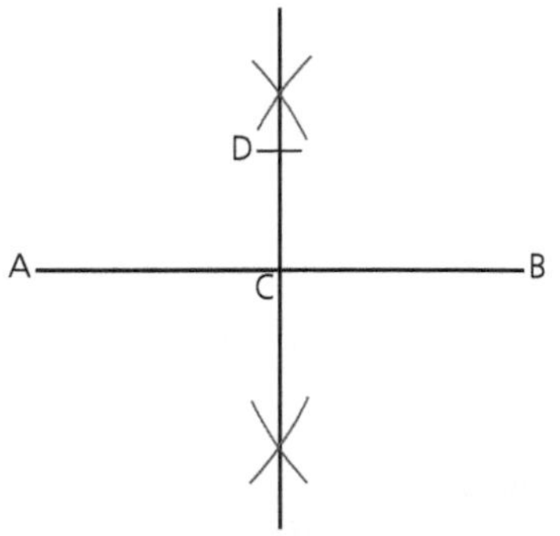
Step 2	Drive in temporary nails at points A, B and D.	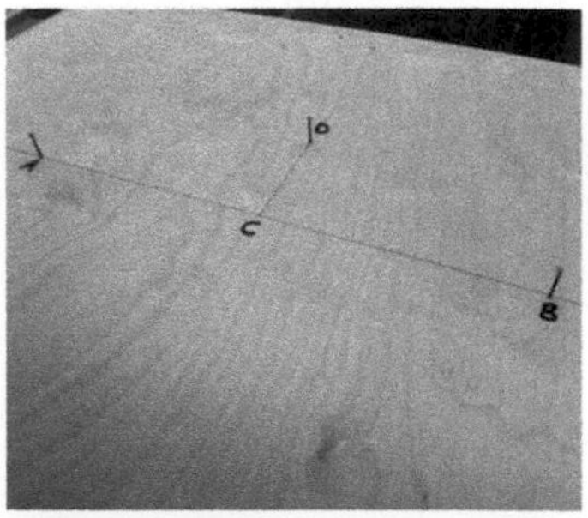
Step 3	Rest ply or MDF laths against the nails at lines A–D and B–D, and secure the two laths together at point D.	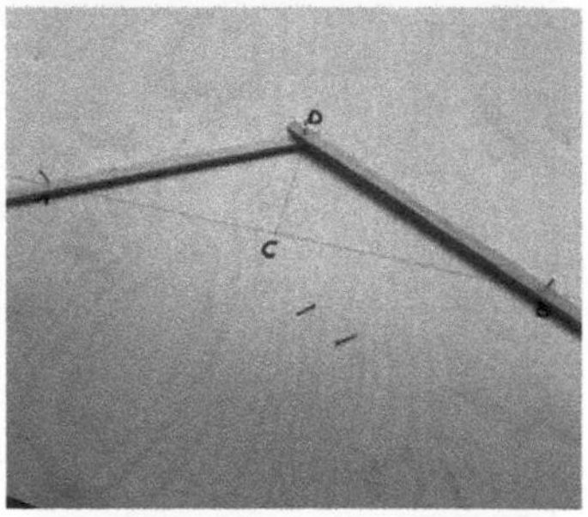
Step 4	Tack a third lath between the first two to create a triangular frame.	
Step 5	Remove the nail at point D and replace it with a pencil or pen. Slide the triangular frame to describe the segmental curve, being careful to keep the laths against points A and B.	

This method is practical only where the span is so great that a trammel is difficult or impossible to use. It was commonly used to set out a segmental **centre** or turning piece for arches to be formed on site. A turning piece is made from a solid piece of stout timber and is used as a former for the bricklayer to turn and lay the bricks around its shape. It can be made on site but can be cut more easily in the joinery shop. A turning piece is used only for low-rise segmental arches where the shape can be cut from solid timber.

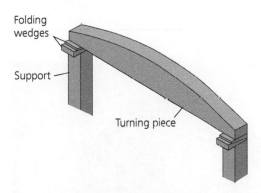

▲ Figure 6.9 A turning piece used for low-rise segmental arches

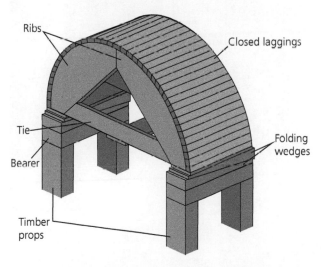

▲ Figure 6.10 A semi-circular arch centre and its parts

ACTIVITY

Draw a full-size templet for a segmental arch with a rise of 200 mm and a span of 900 mm.

Gothic arch

This arch shape is most recognised from ecclesiastical (church) buildings. Gothic architecture began in France in the 12th century and was commonly used in Western Europe until the 16th century. It is also the architecture of many old castles, palaces, town halls, universities and some houses. In the 19th century, the Gothic style became popular again, particularly for building churches and universities. This style is called 'Gothic Revival' architecture.

There are three commonly used Gothic shapes:
1 drop Gothic – where the radius is less than the span
2 equilateral Gothic – where the radius of the arch is equal to the span
3 lancet Gothic – where the radius is greater than the span.

In each case the radius origin is on the springing line.

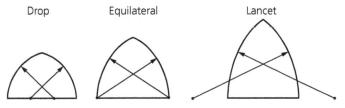

▲ Figure 6.11 Types of Gothic arch shape

Drawing an equilateral Gothic arch

▼ Table 6.5 Drawing an equilateral Gothic arch: step by step

Step 1	Draw the span, line A–B.	
Step 2	Using the span as the radius, draw arcs A–B and B–A to intersect at point C.	

The procedure required to draw drop and lancet arches is very similar in that both radii will be struck from the base line A–B. The radius will generally be taken from the architect's drawings. If the arch has been built, we will have to take the rise and span dimensions from the survey. There are no specific proportions used for either; these will be decided by the architect.

INDUSTRY TIP

You should never take manufacturing sizes from a drawing as they may not accurately represent what has been built. To be safe, a site survey is required to establish accurate information for manufacturing purposes.

ACTIVITY

Draw an equilateral arch shape with a span of 150 mm.

▲ Figure 6.12 A pair of glazed drop Gothic doors

Drawing a drop Gothic arch

▼ Table 6.6 Drawing a drop Gothic arch: step by step

Step 1	Bisect span A–B to produce point C.	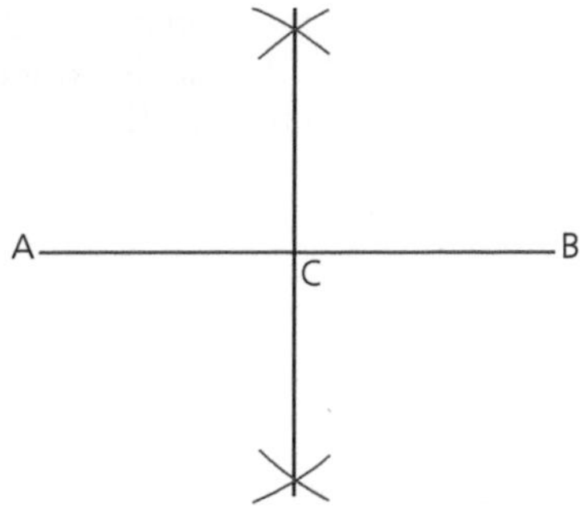
Step 2	From point C, measure the rise to produce point D. Draw lines A–D and D–B.	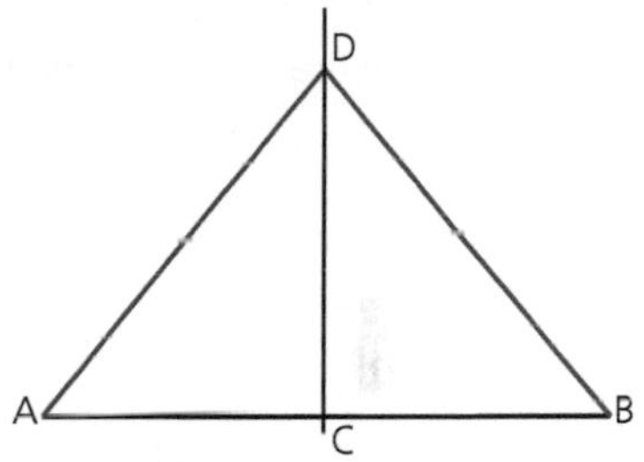
Step 3	Bisect lines A–D and D–B to produce points E and F on the span A–B.	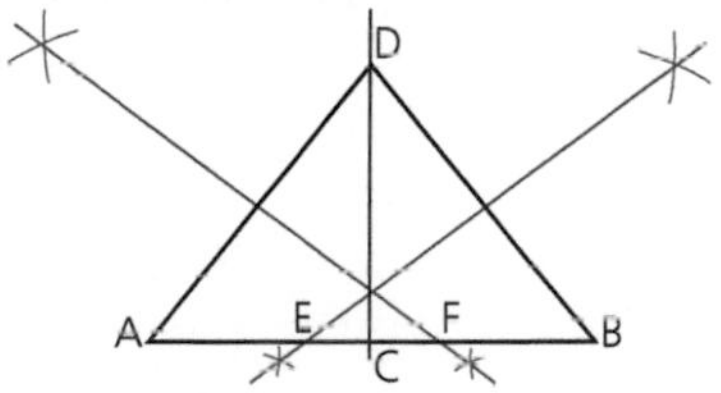
Step 4	Draw arcs with radius A–F and E–B from points E and F to intersect at point D.	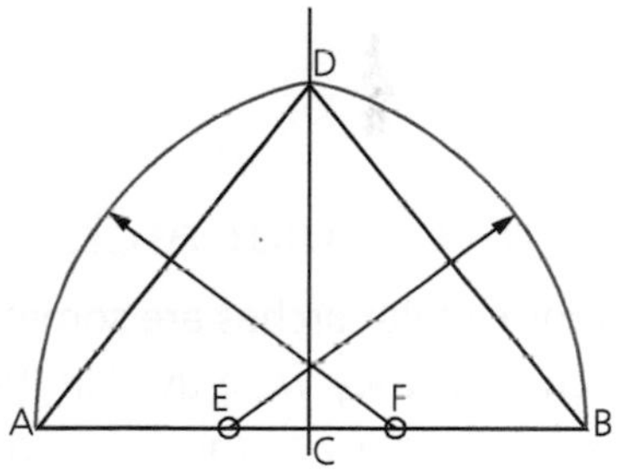

IMPROVE YOUR ENGLISH

Research the arch shape called 'pointed segmental'. Write an informational PowerPoint slide describing this shaped arch and how it differs from a segmental shaped arch.

ACTIVITY

Draw a lancet gothic arch with a span of 600 mm and a rise of 800 mm.

Drawing a lancet Gothic arch

▼ Table 6.7 Drawing a lancet Gothic arch: step by step

Step 1	Draw the span A–B and extend along the springing line beyond the span. Bisect line A–B. Mark the rise as C, using the measurement taken from the drawing. Draw chord B–C.	
Step 2	Bisect chord B–C, where the bisecting line falls on the springing line past point A. Mark this as point D (the striking point). To obtain the opposite striking point, measure equal distances from point O–D to point O–E.	
Step 3	Using a radius E–A and D–B, draw arcs to intersect at point C.	

Semi-circular arch

Semi-circular arches are sometimes known as Roman arches, as they were used in that era. They are the easiest to draw as the rise of the arch is half the span. In other words, the rise is the same as the radius.

▲ Figure 6.13 Semi-circular window

Elliptical arch

An elliptical arch is a flattened version of a semi-circular arch. Elliptical arches have a significant advantage over semi-circular arches as they allow much more headroom over a greater width. Therefore, an elliptical arch can be lower.

True ellipses are rarely used in construction unless the shape is standalone (nothing has to fit into it), such as a railway arch or the arched opening shown in Figure 6.14. The problem lies mainly with the fact that no two ellipses can be drawn exactly parallel to one another, so in joinery, where you have to draw many parallel lines, it becomes a considerable challenge. To overcome this, a pseudo (false) ellipse is drawn with the aid of a compass or trammel. As you will see later, on page 303, in the construction of a three-centre arch any lines drawn from any of the three centred will be parallel to one another, overcoming this problem.

Drawing a semi-circular arch

▼ Table 6.8 Drawing a semi-circular arch: step by step

Step 1	Draw span A–B.	
Step 2	Bisect line A–B to produce the radius centre, point O.	
Step 3	Draw radius O–A to produce the semi-circular shape.	

▲ Figure 6.14 An elliptical arched opening

Before drawing an ellipse, we need to know some additional terminology:

- major axis – the widest part of the ellipse
- minor axis – the narrowest part of the ellipse focal points – the ellipse is a curve surrounding two focal points where a straight line drawn from either of the focal points to any point on the curve and then back to the other focal point will have the same combined length, regardless of which point on the curve it touches (see Figure 6.15, which shows ellipse terms).

Drawing a true ellipse

There are several ways of drawing a true ellipse; unfortunately, they are not practical to use for setting-out purposes, as you will see, particularly with the first two methods shown.

Pin and string method

This is a common method but can be very fiddly. If care is not taken it won't produce an accurate outline.

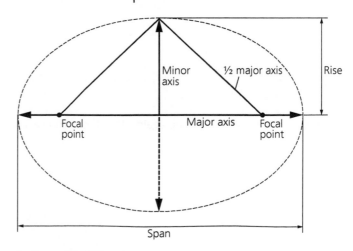

▲ Figure 6.15 Ellipse terms

▼ Table 6.9 Drawing a true ellipse (pin and string method): step by step

Step 1	Draw the major and minor axes, lines A–B and C–D.	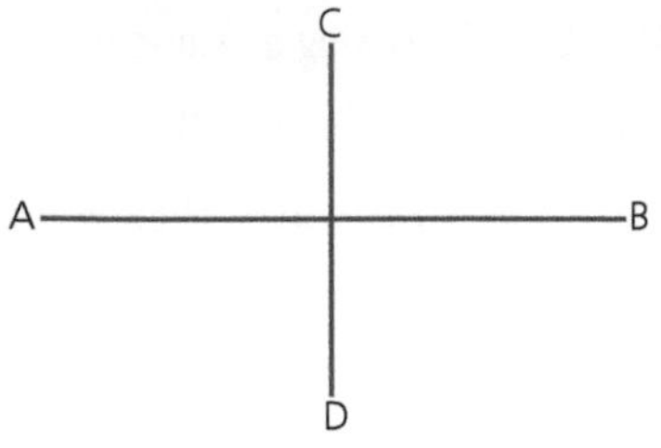
Step 2	With a centre at C and a radius of A–E, mark focal points F1 and F2 on the major axis.	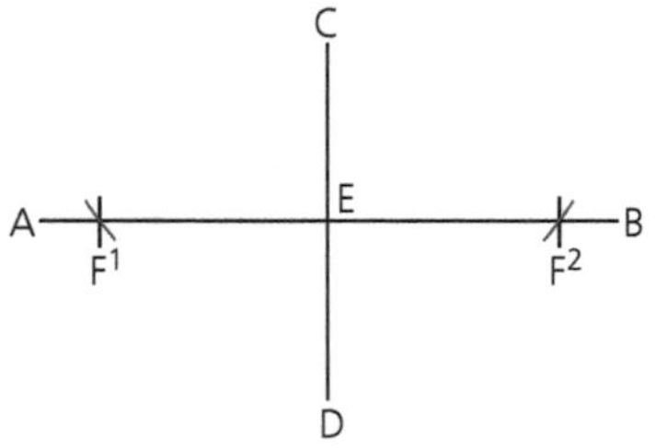
Step 3	Place pins at points F1, F2 and C.	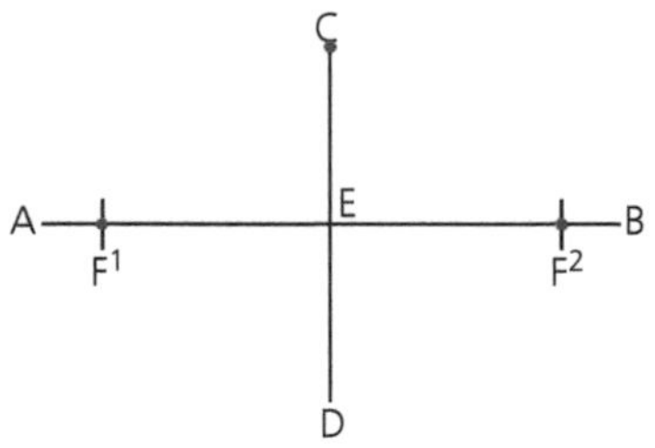
Step 4	Place taut string around the pins.	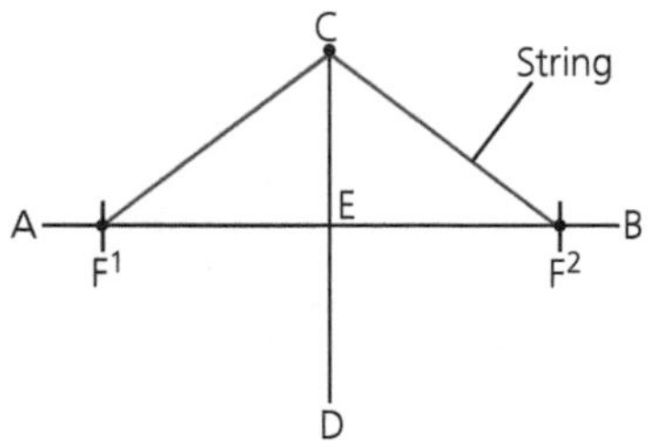
Step 5	Remove the pin at point C.	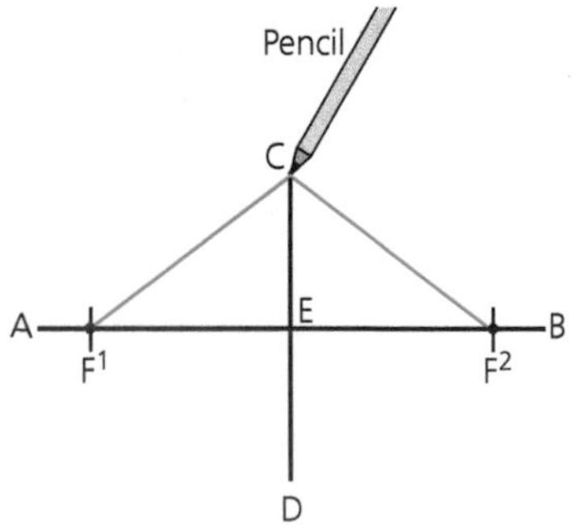
Step 6	Place a pencil tight up against the string and use it as a guide to draw the ellipse.	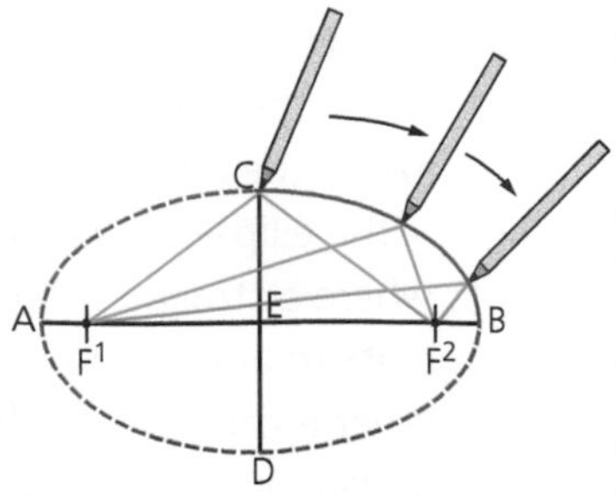

Concentric circle method

This method is usually used for scale drawing due to the difficulty of obtaining and using large set squares.

▼ Table 6.10 Drawing a true ellipse (concentric circle method): step by step

Step 1	Draw the major axis (line X–X), then bisect this to produce the minor axis (line Y–Y).	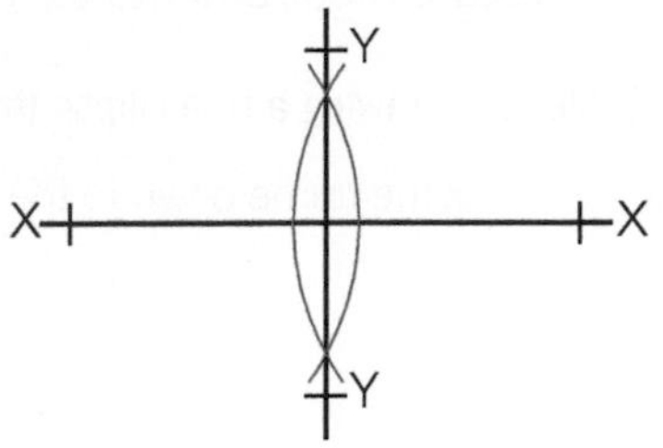
Step 2	From the centre, draw circles with radii that are half the minor axis and half the major axis.	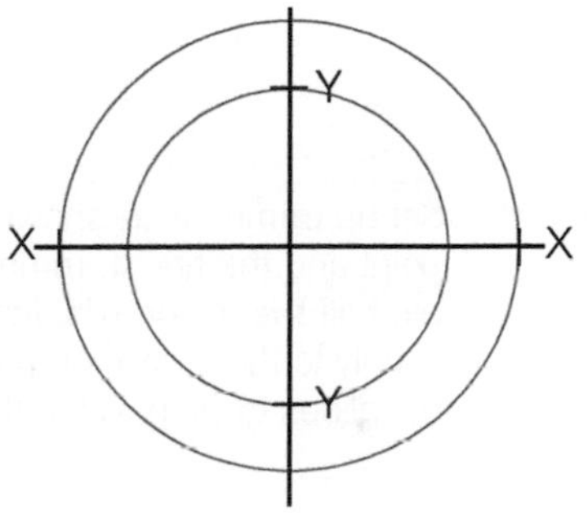
Step 3	Using a set square, draw a number of radiating lines.	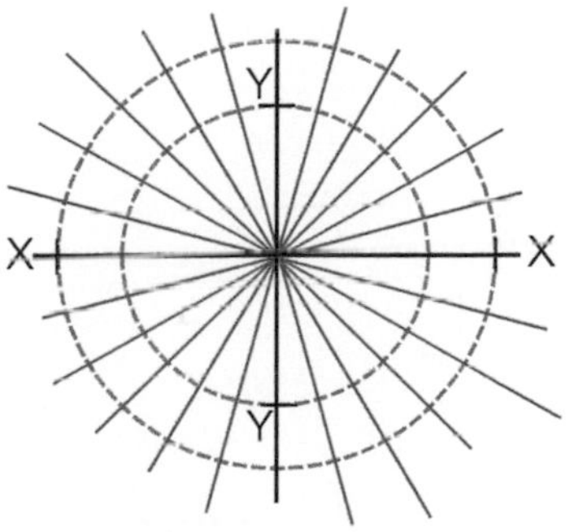
Step 4	From the points where the radiating lines cross the smaller circle, draw horizontal lines towards the larger circle, and vertical lines to intersect with the horizontal lines from the larger circle. From the points where the radiating lines cross the larger circle, draw vertical lines down towards the smaller circle to intersect with the horizontal lines.	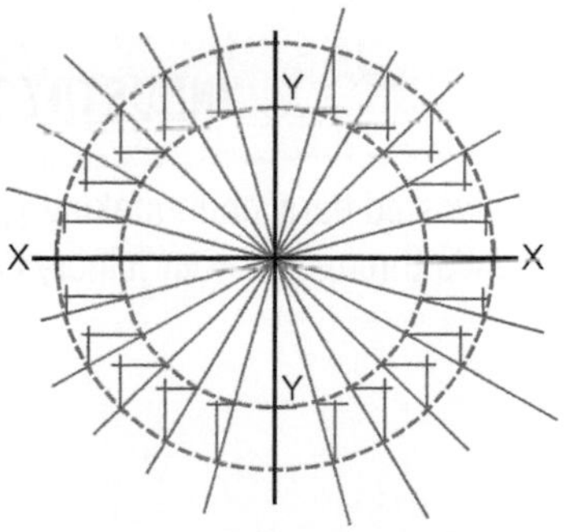
Step 5	Carefully draw a freehand curve that joins all these intersection points to produce the elliptical shape.	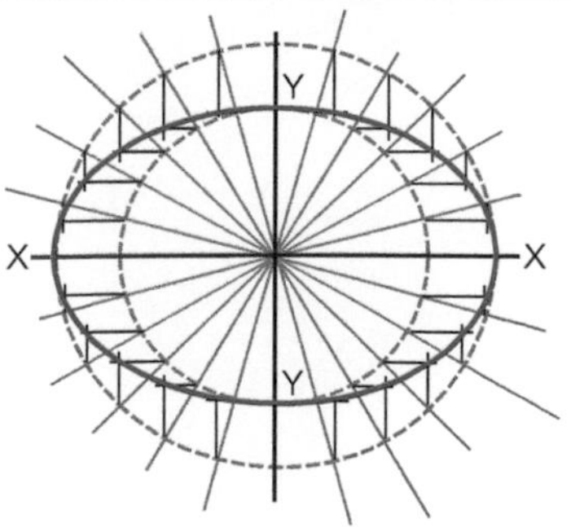

Trammel frame method

This is by far the most practical workshop method. Once an accurate frame has been made it can be kept for any future requirements as it is not size specific.

Furthermore, MDF board versions do not need to be braced. The frame is used in conjunction with a set of trammel heads and a trammel beam.

▼ Table 6.11 Drawing a true ellipse (trammel frame method): step by step

Step 1	Fix the frame down to the setting-out board.	
Step 2	Set up trammels as shown in the images, with the distance between the pencil point and the first trammel pin half the minor axis and the distance to the second pin half the major axis. Insert the trammel points into short wooden glides that fit snugly to the groove on the trammel frame. Gently slide the trammel frame to allow the glides to slide within the grooves and draw the outline of the elliptical curve.	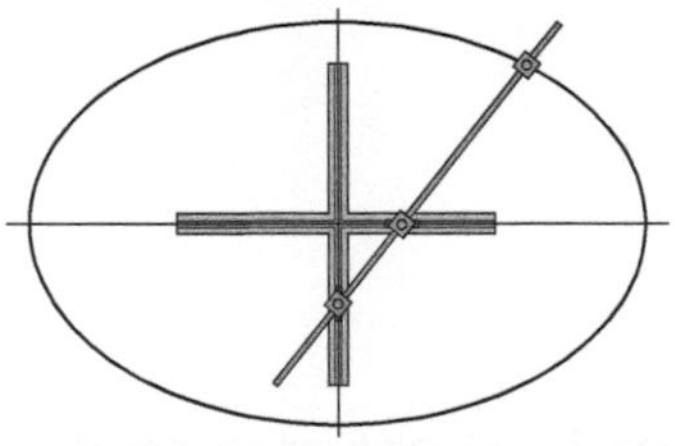 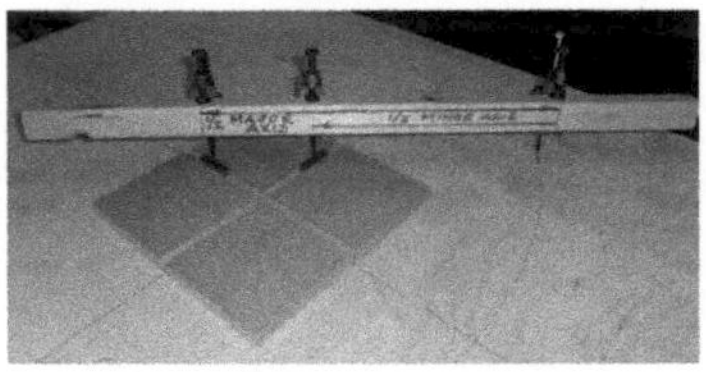 ▲ Figure 6.16 Trammel frame aligned with the major and minor axes

---

INDUSTRY TIP

Time invested early on in making jigs and aids will pay dividends throughout your joinery career.

Three-centred arch

Also known as a pseudo ellipse, this approximate ellipse shape is often used when constructing joinery that requires an elliptical shape. Being drawn from three fixed centres, the drawing and construction are made simpler than is the case for true ellipse shapes.

▼ Table 6.12 Drawing a three-centred arch: step by step

Step 1	Draw a rectangle A–B–C–D, where line A–B is the major axis and line A–C is half the minor axis. Next bisect line A–B to produce points E and F.	
Step 2	Draw a line between points A and E. Using point C as the centre and line C–A as the radius, draw an arc to cut line C–D at point G. Using point E as the centre and line E–G as the radius, draw an arc to cut line A–E at point H.	
Step 3	Bisect line A–H to produce point 1 cutting line A–B, and point 3 crossing extended line E–F. Using point F as the centre, draw an arc with radius F–1 to produce point 2 on line A–B. Draw a line connecting points 3 and 2 and extending up to cross line C–D as shown.	
Step 4	Using point 1 as the centre, draw an arc with radius 1–A to cut the bisector of line A–E at point J. Using point 2 as the centre, draw an arc with radius 2–B to cut the line through points 3 and 2 at point K. Using point 3 as the centre, draw an arc with radius 3–J through point E to point K.	

ACTIVITY

Draw a three-centred false ellipse with a span of 600 mm and a rise of 200 mm.

Lines 3–J and 3–K are called 'common normals' and provide the ideal jointing place when producing the curved component.

Setting out

With CAD drawings (further information on CAD can be found in Chapter 2, pages 85–86) becoming more common, joinery shops are making good use of its benefits in setting out joinery work. Unfortunately, while large sheets and rolls of drawing paper can be used and printed on, generally on a plotter, paper will shrink, expand and distort, and can become damaged in use. For these reasons all shaped work should be set out full size on a flat and light-coloured board.

Birch ply is the best material to set out on. If MDF is used, lightly sand it first with a very fine abrasive, as the surface as supplied is 'waxy' and cannot be drawn on easily.

It is very important to keep the material flat as the circular part of the construction is marked out, fitted up and assembled over the full-size drawing, to ensure that it will be the correct size and shape.

The main pieces of equipment required when setting out curved work (shown in Table 6.13) are:

- trammel points and beam
- 2 m steel rule
- accurate parallel straight edge
- large Perspex set squares
- line runner
- protractor.

Other equipment you will have used already that may also be required for scale drawing work is:

- drawing board (with tee square or parallel motion)
- scale rule
- dividers
- compasses
- mechanical pencil (2H grade).

When setting out curved joinery, you generally only need to draw the elevation of the shaped part of the product, with just enough straight lines to align the work while marking out, fitting up and assembling. The straight part of the door or frame can be drawn using standard height and width sections, as shown in the images that follow.

INDUSTRY TIP

When drawing circular work, always draw the curved lines first and then match the straight lines up with them rather than the other way around. This will produce a more accurate drawing.

INDUSTRY TIP

A parallel Perspex straight edge allows you to see the drawing below the straight edge and is easier, more efficient and produces more accurate work.

IMPROVE YOUR MATHS

Find the equipment listed in the following table on an appropriate website and find the total cost for all items.

INDUSTRY TIP

Before starting the setting out, take a moment to think about where the setting out will be positioned on the board. There is nothing worse than starting the setting out only to discover that you have run out of board because you started in the wrong place.

▼ Table 6.13 Equipment required for setting out curved work

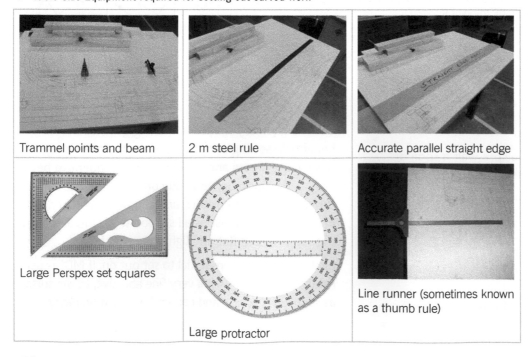

| Trammel points and beam | 2 m steel rule | Accurate parallel straight edge |
| Large Perspex set squares | Large protractor | Line runner (sometimes known as a thumb rule) |

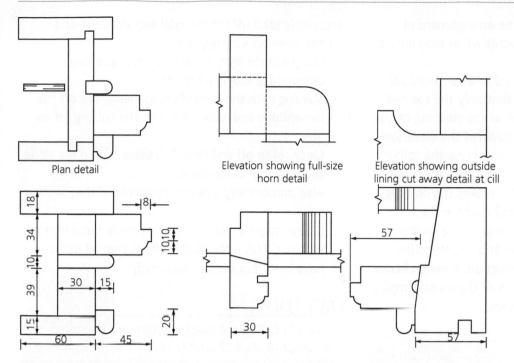

Plan detail

Elevation showing full-size horn detail

Elevation showing outside lining cut away detail at cill

▲ Figure 6.17 Height and width sections of semi-circular headed box frame (dimensions in mm)

These height sections are all that would be required in addition to the full-sized rod (elevation) as shown in Figure 6.17.

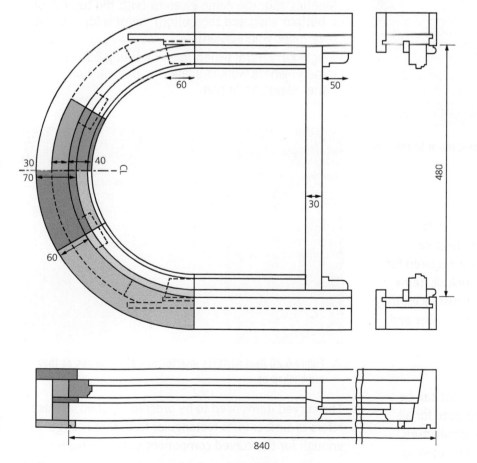

▲ Figure 6.18 Construction details of a semi-circular headed box frame (dimensions in mm)

The construction details show the arrangement of components and the jointing details when they are not standard.

Figure 6.18 shows a completed rod of a semi-circular headed **box frame**. You will see that only the top half of the frame is drawn to just below the **meeting rail**, as this is all that is required to be drawn of the elevation in order to manufacture the window. Note the colour coding of the vertical section and the corresponding components in the elevation. This makes the drawing easier to interpret, and is very useful where there are several components lying on top of one another. In this example, there are seven components to the shaped top sash and head construction. A semi-circular headed box frame is probably one of the most complex examples of **single curvature** work.

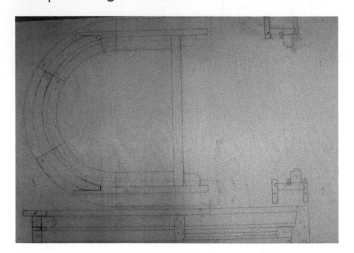

▲ Figure 6.19 Completed full-size rod of semi-circular headed box frame

Order requisitions

Most companies will have standard documents for this purpose. The setter-out is generally responsible for writing up the orders for all the items required for the finished joinery product, and may include items such as glass and ironmongery. In some instances, they may even be responsible for ordering the timber and manufactured boards from the suppliers.

Cutting lists

Once the drawing is complete, a cutting list can be produced. This gives the timber sizes needed for all the components that are required to manufacture the product. The following method can be used to accurately **take off** the material requirements and record them on a cutting list.

1 Using a circle templet, draw a circle in every component shown on the rod.
2 Starting with the longest component, take off the dimensions and record them on the cutting list as 'Item 1'.
3 Once taken off and recorded, place a '1' in the circle showing that component.
4 Also number any other components of that section size.
5 Repeat steps 3 and 4 until every circle has a number within it. This is the double-check that all items have been 'booked up' (recorded).

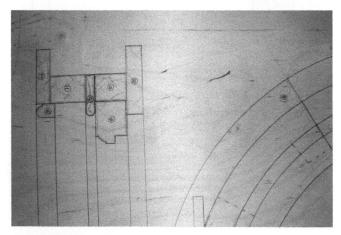

▲ Figure 6.20 Item numbers inserted into blank circles as they are booked up

The curved items need to be ordered as 'blanks'. This refers to timber in its rectangular form that is large enough for the curved component to be cut from.

It is best to group all the curved components together on the cutting list rather than jump around all over the cutting list later, trying to find the individual curved components.

IMPROVE YOUR MATHS

Research the current cost of unsorted European redwood and calculate the cost of the timber required in the example cutting list.

Table 6.14 shows an example cutting list for the semi-circular headed box frame window shown in the images.

▼ Table 6.14 Example cutting list for a semi-circular headed box frame window

Item	Qty	Description	Mat.	Length	Sawn size		Planed size		Instructions
					W	Th	W	Th	
1	2	Pulley stiles	S/W	730			95	20	
2	1	Cill	H/W	620			116	57	
3	3	Staff bead	S/W	600			20	15	
4	2	Parting bead	S/W	600			21	10	
5	2	Inside linings	S/W	650			60	15	
6	2	Outside linings	S/W	650			75	18	
7	1	Curved in. lining	S/W	1100	105	20		15	Cuts 3 in length
8	1	Curved out. lining	S/W	1100	115	23		18	Cuts 3 in length
9	1	Curved parting bead	S/W	1100	90	15		10	Cuts 3 in length
10	1	Curved staff bead	S/W	1100	60	25		20	Cuts 3 in length
11	1	Curved inner head	S/W	1100	75	45		39	Cuts 3 in length
12	1	Curved outer head	S/W	1100	75	38		34	Cuts 3 in length
13	1	Curved top rail	S/W	850	85			33	Cuts 3 in length
14	1	Stiles (top/btm sash)	S/W	1600			45	33	Cuts 4 in length
15	1	Bottom rail	S/W	500			43	33	
16	2	Meeting rails	S/W	500			57	30	

Selecting and preparing timber from a cutting list

Always select the timber carefully before cutting to length. Pull out a number of boards to examine them, rather than cutting as you go. This allows you to select for consistent colour and grain characterisation. This is particularly important for polished hardwood jobs. Try to cut all the blanks for the curved work out of one piece; this ensures consistency of colour and grain, and helps produce a harmonious-looking frame.

The general rule is that the longest lengths on the cutting list should be cut first, leaving the offcuts for the shorter lengths. (If you cut the short pieces first you may not be left with any timber in the rack long enough to cut the longer lengths required.) Try to cut between defects (knots and shakes, etc.) if possible.

The order of machining is as follows and the page numbers afterwards show where these processes were covered in detail in Chapter 3:

1 crosscut to length (pages 105 and 120–123)
2 rip to width (pages 103–104 and 106–109)
3 surface plane face and edge (pages 131–139)
4 plane to width the thickness (pages 141–144).

② MARKING OUT AND MANUFACTURING CURVED JOINERY

In Chapter 3, your learned how to produce components using a range of woodworking machinery. Remember that you must always safely set up and use machines and power tools to comply with the Health and Safety at Work Act, Provision and Use of Work Equipment Regulations and the Approved Code of Practice. Chapter 3 covers the types of machines and tools you are likely to encounter, including:

- radial arm, table ripsaw and panel saws
- surface planers/thicknessers
- narrow bandsaws
- spindle moulder
- router table
- tenoner
- hollow chisel morticer
- dust extraction.

Remember that risk assessments and method statements must be completed for all machines that are used.

Templets

When the timber required for the job is complete, the curved component templets can be produced. Templets are made to allow the blanks to be marked out, cut and brought to shape and profile. Traditionally they were produced by marking the shape with the trammel heads and beam, bandsawn slightly oversize, and brought to shape with a **compass plane** or **spokeshaves**.

▲ Figure 6.21 A templet being trimmed to the exact shape required using a compass plane

A templet will need to be made for each shaped component. They are initially used to mark out the shape of the component onto the wide boards before being bandsawn to shape. Templets are usually made from 12 mm birch ply or MDF. They need to be cut to the exact shape required to make the curved component, but need to be between 75 and 100 mm longer than required to allow for safe 'lead in and out' when spindling later. The best way of producing templets is to use a portable power router and trammel bar. As the router cutter takes the place of a pencil in a compass, it will produce accurate templets very quickly.

▲ Figure 6.22 A templet being produced using a trammel bar and router

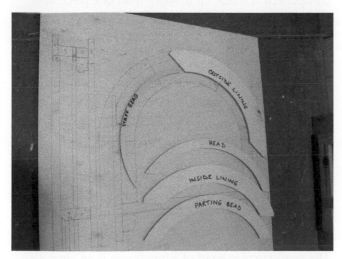

▲ Figure 6.23 The templets required to produce a semi-circular headed box frame

Methods of constructing curved components

The setting out will show the method to be used for constructing the curved components. There are three possible methods:

1 solid
2 built up
3 laminated.

Which of the three construction methods is chosen will depend on a number of factors, including:

- the strength required
- the component type
- whether the finish is painted or polished
- the radius of the component.

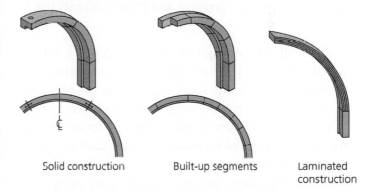

Solid construction   Built-up segments   Laminated construction

▲ Figure 6.24 Methods of forming curved components

Solid construction

With the solid (and the built-up) method of construction, we need to be aware of the problems of 'short grain'. Short grain occurs where the shape is made in too few segments or where the templet is not applied 'with the grain' of the timber.

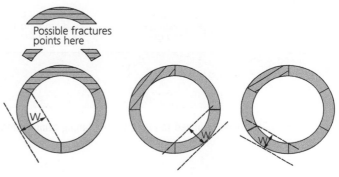

W = width of timber required for blank

▲ Figure 6.25 Frames made in three, four and six parts; the short grain is reduced as more parts are used

As you can see in Figure 6.24, using fewer parts to make the circular frame will lead to more short grain being encountered. This will increase the chance of the grain splitting and of breakages due to **pick-up** on the spindle or router where the timber is weakest. The other advantage of using more parts is that the timber section is made from narrower material, minimising any shrinkage and therefore distortion of the shape of the finished components.

For a circle, a minimum of four parts should be used, with six being good practice and eight being the best. Obviously, the more parts are used, the costlier the circle will be to construct due to the extra labour required to joint and manufacture the sections of the frame. The main advantage of using four parts is that it is the simplest to produce and therefore costs the least.

Traditionally, the templet would be applied to a full-width board that had been planed to finished thickness to minimise waste and short grain. It can still be done in this way if the components are to be shaped by hand using a compass plane or spokeshaves – for example, if the frame was to be a 'one-off' or a job does not need heavy rebates to be produced on the spindle. In this case, the mouldings, grooves, etc. can quite safely be **stuck** using a portable power router.

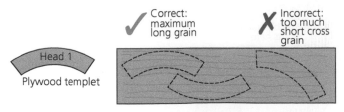

▲ Figure 6.26 Templet applied to board when planing components by hand

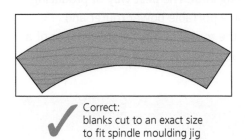

Correct:
blanks cut to an exact size
to fit spindle moulding jig

▲ Figure 6.27 Rectangular blanks required when using a spindle moulding machine

It is a requirement of the wood machining Approved Code of Practice (ACoP) that all circular work is held in secure holding jigs while being machined. This means the blanks must be cut to an exact size so they fit into the jig and can be safely machined on the spindle moulding machine.

Built-up construction

In built-up construction, the thickness of the curved component is made up of more than one piece. It is quite common for two or three layers of segments to be assembled with the joints staggered to create a strong curved component.

This method suits frame construction where the rebate forms a natural break. It is not particularly suitable for shaped top rails to doors or sashes, as when you plane the edges to fit to the frame you will get considerable pick-up, as the grain in each layer is running in a different direction. Another disadvantage is that the timber sections will move (shrink or expand) in different directions, distorting the finish to the face. This can be especially obvious when the work is gloss finished.

The semi-circular head of a box frame is an ideal example of where this type of construction works well due to the construction method required.

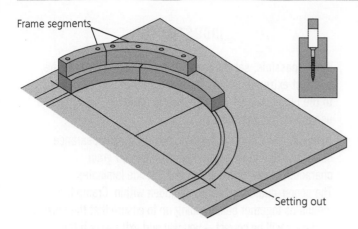

Frame segments

Setting out

▲ Figure 6.28 The curved head is built up over the rod

Laminated construction

Laminated construction (sometimes referred to as **glulam** construction) is the most labour-intensive method but can produce very stable components with no short grain. The section required is made up of a number of thin laminates (thin strips of timber) that are glued and bent around a former. The former is usually made of two parts – one male and one female. The glued laminates are placed between the two with cramps applied until the adhesive has thoroughly cured.

KEY TERM

Glulam: a common abbreviation for 'glue laminated'.

▲ Figure 6.29 The laminates are cramped between a male and female former

INDUSTRY TIP

Use a sheet of clear polythene under the former to prevent the glue that is squeezed out from between the laminates from ruining the bench or becoming stuck to it.

A synthetic resin adhesive should be used rather than a PVA to prevent **spring-back**. As there is considerable tension involved in bending the laminates around the former, the former needs to be very robust in construction. The thinner the laminates, the easier they are to bend around the former.

There needs to be a balance between the number of laminates used and the extra material required to produce the laminates. Each laminate has to be planed up so will need to be sawn with a planing-up allowance, adding to the cost of the material used. The thickness of laminate depends on the species of timber. Generally, hardwood laminates have to be thinner than softwood laminates as they are more resistant to bending.

Working out the thickness of laminate required

As a rule of thumb for softwood (European redwood), divide the radius by 200. For example, if the radius was 1000 mm the thickness of laminates would be 1000 mm ÷ 200 = 5 mm.

For hardwood, divide by 250. For example, for a radius of 1000 mm in mahogany the thickness of the laminates would be 1000 mm ÷ 250 = 4 mm.

It is not practical to use laminates less than 3 mm thick as this is generally the thinnest that a thicknessing machine will plane to. This determines the minimum radius that laminated construction can be used for. Laminates can be sawn thinner but they will not be of a consistent thickness and the quality of the **joint line** will be poor.

KEY TERMS

Joint line: the fit of a joint where two pieces of material are bonded to each other.

Spring-back: where the tension in the timber wants to pull the curve flat and it loses its intended shape.

Once the former has been made, it is advisable to do a trial run for the thickness of the laminates, particularly when using some dense hardwoods, to ensure they are sufficiently pliable to bend around the former. When preparing softwood laminates, prepare about 20 per cent more than required. This is because many of the laminates will have shakes or knots that will break or split during the bending process.

A vacuum bag can be used for laminating certain components, as long as they are small enough to fit in the bag. This method is best when laminating plywood structures. A male former is constructed and placed in the bag. The laminates have adhesive applied to each bonded face and are then put on top of the former, held with two light tacks (punched so as not to puncture the bag) that hold them to the top of the former to prevent slipping. The compressor is then started, which will slowly suck the air out of the bag. The polythene will gradually pull the laminates tight to the former. The compressor remains switched on until the adhesive has cured.

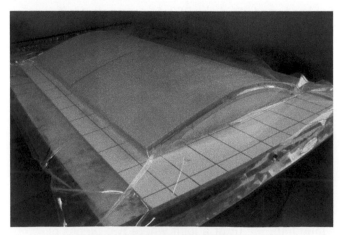

▲ Figure 6.30 A laminated component being formed in a vacuum bag

Jointing methods

The method of joint chosen will depend on the nature of the work, whether the joints will be seen and how the components are arranged within the structure. The jointing arrangement will, however, need to be decided at the setting-out stage and these details shown on the rod. Typically, the straight parts of the work will employ variations of machine-cut mortice and tenon or bridle joints, as described in *Level 2 Site Carpentry and Architectural Joinery*, Chapter 7, pages 342–3. These are adapted as shown with the springing line joints below.

Heading joints

The **heading joints** used between parts of a continuous curved component could be formed using any of the following jointing methods:

- hammer-headed keys
- dovetail keys
- handrail bolts
- loose tenons
- kitchen worktop connectors.

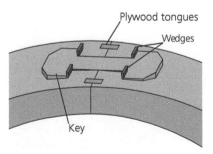

▲ Figure 6.31 Hammer-headed key

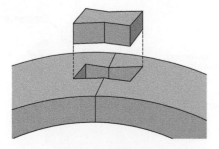

▲ Figure 6.32 Dovetail key

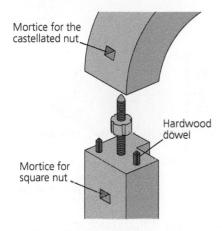

Mortice for the castellated nut

Hardwood dowel

Mortice for square nut

▲ Figure 6.33 Handrail bolt

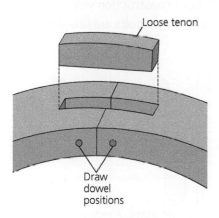

Loose tenon

Draw dowel positions

▲ Figure 6.34 Loose tenon

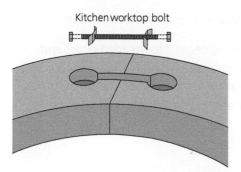

Kitchen worktop bolt

▲ Figure 6.35 Kitchen worktop connector

Hammer-headed and dovetail keys are both labour-intensive joints to produce. The hammer-headed key is the better of the two as it uses wedges to pull the joint up tight.

Handrail bolts are very efficient as they also pull the joint up, but they can be quite fiddly to fit. A pair of dowels or loose tongues should be incorporated into the joint to prevent it twisting. (These bolts are now difficult to obtain.) Figure 6.35 shows a traditional handrail bolt with its modern replacement, a Zipbolt™. The handrail bolt has two nuts, a captured (square) nut and a castellated (grooved) nut. The bolt is tightened on the castellated nut side. The threaded side of the Zipbolt™ is turned into one side of the joint and is tightened with an Allen key on the other side.

▲ Figure 6.36 Handrail bolt with a captured (square) and castellated (grooved) nut (above) and Zipbolt™ (below)

Loose tenons are now probably the most commonly used joint as they are easy to produce. They require pinning through the face as, unlike the other jointing methods listed, they are not **mechanical joints**.

KEY TERM

Mechanical joints: joints that, due to their design, hold or pull themselves together.

Kitchen worktop bolts can be used as a substitute on large frames but do require large holes to be bored, which weaken the component, and should be used only on the back of a frame.

▲ Figure 6.37 Worktop connecting bolt

313

Jointing on springing lines

Usually a variation of a mortice and tenon will be used when jointing on **springing lines**, with the tenon extending from the jamb of the frame or stile of the door. This joint is very easy to produce and can be draw pinned to cramp the joint together. If there is more than one component coming together at the springing line, the joints of the different components will have to be offset to avoid clashing. A twin tenon is commonly used between the transom and the jamb or the rail and the stile, with a single tenon/bridle on the jamb to the head or stile to the top rail.

A more complex and time-consuming joint could be used if specified, such as a hammer-headed tenon joint or handrail bolt. Alternatively, the joint position could be moved further towards the **crown** of the frame or door; this avoids the grain running in three directions at one point, which can look ugly on polished work and weakens the frame at this point.

KEY TERMS

Springing line: where a curved section starts to 'spring away' from a straight line.

Crown: the uppermost part of a shaped headed door or frame.

▲ Figure 6.38 Springing joint and transom intersection using a twin tenon and hammer-headed tenon joint

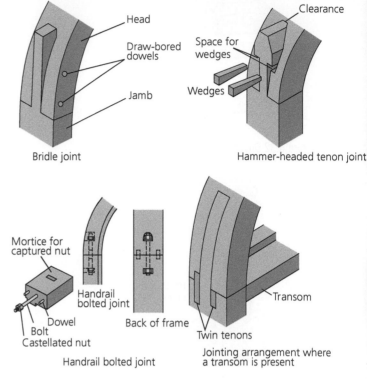

▲ Figure 6.39 Springing joints used in frame construction

Springing joints used in door construction will generally be limited to tenons, as bolts are unsightly even when plugged.

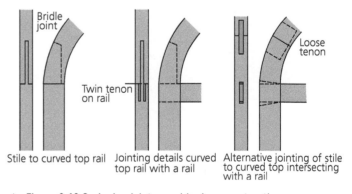

▲ Figure 6.40 Springing joints used in door construction

The springing joints of segmental headed doors and frames are different, as the jamb/stile does not run at a tangent to the head/rail. The most common joints used in this situation are shown in Figures 6.41 and 6.42.

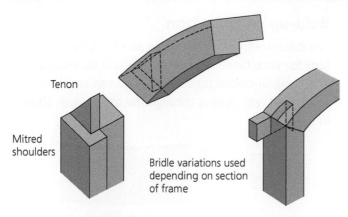

▲ Figure 6.41 Springing joints used for segmental headed frames

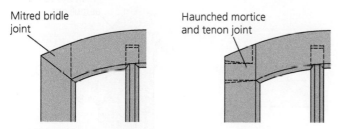

▲ Figure 6.42 Joints between curved top rail, stile and bar

Crown joints

For the frame, any of the heading joints already mentioned can be used. Joints used for single-width doors can be similar.

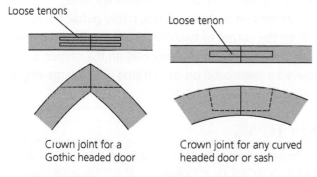

▲ Figure 6.43 Crown joints suitable for single doors

ACTIVITY

Draw an isometric view of a suitable crown joint for a semi-circular headed door.

Where the frame is wide enough to require a pair of doors, a different jointing arrangement is needed where the curved rail meets the meeting stiles on the door. Inserting false tenons removes the problem of the short grain that would occur if a tenon was

machined on the crown end of the rail. The false tenon should be inserted and glued to the curved rail, then allowed to dry. The joint can then be fitted up in the usual way.

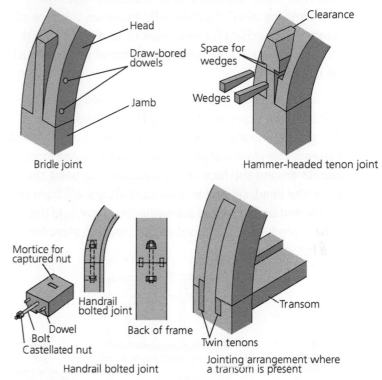

▲ Figure 6.44 Pair of glazed Gothic doors

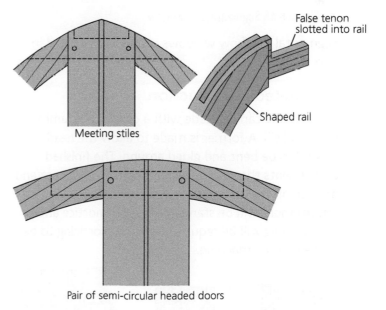

▲ Figure 6.45 Using false tenons at a pair of drop Gothic doors

Segmental bay/bow windows

Work on segmental bay/bow windows is fairly rare due to the expense of building curved brickwork and the fact that most replacement windows are now high-performance uPVC. Occasionally, however, this type of window is specified for new work or as a replacement in a conservation area where an identical window must be installed.

Technically, a true bow window has curved glass but – due to the expense of making the formers to enable the glass company to bend it – flat glass is usually used instead. The glass will sit in straight rebates and form **facets** around the face of the window. This being the case, the head, cill and any horizontal bars will have the inside and outside faces following the curve, and the rebate produced as a chord to the shape to allow for the flat glass.

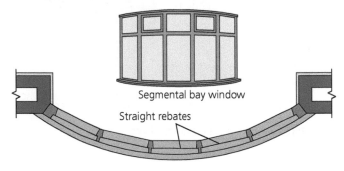

Segmental bay window

Straight rebates

▲ Figure 6.46 Segmental bay window

Segmental bay/bow windows can be constructed in two ways.

Laminated construction

The window can be made with a continuous laminated head and cill. A former is made to allow the head and cill to be bent and glued around. The finished components then need to be brought to thickness and marked off the plan in the usual manner. The jointing arrangement will be standard through mortice and tenons. A jig will be required to allow morticing to be carried out by machine.

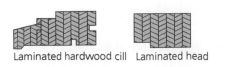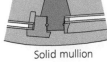

Laminated hardwood cill   Laminated head   Solid mullion

▲ Figure 6.47 Laminated construction with solid mullions

Built-up construction

The other method involves a series of flat frames (one for each facet). The ends of the cill are mitred and handrail-bolted together. The frames are then connected with screws through the mullions, as shown in Figure 6.47.

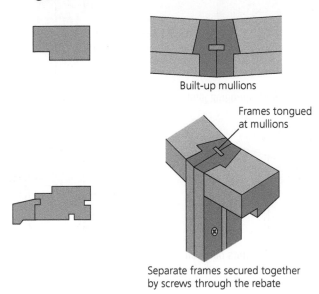

Built-up mullions

Frames tongued at mullions

Separate frames secured together by screws through the rebate

▲ Figure 6.48 Built-up segmental bay construction

Doors and frames shaped on plan

Curved segmental doors and frames are rarely encountered. In Victorian times, many public buildings built on the corner of two roads had a **radiused corner**. If the entrance to the building was on the corner, it required a segmental on-plan frame with a large single or pair of double curved doors.

> **KEY TERMS**
>
> **Radiused corner:** any corner whose sharp point has been softened by a radius.
>
> **Facet:** a section forming a flat face – for example, the panes of glass within a segmental bay window.

Construction of the frames follows the methods described in the previous section. The plan section of the jambs, however, can take two forms. These are known as 'radiating' or 'parallel' jambs. Whichever type is detailed by the architect, the rebates must be parallel; if they are not, the door will be trapped and will not clear the frame as it opens.

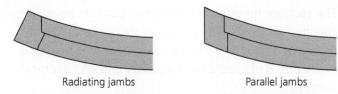

Radiating jambs   Parallel jambs

▲ Figure 6.49 Jamb details for segmental on-plan door frames

The rails in the curved door construction should be built up in 50–75 mm layers to make up the width required. The tenons can be produced by hand or by using a false bed on the tenoning machine. This will lift the rail to bring the tenon parallel to the cutting circle of the tenoning heads.

The panel will be constructed by laminating thin short-grained ply to make the thickness required. If the door is to be polished, before they are bent, the faces of the ply will need to be veneered to match the species of timber.

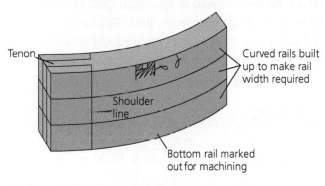

Tenon   Curved rails built up to make rail width required

Shoulder line

Bottom rail marked out for machining

▲ Figure 6.50 Bottom rail marked out for machining

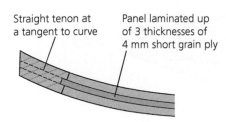

Straight tenon at a tangent to curve   Panel laminated up of 3 thicknesses of 4 mm short grain ply

▲ Figure 6.51 Plan view of curved door

Section profiles

Section profiles can vary considerably depending on whether they are for new or restoration work. This applies particularly to window construction, where high-performance windows are required to conform to Part L of the Building Regulations (covering conservation of fuel and power). High-performance windows require much more complex sections, which can sometimes be difficult to machine, particularly on curved sections.

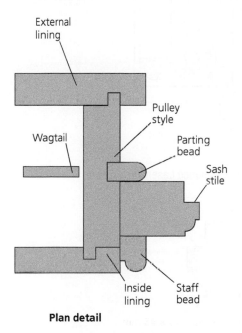

External lining

Pulley style

Wagtail

Parting bead

Sash stile

Inside lining

Staff bead

Plan detail

▲ Figure 6.52 Traditional sections for a box frame window

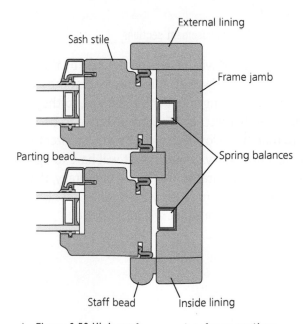

Sash stile

External lining

Frame jamb

Parting bead

Spring balances

Staff bead   Inside lining

▲ Figure 6.53 High-performance box frame sections

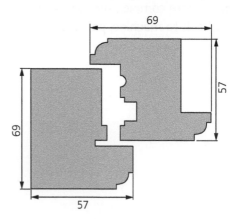

▲ Figure 6.54 High-performance casement sections

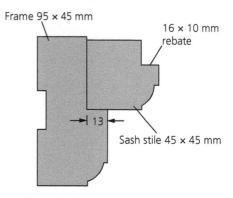

Frame 95 × 45 mm

16 × 10 mm rebate

13

Sash stile 45 × 45 mm

▲ Figure 6.55 Traditional casement sections

The sections required will have been detailed by the architect on her/his drawings. It is unlikely that they will be drawn from a tooling supplier's catalogue, so they will not be exactly matched. Tooling will have to be sourced to match as closely as possible the architect's details and, if necessary, the architect will have to agree that these will be a suitable alternative. Generally, the architect will not be concerned with the fine detail, only that the finished product will look correct.

The sections for internal joinery components have not changed much over the years and should not pose any problems. (More information about them can be found in Chapter 8 of *Level 2 Site Carpentry and Architectural Joinery*.)

Marking out curved joinery products

Once all the materials have been machined to size, they can be marked out ready for jointing and profiling. The first process, as ever, will be to select the face side and edge for each of the components. You should look to remove any defects in the profile section where possible, such as machining out knots and shakes in rebates or positioning them to the back of the work. Components with large dead knots, or that are twisted, bowed or sprung, should be discarded and replaced. In painted work, **piecing in** avoids the need to replace components and reduces wastage. This should be for minor damage, or to replace loose, dead, face or edge knots.

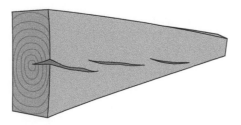

▲ Figure 6.56 Shakes can be placed to the back of the component

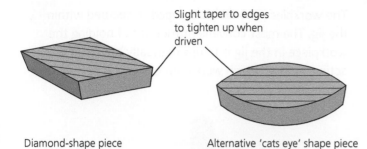

Slight taper to edges to tighten up when driven

Diamond-shape piece

Alternative 'cats eye' shape piece

▲ Figure 6.57 Piecing up dead knots or damaged timber in hardwood

Before the face marks are applied, the curved blanks should be examined carefully for defects, particularly knots and shakes. These can cause pick-up which could lead to snatching when the blanks are being spindled to shape; this is hazardous (and is one of the reasons a robust jig is required). Wherever possible, try to lose knots by ensuring they are on the part of the timber blank that will be bandsawn away as waste.

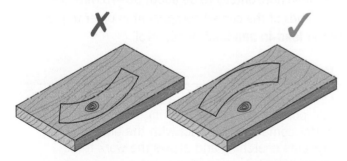

✗   ✓

▲ Figure 6.58 Placing the templet on the blank to avoid knots and defects

Manufacturing curved joinery products

The machining processes for straight work have been described in both Chapter 6 of *Level 2 Site Carpentry and Architectural Joinery* and Chapter 3 of this book. This chapter will cover only the additional processes required for curved work.

Marking out and machining curved components for a replacement semi-circular headed box frame

In this example, curved components are marked out and machined for a replacement semi-circular headed box frame in a heritage building. This example has been selected as it is probably the most complex of curved constructions. The rod for this box frame is as shown on page 306.

As previously mentioned, the straight frame components can be marked out from the full-size height and width sections on the rod. The machining of straight components is covered in detail in Chapter 3. The circular frame components will have to be machined to the curved shape (including the section shapes) before marking out from the full-size elevation. Note in Figure 6.59 the use of a Perspex bridge. This allows the lines to be lifted from the rod and marked exactly on the face of the timber.

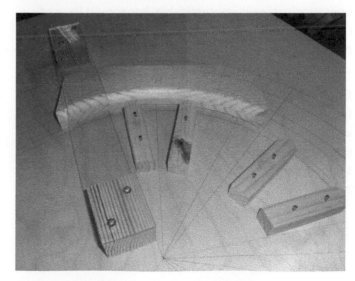

▲ Figure 6.59 Circular components marked out from full-size rod

Machining jigs for shaped components

Using jigs and work holders is necessary for all stopped and curved work, unless the nature of the operation makes it impracticable. Jigs should be used even if workpieces are irregularly shaped and limited production runs are involved.

The design of jigs and work holders is determined by the work to be done. They must be robust, and are typically made of hardwood and plywood. They should allow quick and accurate location of the workpiece, which should be held firmly in position. Jigs should have secure handles and wide bases so that machinists have a firm grasp at a safe distance from the cutters.

The workpiece should be clamped or secured within the jig. The most convenient method of holding the workpiece in the jig is to use manually operated quick-acting clamps, which work with either a toggle or a cam action.

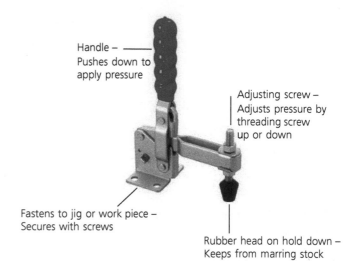

Handle –
Pushes down to apply pressure

Adjusting screw –
Adjusts pressure by threading screw up or down

Fastens to jig or work piece –
Secures with screws

Rubber head on hold down –
Keeps from marring stock

▲ Figure 6.60 A typical toggle cramp

A combined templet and jig helps to ensure that curved work is held firmly and correctly, to produce the required shape and finish. Jigs can be one or two sided, allowing both the internal and external curves to be produced using one jig. The templet should also be extended horizontally to be about 50–75 mm longer at both ends of the curved component in order to provide better **lead-in and lead-out** control.

Producing circular components by machine

▼ Table 6.15 Producing circular components by machine: step by step

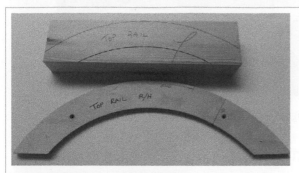

Step 1 Gather the templets that will be used to mark out the shape on the blank.

Step 2 Bandsaw the blanks to shape, leaving about 2 mm to be removed on the spindle to bring them to the correct width.

Step 3 Remove the straight fences from the spindle, and set up the ring fence and bonnet guard.

Step 4 Set a planer or profile block flush to the ring fence.

Step 5 Pin the templet onto the blank and cramp securely in the spindle jig.

Step 6 Carry out a trial run to ensure the block is cutting flush to the templet.

Step 7 Machine all blanks for each templet.

Step 8 Machine the profiles to the curved components.

Figures 6.61–6.63 show different types of profile block commonly used on the spindle moulder. Limited cutter projection tooling (LCPT, also known as chip or chip limited tooling) must be used if machines are hand fed. For more information about this, see Chapter 3.

▲ Figure 6.61 Moulding block

▲ Figure 6.62 Expanding grooving block

▲ Figure 6.63 Rebating block

Router table

For small and lightweight sections, a router table could be used to profile and shape the section profiles. It must be remembered that the ACoP still applies when using this method.

▲ Figure 6.64 Router table used for light work

Fitting up component parts

Once the components have had their **secondary machining** carried out, the various parts can be jointed using one of the methods shown and then fitted up (sometimes referred to as dry fitting). It is essential that the curved components are fitted up over the full-size rod to ensure the joints fit up tight, and that it conforms to the shape and size required. Note the use of the offset temporary pointed blocks to correctly position the component parts while they are being assembled (in this case, the curved head of a box frame).

KEY TERMS

Secondary machining: jointing and profiling of planed components.

Draw-bored: a method of creating joints that allows them to be pulled together without the need for cramps.

▲ Figure 6.65 Curved head fitted up over rod

Any dowelled joints can now be **draw-bored** ready for pinning in the assembly process. Draw pinning allows assembly to be carried out without the use of cramps. This method is suitable for shaped work because it is difficult to cramp positively against a curve's surface. See the step-by-step method presented in Table 6.16 for this process.

▼ Table 6.16 Draw-boring ready for pinning: step by step

Step 1	Bore a hole through the mortice.	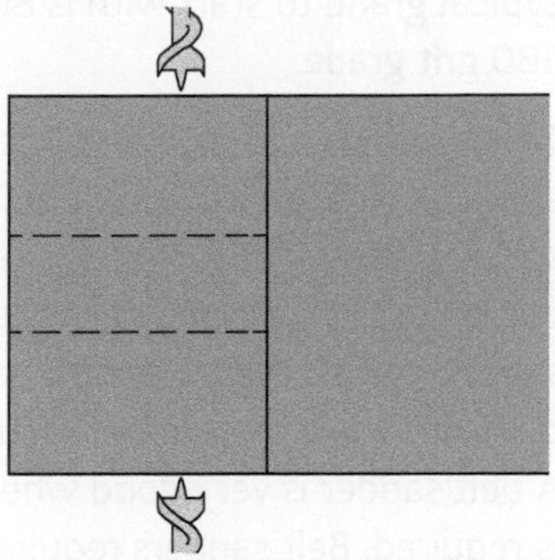
Step 2	Push the tenon into the mortice and mark the centre of the hole on the tenon using a boring bit.	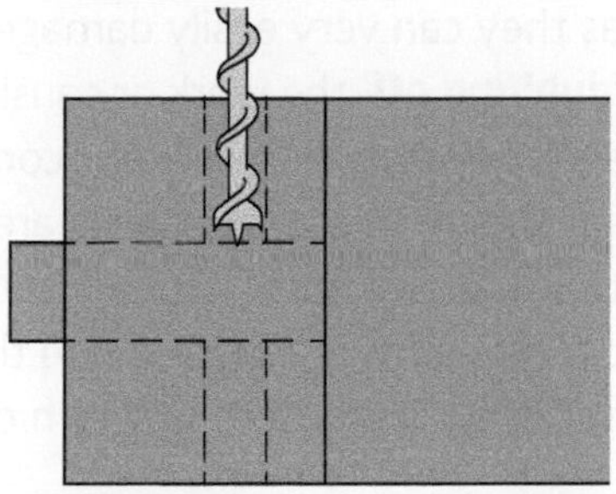
Step 3	Remove the tenon and bore a hole 1.5 mm closer to the shoulder.	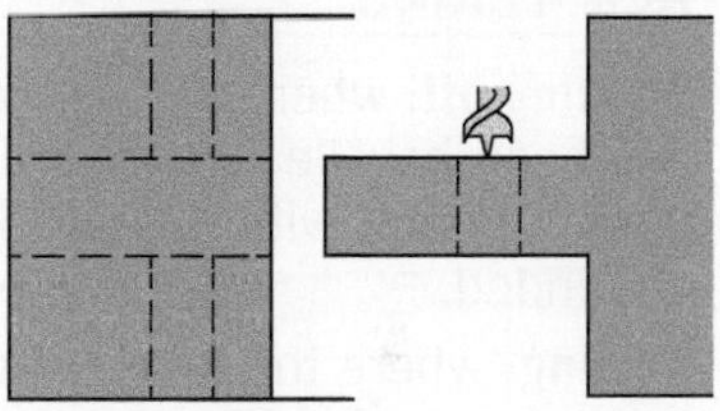
Step 4	As the dowel is driven in, the shoulders of the tenon will be pulled up tight.	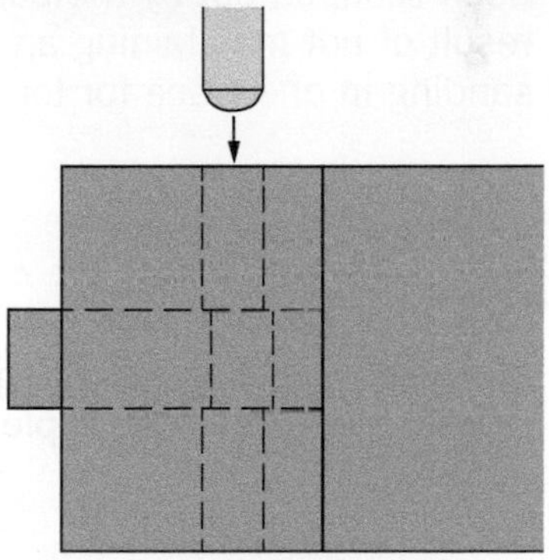

Papering up

When you have made any adjustments required and are happy that everything is correct, all the faces that will be inaccessible after assembly can be papered up.

Sometimes called 'cleaning up', this is the process of using abrasive sheeting to remove any slight blemishes and machine or pencil marks, ready for the surface finish. This is essential in order to achieve a good finished appearance, otherwise any defects will appear magnified, particularly by a high-gloss finish.

The grade of abrasive paper used will depend on the species of timber and the surface finish to be applied. Softwood requiring a painted finish requires a grit grade of between 80 and 100 – any coarser and the scratches would show through the paint, and any finer would clog and not remove the machine marks.

Hardwood timber with a clear finish (polish or varnish) requires working through several grades of paper. Starting with coarse to remove defects, work through to finer papers to gradually lose the deeper scratches and provide a very fine finish suitable for polishing. A typical grade to start with is 80, then 100, 120, 150 and 180 grit grade.

The sanding can be carried out by hand on narrow surfaces where a sander would topple and ruin the surface. On wider surfaces, a random orbital sander can be used. This type of sander produces a far superior finish than a traditional orbital sander and does not leave discernible circular (orbital) scratches. A belt sander is very good where fast stock removal is required. Belt sanders require significant skill to use as they can very easily damage the finished work, '**dubbing off**' the work or causing '**dishing**' in the surface of the timber if not controlled properly. Belt sanders with a framed base are superior and minimise this risk. The frame around the revolving belt prevents contact with the surface and the height can be adjusted to control the depth of cut.

▲ Figure 6.67 Orbital sander

▲ Figure 6.68 Belt sander

KEY TERMS

Dubbing off: where the face or edge of the timber has been bevelled as a result of not maintaining direct contact with the surface while it was planed or sanded.

Dishing: where the face or edge of the timber has been scooped out by the base of a sander as a result of not maintaining an even pressure, or sanding in one place for too long.

ACTIVITY

Carry out research into open and closed coat abrasives, and state which type is suitable for papering up softwood. Explain the reasons for your choice.

All rebates and moulding profiles will also need papering up. The rebates can be papered up using a cork rubber with a sharp corner to make sure that all the machine marks are removed and the internal corner area is not missed. Moulded profiles need to be papered up using a purpose-made cork or timber 'rubber' block, to the opposite profile of the moulding. A supple abrasive sheet is laid face down over the moulded profile and the shaped rubber is pushed into it. This ensures that the abrasive fits the profile and does not lose its sharpness while the machine marks are being removed.

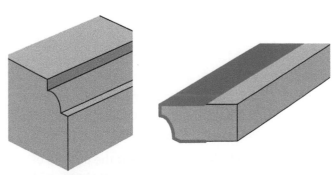
▲ Figure 6.69 Abrasive paper wrapped around a moulded rubber block shaped to the opposite profile of the moulding

▲ Figure 6.66 Random orbital sander

Assembling curved joinery

Successful assembly depends on good preparation. Everything required should be collected and set up ready for assembly. Bench bearers will need to be levelled, the job dry fitted and laid on the bearers, and cramps with any necessary protection blocks should be set to the required size. The adhesive also needs to be to hand, along with a glue brush. A rule or squaring rod is required for carrying out the quality check (checking the diagonals for square), along with any stretchers required. Nails or screws and fixing tools will also be required for fixing the stretchers. Gathering these things in advance means they are available when the adhesive is **going off** and time is tight.

KEY TERM

Going off: the part-drying of an adhesive.

When all the equipment required has been collected, the assembly process can start. Adhesive is applied, the component parts assembled and the cramps applied. The frame should then be checked for square and wind before any wedges are driven.

Adhesives

The type of adhesive used will be dependent on the specification and the final position of the finished item. More information about adhesives can be found on page 407 of *Level 2 Site Carpentry and Architectural Joinery*. The most common adhesive used is PVA as it is relatively inexpensive and meets most needs.

ACTIVITY

Carry out research into cyanoacrylate, polyurethane, PVA and synthetic resin adhesives. Produce a poster with a table showing the pros and cons for each type of adhesive and suitable uses for them.

▲ Figure 6.70 Semi-circular headed box frame part assembled

INDUSTRY TIP

Have a clean, damp cloth available to clear any excess adhesive off the work.

It is good planning, where possible, to time assembly to be just before lunch or at the end of the day. This allows the adhesive to dry thoroughly before the cramps are removed and faces are cleaned, minimising the loss of production time.

Cramping techniques for assembling curved work

Sometimes the shape of a job will require special cramping techniques.

Ratchet, band/web/strap cramps

A useful addition to any joiner's toolbox, web is available in various lengths and is made from a strong nylon. The cramp can be used to pull up many types of odd shape when assembling. It can apply pressure in ways that traditional cramps cannot and can reduce the number of cramps required as they can surround the work.

▲ Figure 6.71 Band cramp

▲ Figure 6.72 Band cramps in use

Joiner's dogs

These are invaluable when assembling curved work. They can be driven into the back of a frame to pull the joints up. As they leave a hole, they are not suitable for face work unless it is to be painted (because then the holes can be filled).

▲ Figure 6.73 Joiner's dogs in use

Bar cramps

Any straight parts of the curved joinery can be assembled with sash or bar cramps, as described in *Level 2 Site Carpentry and Architectural Joinery*, pages 427–428.

Cleaning off assembled work

Once the adhesive has dried, the cramps can be removed and any wedges trimmed off. It is good practice to have a pair of soft bearers (covered in carpet or felt) to replace the cramping-up bearers, to minimise the possibility of surface scratching during the papering up. Traditionally, the back side of the frame is cleaned up first and the face last. The shoulders of joints and the heading joints should be made flush (flushed) with a sharp and finely set smoothing plane before being sanded.

Some hardwoods, particularly those of tropical origin, feature interlocking grain. Practically this means that the grain will lift and tear in whichever direction it is planed. If **interlocking grain** is encountered when using the smoothing plane, a cabinet scraper can be used. The cutting action scrapes and will not lift or tear up the grain as with planning. A coarse grit grade abrasive paper will also help clean out the tearing up of the grain. Always look to see which way the grain is running before planing. This will eliminate most grain tearing.

KEY TERM

Interlocking grain: also known as 'refractory grain'; grain that spirals around the axis of a tree but reverses its direction regularly, causing a poor finish whichever way it is planed.

HEALTH AND SAFETY

A cabinet scraper can generate a lot of friction-generated heat, so wear gripper gloves with latex-coated palms, or similar, to protect your fingers from being burned.

The scraper is a piece of silver steel between 16 and 18 gauge thick. Each of the four edges can have a burr turned and, when correctly sharpened, will produce a shaving rather than a scraping. Sharpening a scraper is a skilled task and can be carried out as shown in Figures 6.74–6.77.

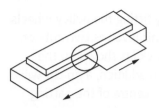

▲ Figure 6.74 Re-squaring the edge

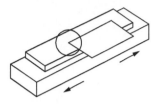

This is the cutting edge

▲ Figure 6.75 Removing the burrs

Edge square

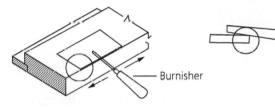

— Burnisher

▲ Figure 6.76 Drawing out the burr

Edge now ready for use

▲ Figure 6.77 Turning the burr

ACTIVITY

Following the method given, re-sharpen a scraper and try it out on a piece of hardwood to see if you can produce shavings.

▲ Figure 6.78 A cabinet scraper in use

When the job is complete, the delivery address and any reference number should be placed on the back for identification purposes. It should then be stored in a protected, dry and stable environment until it is ready to be delivered.

Sanding machinery

Most joinery manufacturers will have some form of sanding machine, either a hand pad belt sander, or a wide belt or drum speed sander. These can be used to save considerable hand work in the cleaning-off process.

▲ Figure 6.79 Wide belt sander

▲ Figure 6.80 Hand pad belt sander

INDUSTRY TIP

An orbital sander should not be used to level joints as it will not flatten the face and the undulations will show after the surface finish has been applied.

Typical assembly method for curved joinery

This section outlines, step by step, the assembly of a semi-circular headed box frame.

1 Fit up and assemble the top sash over the full-size rod. The semi-circular headed sash and frame head then need to be built up directly over the rod to ensure conformity of shape and that it is constructed on a flat surface.

▲ Figure 6.81 Top sash fitted up

2 Assemble the sashes, flush and leave to one side to fit to the box frame later.

3 Assemble the head to create one jointed curved component. It will need to have a stretcher tacked across the open end to maintain its size.

INDUSTRY TIP

Traditionally, where a number of box frames are being manufactured, the sashes would be assembled first. These would then be returned to the mill to be sanded and have the grooves ploughed for the sash cord. Meanwhile, the joiner would fit the pulley wheels and cut the pockets to the pulley stiles. By the time the first box frame was ready for assembly, the sashes would have been returned. These could then be shot into each box frame as it was assembled.

▲ Figure 6.82 Circular head built up

4 Prepare the pulley stiles by fitting the pulley wheels and cutting the pockets. This involves two cuts on the face and two on the back (as shown on the marked-out component). In addition, a rip cut will need to be made down the centre of the parting bead groove between the top and bottom cuts.

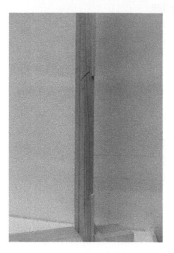

▲ Figure 6.83 Pocket cuts and removal of tongue to the inside face of the pulley stile

INDUSTRY TIP

Remember to remove the tongue along the full length of the pocket, otherwise it will be trapped and the weights will not be accessible.

5 Glue and wedge the pulley stiles into the cill and check for wind.

▲ Figure 6.84 Pulley stiles glued and wedged into the cill

6 Assemble the frame (with the inside of the window facing downwards) across a pair of levelled bearers on the bench, directly over the rod.

7 Connect the shaped head by screwing through the back of the pulley stiles into the shaped head. The diagonals will need to be checked for square, then braced temporarily.

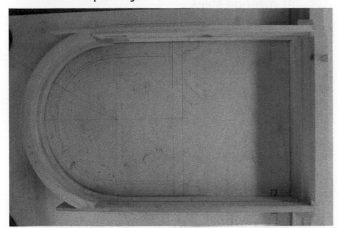

▲ Figure 6.85 Frame assembled and ready for the fitting of the linings

8 Cut and fix the inside linings, flush the joints and sand the surface. Turn the frame over and repeat this for the outside linings.

9 **Shoot in** the top sash and fit the parting bead. Knock through the pockets from the back of the frame. Shoot in the bottom sash and fix the staff beads.

KEY TERM

Shoot in: the process of fitting a door or sash to an opening with a parallel gap that allows for fitting and finish clearance. Allow 2–3 mm for a painted varnish and 1–2 mm for a polished finish.

▲ Figure 6.86 Completed box frame

Case Study: Eleanor

▲ Figure 6.87 Eleanor

Eleanor's supervisor has asked her to make a replacement semi-circular headed box frame. Eleanor has never made one before but remembers learning about them in college. Her supervisor, Clare, has not made one either, but is keen to encourage Eleanor to develop. Clare has said that Eleanor can make the frame in the way she feels is most appropriate so long as it is an accepted form of construction, and no special equipment needs to be bought to make it.

Put yourself in Eleanor's position. Remember that, without planning and research, you'll encounter problems later, so thorough knowledge is required before any setting out takes place. The following list gives some of the key knowledge areas required before you start. See how you would get on.

- Research three different ways of constructing the head to the box frame.
- What special provision is made to prevent the crown of the sash from hitting the underside of the head, to minimise the chance of the glass in the top sash breaking?
- Draw the construction details of the top sash showing the joint you would use. Explain why you have chosen this joint.

Test your knowledge

1 What is the term given to the process of collecting dimensions from site for the manufacture of a shaped window?

A A check   C A survey

B An audit   D An appraisal

2 How is the shape of an existing segmental opening best obtained?

A Drawing a sketch

B Making a templet

C Taking a photograph

D Using a CAD program

3 What piece of equipment is used to draw a large radius?

A Flexi-curve

B French curve

C Trammel and beam

D Springbow compass

4 Which Gothic arch is the most pointed in shape?

A Drop

B Lancet

C Segmental

D Equilateral

5 Which method is most practical when drawing an ellipse in the workshop?

A Conic section

B Pin and string

C Auxiliary circle

D Trammel frame

6 What is the term for a line bounding a circle?

A Chord   C Tangent

B Radius   D Circumference

7 Which is a three-centred arch closest to in shape?

A Elliptical   C Drop Gothic

B Segmental   D Semi-circular

8 Where are false tenons commonly used?

A At heading joints

B At crown joints

C At meeting stiles

D At meeting rails

9 What document outlines the component parts of a product and their sizes?

A Specification

B Cutting list

C Invoice

D Quote

10 Who should any discrepancies discovered during a site survey be reported to?

A Estimator   C Supervisor

B Site agent   D Marker out

11 A State two arch shapes drawn from one centre.

B Describe how the focal points of a true ellipse are obtained.

12 A List the equipment required when setting out a circular-headed door frame.

B Describe the purpose of templets.

C State the required properties of a templet.

13 A Describe how short grain can be minimised when constructing curved joinery.

B Describe built-up construction in relation to curved joinery.

14 A What thickness of laminate is required when producing a curved head in mahogany for a door frame with a radius of 1500 mm?

B What is the minimum practical thickness of laminate possible when using a thicknessing machine?

15 A one-off sapele traditional box frame window has been ordered. Only the approximate size has been obtained for pricing purposes. An operative has been asked to produce the window to match the existing windows. The timber has been ordered in and delivered. Discuss the planning and machining requirements to manufacture the window.

CHAPTER 7

MANUFACTURING STAIRS WITH TURNS

INTRODUCTION

Manufacturing stairs with turns will provide you with some of the most enjoyable and rewarding experiences of your joinery career. Stairs with turns are normally the responsibility of the most experienced joiners in the workshop, so when you are given one to make you will know you are well thought of by your supervisor.

Stairs with turns can be as simple as two flights connected by a quarter- or half-space landing, or as complex as a geometrical stair. This chapter will prepare you to take on most types of turning stair that you may encounter.

It is often not the ability of a joiner that poses the challenge when working on complex stairs but actually finding an opportunity to work on them, as they form only a very small proportion of staircases manufactured.

The construction of straight flights and stair terminology were covered in *Level 2 Site Carpentry and Architectural Joinery*, in Chapters 4 and 8. We recommend that you read those chapters to refresh your memory before reading on.

LEARNING OBJECTIVES

By reading this chapter, you will learn:
1 types of stairs with turns
2 types of string construction
3 marking out cut string stairs
4 assembling a cut string stair
5 shaped entry steps
6 winder step construction
7 handrails
8 newels
9 balusters
10 setting out stairs with turns
11 marking out components
12 setting out, surface developments and constructing geometrical stairs.

1 TYPES OF STAIRS WITH TURNS

The planned arrangement of a staircase will have been determined by several considerations, including:
● the need to divide the stair into more than one flight
● the need to change direction due to the shape of the building
● the demands of floor space
● the space available for the stair to fit in.

Stairs are often named or described according to their plan shape or string construction type, and they fall into two classes:
1 newel stairs
2 non-newel stairs (commonly called geometrical).

Newels are generally used at both the bottom and the top of the stair. They allow the string and the handrail to be jointed into them, and provide a means of supporting the stair by fixing it to the landing trimmer. (See Chapter 4 of *Level 2 Site Carpentry and Architectural Joinery* and Chapter 8 of this book for further information on how stairs are fixed.)

Turning stair arrangements

Stairs will be positioned as shown on the architect's drawings. In small residential properties, consideration is often given to minimising the space the stairs take up.

In the examples illustrated in Figures 7.1–7.4, each set of stairs contains 13 risers; this gives an idea of how much floor space each stair type will take. In examples A, B and C, note that adding winders reduces the **total going** required. Example D shows winders at both ends of the stair and allows access to the stairs where the going is restricted by walls.

Stairs can turn a corner by using either **winders** or a landing. The turn is described by the method used – for example, 'quarter-space landing' (as in example A) or 'quarter-space of three tapered steps' (as in example C).

KEY TERMS

Total going: the horizontal distance between the first and last riser in a straight flight.

Winders: tapered steps used to save space by allowing extra risers to be incorporated. They generally turn through 90° or 180°.

ACTIVITY

Carry out online research to find the maximum number of risers allowed in one flight.

INDUSTRY TIP

It is important to recognise that, while the use of winders can reduce the total going, this has to be balanced with the complexity of the manufacture and fitting, and therefore the cost.

Figures 7.1–7.4 show examples of how much space a stair with 13 risers will occupy.

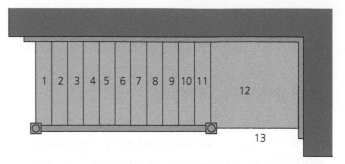

▲ Figure 7.1 Example A: quarter-space landing

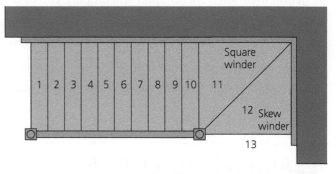

▲ Figure 7.2 Example B: quarter-space of two tapered steps

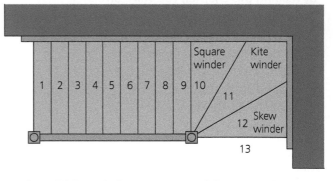

▲ Figure 7.3 Example C: quarter-space of three tapered steps

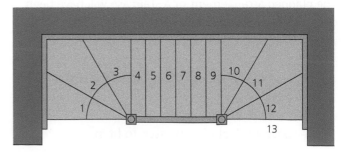

▲ Figure 7.4 Example D: two quarter-space turns each with three tapered steps

Newel stairs

Newels are an integral part for the following stair types:

- straight flight
- quarter-space
- half-space
- winding.

Straight-flight stairs

These have been dealt with in *Level 2 Diploma in Site Carpentry and Bench Joinery*. Example A in Figure 7.1 shows a straight flight with a quarter-space landing. This is simply a straight flight with an additional riser turned through 90° and should present few difficulties to manufacture. This is a very common turning stair type that allows access to the first-floor landing.

Quarter-space stairs

The addition of a quarter-space landing at the top of the flight saves one going space at the bottom of the stairs and gives more circulation space at the foot. The Building Regulations state that the landing depth must be at least the width of the flight (Approved Document K Regulation 1.20).

Half-space stairs

A half-space landing can take two forms:

1 Dog-leg stair: where width is restricted, a dog-leg stair allows both the lower and upper flight to be as wide as possible as the string of the upper flight is directly over the lower, both being jointed centrally to the newel.

2 Open-well stair: if the opening is wider than the combined width of the two required flights, the gap between them is termed the **well**. This type of stair is sometimes referred to as an 'open-newel stair' because two newels are used at landing levels.

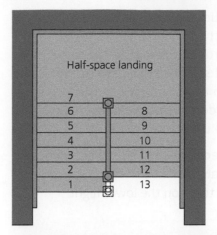

▲ Figure 7.5 Dog-leg stair with half-space landing

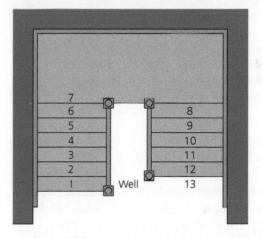

▲ Figure 7.6 Open-well stair with half-space landing

Care must be taken when designing half-space stairs, to ensure that there is enough headroom. Remember that the landing trimmer will run across the trimmed opening for the stair and therefore can cause a headroom obstruction. The Building Regulations require a minimum of 2000 mm headroom (Approved Document K Regulation 1.12). This means that there will need to be about 11 risers above the first step, depending on the step rise.

Winding stairs

Where winders are used, it is most common to have two or three per quarter turn; in half a turn, four or six winders are common. (Any number is possible as long as the stair conforms to the Building Regulations.)

In straight flights, the riser face is positioned at the centre of the newel. This is not possible with winding risers as the Building Regulations (Approved

Document K Regulations 1.25–29) state that the minimum going must be 50 mm. This spreads the riser faces out around the face of the newel. The minimum newel size that will allow for this is 95 mm × 95 mm. To avoid complex construction problems where winders turn around half a turn, a double newel can be used.

A tongue and trenched joint will be used between wall strings. As stairs are always fixed from the top down, the tongue will need to be on the lower flight.

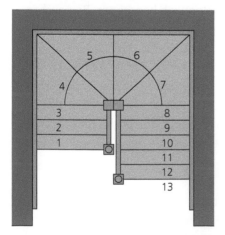

▲ Figure 7.7 Half-space of four winders with a double newel

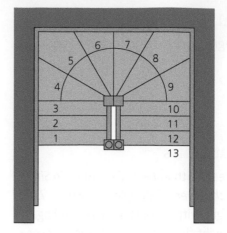

▲ Figure 7.8 Half-space of six winders with a double newel

Non-newel stairs

This type of stair is often termed a '**geometrical stair**' and covers a multitude of varieties. The newels are replaced by a continuous string running from the bottom to the top of the flight, generally with some part of it being curved. The construction of this type of stair is far more complex, with the setting out requiring a degree of geometrical development to determine the 'stretchout' (true shape) of the string. This type of stair is used only for prestigious work as the time taken to construct it, due to its complexity, means that it is very expensive to produce.

Two examples are shown in Figures 7.9 and 7.10. The first (Figure 7.9) has a continuous well string but straight wall strings. The complexity here over and above winding stairs is limited to the well string, which replaces the newel as a means of turning the steps. The wall strings will be constructed in the same way as for winding stairs.

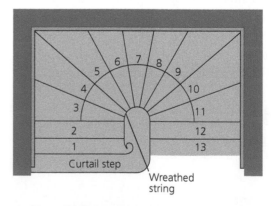

▲ Figure 7.9 Geometrical stair with a continuous well string

The second illustration (Figure 7.10) shows both the wall and the well strings radiating from a centre point. Each tapered step would be the same size and shape. Both strings here require building up.

Stair terms

Figure 7.11 and Table 7.1 describe terms and components you are likely to come across while fitting stairs.

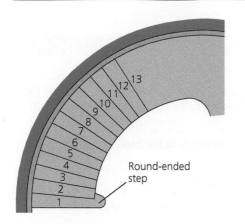

▲ Figure 7.10 Circular or helical stair

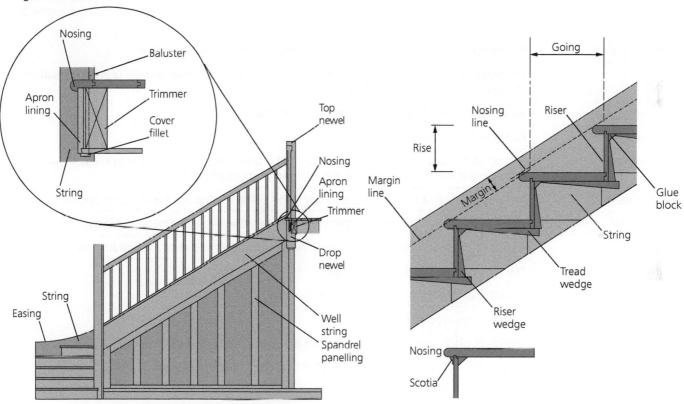

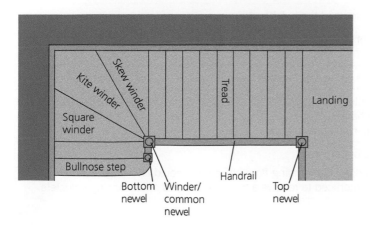

▲ Figure 7.11 Stair terms

KEY TERM

Geometrical stair: a staircase that turns and rises without the use of landings. The stair strings are formed to rise and turn and are also known as 'wreathed strings'.

▼ Table 7.1 Common stair terminology

Term	Definition
Apron lining	A thin board (timber or MDF) that faces the sawn joists in the stairwell.
Balusters	Vertical components fixed between the string (or its capping) and the underside of the handrail on the stair, and the handrail and nosing on the landing. Provide guarding to the open side of the stairs and landing areas.
Baluster spacers	Thin spacers that fill the groove in the string capping and the underside of the handrail between the balusters to ensure consistent gaps.
Balustrade	A collective term for the area between the handrail, string and newels.
Brackets	1 Timber used to provide interconnection between the carriage and the underside of the steps. 2 A decorative step end to cut string staircases.
Bulkhead	The intersection between the wall and the trimmer above a staircase. The headroom is measured to this.
Bull-nose step	A common entry step with a radiused end, found at the beginning of a flight.
Cap	The name given to the shaped top of the newel post. It can be worked on solid (a finial) or fixed on. The detail can also be applied to the newel drop.
Carriage piece	An inclined bearer that acts as a central support to wide stairs.
Commode step	A step with a curved riser.
Cover fillet	A small section of timber used to cover the joint between two parts of the structure, e.g. between the spandrel and the string, or between the apron and the plaster finish.
Curtail step	An entry step with a scrolled end, generally found with a handrail scroll and cage of balusters over.
Drop newel	The portion of newel that projects below the landing, often with a decorative end (finial).
Easing	A gentle radiused section used to connect two inclinations on a level to a raking string or handrail.
Finial	The term given to the decorative shaped end to newel posts.
Flier	A traditional term for a parallel tread in a staircase.
Flight	A set of steps running from floor to floor.
Glue blocks	Triangular blocks used to reinforce the joint between the tread and the riser.
Going	1 Step: the horizontal distance between the face of one riser and the face of the next. 2 Total: the distance between the first and last riser face in the flight.
Handrail	A rail positioned at waist height to provide support and guarding to the stair. It may be required on one or two sides, depending on the width of the flight.
Headroom	The vertical distance between the step and the bulkhead above.
Landing	A platform at the top of the stairs, or a resting place between one or more flights. May be quarter- or half-space.
Margin	The distance between the top of the string and the intersection of the tread and riser face.
Newel post	Top, bottom and common (to two interconnecting flights) newels are stout vertical components used at the top and bottom and at any change of direction. They provide a means of jointing the string and handrail and provide support to the flight.
Nosing	The moulded projecting front edge of the tread. The top reduced-width tread that adjoins the landing floor boarding. It could also run around the stairwell, in which case it would be morticed to receive a tenon on the end of the landing balusters.

➡

▼ Table 7.1 Common stair terminology (continued)

Term	Definition
Nosing line	An imaginary line that touches each nosing in the flight. It is referred to in the Building Regulations but has no practical function.
Pitch	The angle of the staircase.
Pitch line	An imaginary line that touches all the nosings in the flight.
Rise	1 Step: the vertical distance between the top of one tread and the top of the next. 2 Total: the vertical distance between finished floor line (FFL) at the bottom of the stairs and FFL at the top of the stairs.
Riser	The vertical component of a step.
Scotia	The shape of a small moulding used on the underside of a nosing in prestigious stair construction.
Shaped bottom steps	These are intended to give the user more access to the bottom of the flight by providing a decoratively shaped bottom step that projects beyond the face of the bottom newel. Typical examples are bull-nose and curtail.
Spandrel	The triangular area below the string on the first flight of a staircase. It can be panelled in or made up of a series of doors, allowing the area below the stair to be used for storage.
Splayed step	An entry step with a splayed end.
Staircase	This is the complete stair structure including the flight, landings and balustrade.
Stairwell	This is the opening formed in the floor layout to accommodate the flight (or flights) of stairs.
Step	The name given to an assembled tread and riser; parallel steps are often termed 'fliers'.
Storey rod	Traditionally, a lath of timber onto which the position of the landings was marked on site. The rod was taken back to the joinery shop and divided up to find the step rise for the stair. This is very seldom used in modern practice but the term is still used and shown on drawings.
String capping	This component is planted on the top of the string to increase its thickness, allowing balusters to be fixed where the balusters are thicker than the string.
Strings	1 Wall: the string fixed against the wall. 2 Well: the string on the open side of the staircase, sometimes called an 'open' or 'outer string'. 3 Geometrical: a string that runs continuously from the top to the bottom of a flight with part of it curved on plan. Used to change direction on stairs with no newels.
Tread	The horizontal surface of a step.
Well	The gap between two strings on a turning stair.
Winders	1 Square: the first of the two or three winders encountered in a quarter turn. The nosing is square to the wall string. 2 Kite: the kite-shaped second of the three winders encountered on a quarter turn. 3 Skew: the last of the three winders encountered in a quarter turn of two or three winders. The nosing is at a skewed angle to the wall string.
Wreathed handrail	Normally made in 90° sections, these are handrails that follow the plan shape of the string below. The handrail rises and turns at the same time.
Wreathed string	Found on geometrical stairs, these are curved strings that replace newels at a change of direction.

ACTIVITY

1 Draw two finial details commonly found on newels.
2 Research online to find three baluster types.

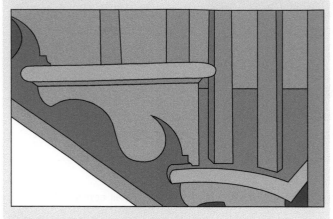

▲ Figure 7.12 Decorative step bracket

Sketch a detail of the next ornamental riser bracket you come across when visiting a building of interest. Research the best way of producing them, using the internet and other sources such as your trainer/lecturer, and explain how they would be manufactured. Then make a templet in MDF for their manufacture.

Component sizes

Table 7.2 shows common component section sizes in a typical residential staircase. They may vary depending on the architect's details.

▼ Table 7.2 Common component section sizes in a typical residential staircase

Component	Size	Material
Treads and winders	19–32 mm thick, depending on requirements. The larger the stair, the thicker the material required. Cut string stairs require thicker treads.	European whitewood/redwood/MDF Hardwood as specified by the architect or client
Risers	9–18 mm thick	Ply/MDF Hardwood
Newels	70–120 mm square, depending on design and winder requirements	European whitewood/redwood Hardwood
Handrails	70 mm × 45 mm, 95 mm × 45 mm, depending on section	
Balusters	22–45 mm square, depending whether plain or turned	
Strings	26–45 mm thick, depending on type	

Machining components to size

Refer to Chapter 3 for information about machining operations.

HEALTH AND SAFETY

The strings of a staircase are long and can be very heavy, so you need to take care when manoeuvring them during the preparation and manufacturing processes. Mishandling could injure you and/or your colleagues, and damage the components and the surrounding area.

Produce a cutting list for a straight flight of stairs with five steps. The flight has a rise of 180 mm and a going of 240 mm, with a newel at the top and bottom.

② TYPES OF STRING CONSTRUCTION

Chapter 7 of *Level 2 Diploma in Site Carpentry and Bench Joinery* concentrated on closed string stair construction, as shown in Figure 7.13. But as stairs become more decorative, cut string stair construction is used more often for the well string. This is where the top edge of the string is cut to the shape of the step profile. The finished effect is designed to be aesthetically pleasing (good to look at).

▲ Figure 7.13 Closed string stair

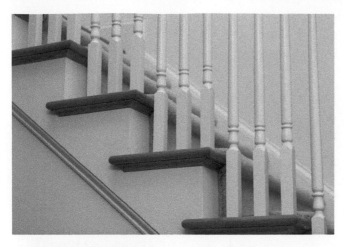

▲ Figure 7.14 Cut string stair

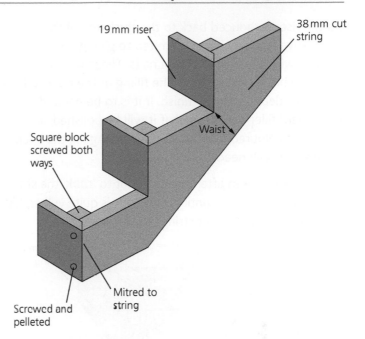

▲ Figure 7.15 Cut and mitred string construction

Cut string construction

Considerable work is involved in this type of string construction. Much of it relies on having a high standard of hand-tool skills and requires a great deal of fitting up on the bench. Badly fitting joints will be visible and spoil the appearance of the finished product, especially on a polished hardwood stair.

Cutting away the step profile to the top edge of the string will obviously reduce its strength. Thicker strings are required in order to compensate for this, typically between 38 and 45 mm thick. The amount of parallel timber below the step profile (known as the 'waist') should be at least 125 mm. Treads are also generally thicker than on closed string stairs, typically 28 to 35 mm thick. There are two types of cut string stair:

1 cut and mitred
2 cut, mitred and bracketed.

Cut and mitred

In a cut and mitred string, the top edge is cut to the profile of the step and the riser is jointed flush with the outside edge of the string. A square shoulder is produced on the riser line of the string in order to provide a positive location for the riser. This joint can then be glued, screwed and pelleted. The back of the joint can be reinforced with triangular glue blocks or square blocks screwed in both directions.

> ### INDUSTRY TIP
>
> If the stair is to be painted and the natural features of the timber will not be seen, make the risers out of manufactured board such as ply or MDF, as this will remain stable, the mitre line will be less likely to open up and it will be more economical.

IMPROVE YOUR ENGLISH

Research two types of hardwood that could be used to construct a stair where the client requests a polished finish. Write a short report explaining reasons for your choice.

Cut, mitred and bracketed

The cut, mitred and bracketed method is a more decorative version of the plain cut and mitred string. In this type of cut string construction, the strings are easier to cut as the riser that is cut on the step profile to the string remains square. This leaves the riser lapping the face of the string and projecting by the thickness of the decorative bracket. This projecting part is mitred to the end of the bracket.

As the riser is reduced back to the thickness of the decorative bracket, fixing is limited to gluing and neatly pinning with three or four oval nails. These will be punched below the surface. The filling of the punched holes will depend on the finish. If it is to be painted, a standard filler can be used. If it will be polished, a matching coloured filler or wax is used. Again, the back of this joint will need to be reinforced.

Care must be taken after assembly not to '**rack**' the stair during transportation: until it is fixed, it is quite vulnerable and the joints on the cut string could fracture.

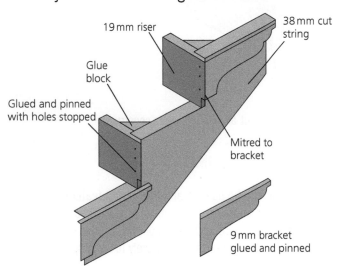

▲ Figure 7.16 Cut, mitred and bracketed string construction

19 mm riser

Glue block

Glued and pinned with holes stopped

38 mm cut string

Mitred to bracket

9 mm bracket glued and pinned

INDUSTRY TIP

The timber used for this type of stair needs to be **second seasoned** to ensure it is stable before use. Considerable problems will be faced if the strings, risers and brackets start to 'curl' before assembly.

KEY TERMS

Rack: a stair or other manufactured component that has been pushed out of shape, allowing the glue line to fracture and the joints to weaken and break.

Second seasoning: where timber is sawn to the sizes required on the cutting list and left '**in stick**' for as many days as time will allow. The freshly sawn faces will acclimatise to the atmospheric conditions in the joiners' shop, minimising the amount the timber will distort after planing.

In stick: where the timber is stacked with thin laths ('sticks') separating the boards, which allows air to circulate and the moisture content to acclimatise.

③ MARKING OUT CUT STRING STAIRS

Marking out cut string stairs is very similar to marking out a closed string stair – the main difference is that there is no margin.

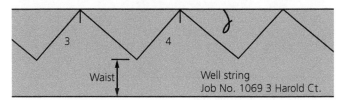

3   4

Waist

Well string
Job No. 1069 3 Harold Ct.

▲ Figure 7.17 Tread and riser lines marked out with a pitch board or steel square

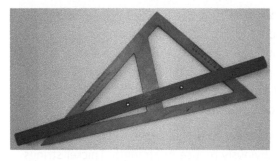

▲ Figure 7.18 Adjustable pitch board made by the author

▲ Figure 7.19 Steel square and fence made by the author

INDUSTRY TIP

To avoid breakout when routing strings, start from the bottom and work up on the left-hand string and then down on the right-hand string.

ACTIVITY

Research online any proprietary jigs suitable for routing cut string stairs. Based on your research, manufacture a half-scale version of each jig. Try them out and write a review of each one.

Shaping the string

The string can be shaped using a router jig made for the purpose. The bulk of the material can be cut away first using a jigsaw. The jig can also be used to produce the shoulder and the mitre, if the risers fit flush with the string.

HEALTH AND SAFETY

Check power tools for safe condition every time you use them. Frequent, careful checking doesn't just protect you from injury – it also shows consideration for anyone else who might use the power tool after you and helps to keep everyone on site safe. If there's a fault or damage, label the tool as unserviceable, report the problem to your line manager and put the tool securely out of service.

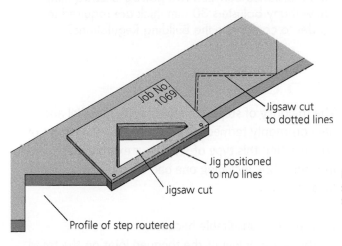

▲ Figure 7.20 Router jig for cut string stair

INDUSTRY TIP

The term 'marking out' can be abbreviated to 'm/o'.

Jointing balusters to the cut string

Cut string stairs usually have the balusters jointed into the top of the tread face. Two methods are commonly used: dovetailed and stub tenoned.

Dovetailed

This method probably provides the best interconnecting joint to the tread. Commonly, a barefaced dovetail is screwed into the end of the tread. The disadvantage is that the return nosings have to be fitted to each step, numbered and sent to the site loose so that they can be fixed after the balusters have been installed. Typically, the return nosing will be slot-screwed to the end of the tread.

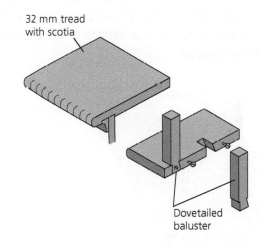

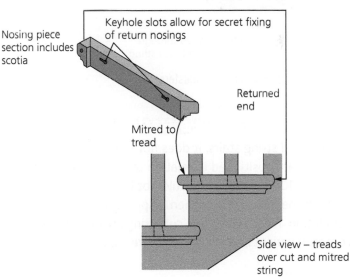

▲ Figure 7.21 Dovetail-jointed balusters

Stub tenoned

This method offers two main advantages. First, the tread can be screwed to the string during assembly. Second, the return nosings can be jointed, glued and flushed to the tread in the joiners' shop and do not have to be sent to the site loose. The return nosing can be loose-tongued or biscuit-jointed to the tread. If the stair is specified to have a painted finish, MDF is ideal for the treads as the return nosing can be worked on the solid without the need for jointing the return nosing to the end of the tread. With this jointing method, the tenon on the end of the baluster can be secured on site by skewing a screw from the underside of the tread into the end of the baluster.

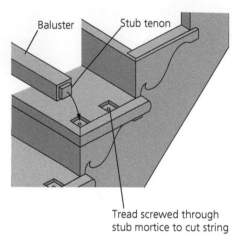

▲ Figure 7.22 Stub-tenoned balusters

Baluster arrangements

With closed string stairs, it does not matter where the balusters are positioned in relation to the riser face, as long as a 100 mm sphere would not be able to pass through the gap between them (Approved Document K Regulation 1.39a). However, cut string stairs have two common variants, as shown in Figure 7.23.

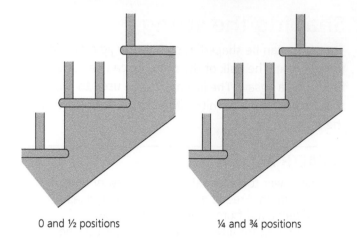

0 and ½ positions   ¼ and ¾ positions

▲ Figure 7.23 Baluster arrangements for cut string stairs

Step construction

The majority of steps in a staircase are parallel and are commonly termed **fliers**. There are many ways of constructing this type of step. Four examples are shown in Figure 7.24, including one bad example that should be avoided.

Example A is unsuitable because:
- the top shoulder of the tongued joint on the tread will open up and show a gap on the face if shrinkage occurs in the riser
- a wider tread is required
- tread shrinkage will cause the riser to split
- the back of the tread will interrupt the riser wedge from driving the face home tight.

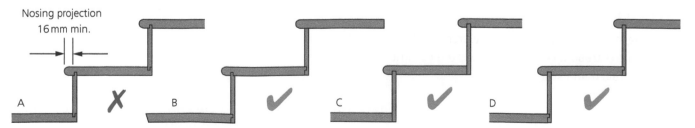

▲ Figure 7.24 Step construction

Example B is a traditional joint used between a solid timber tread and riser.

Example C is the most common method in modern construction, using a solid timber tread and an MDF or ply riser.

Example D is an improved version of example C used when jointing ply or MDF risers, but is more expensive to produce.

Where treads are made from solid timber, they may require jointing in their width. If so, it is good practice for the joint to be made at the back of the tread. In modern construction, where a painted finish is specified or the stair is to be covered in carpet, 25 mm MDF may be used. This is much more economical and stable.

All fliers will be 'boxed up' after the faces have been sanded. Boxing up is the process of assembling treads and risers, checking them for square, and glue blocking them to reinforce the joint. A minimum of three glue blocks should be used (normally made from 50 × 50 mm timber). The fliers are then stacked on a flat surface. Three or four are placed in the first layer, then on the second layer the steps are laid at 90° to the first and so on until they are all assembled. This allows the air to circulate around the faces of the steps, helps keep them stable and prevents them from **casting**.

> **KEY TERMS**
>
> **Flier:** the traditional term for parallel treads/steps.
> **Casting:** the curling and movement of timber.

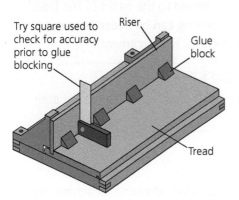

▲ Figure 7.25 A step being squared up and glue blocked

> **INDUSTRY TIP**
>
> Never nail glue blocks — the splinters produced as the nail pushes through the glue block create a gap between it and the tread and riser, reducing the adhesion between the two.

Fitting up cut string stairs

The mitred ends of cut string treads can be profiled using a router templet, as shown in Figure 7.26.

> **KEY TERM**
>
> **Dry fit:** the process of fitting joints one at a time to ensure the shoulder fits on both sides with the shoulder square or at the required angle, that the joint is not twisted, and that it is flush or parallel with the face as required.

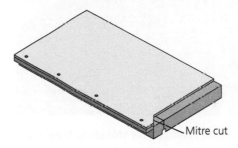

▲ Figure 7.26 Templet used to profile the mitred end of a tread

The purpose of fitting up is to ensure the joints fit, and that the stair is the correct size and shape. Cut string stairs are difficult to fit up and assemble due to the return nosing of the tread and, where used, brackets that project beyond the face of the string. As with all assembly procedures, good preparation is the key to success. The method is as follows.

1 If the stair has newels, **dry fit** them to the strings and handrail first, before any other fitting up. Draw pins can be used as a temporary means of cramping the joints (these will be replaced by dowels on site).

2 Screw a straight bearer to the far side of the bench to support the back edges of the steps.

3 Cramp the well string into the vice, level and at a height equal to the bearers (to ensure the stair is not assembled in wind).

4 Secure the well string at both ends to prevent it from slipping during the assembly operation.

5 Fit the steps to the cut string one at a time, and label them by riser number on the back of the riser. The bottom and top steps will need to be fitted up with the newels on them to ensure they fit on site.

6 Once all the fitting up has been carried out, the parts can be disassembled and all the faces that will be inaccessible after assembly cleaned up with abrasive paper, ready for the surface finish (refer back to page 323 in Chapter 6 for more information about this process).

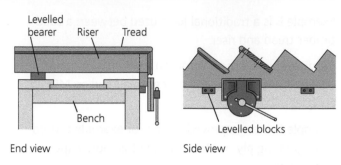

▲ Figure 7.28 Methods of fitting up steps and assembling a cut string stair

▲ Figure 7.27 Draw pin used to temporarily hold string to newels when fitting up

ACTIVITY

Research online to see how many different methods of jointing risers to cut strings you can find. Manufacture a scale version of each type of joint and evaluate the effectiveness and ease of construction of at least two types. You could do this by drawing up a table with columns listing the advantages and disadvantages of each method. Use this information to evaluate which method you think works best. Explain your conclusion.

4 ASSEMBLING A CUT STRING STAIR

A cut string stair is simply reassembled as it was during the fitting-up stage (without the handrail or newels attached), but this time using adhesive and fixings.

The tread is screwed to the string through the stub mortices, and the riser face is pinned with ovals/lost heads or screwed and pelleted. The stair should be left to dry thoroughly overnight. The next day, the joints can be flushed and sanded as required.

The wall string of the stair can now be cramped and wedged in a similar fashion to a closed string stair, but additional care is needed as a packing bearer needs to be cut to the profile of the cut string, slightly thicker than the nosing projection.

Lastly, the risers are screwed to the backs of the treads, and additional glue blocking can be added between the cut string and the steps for reinforcement.

HEALTH AND SAFETY

Write a risk assessment covering all aspects of stair assembly.

Quality checks

For most joinery items, include checking for square, wind, size and shape. For a flight of stairs to be square, the ends of the steps must be cut square. When both strings of the flight are assembled, the front edge of the nosings can be sighted for wind. The size and shape of any winders can be checked during the fitting-up stage.

Preparation for delivery

The stair should be adequately protected with bubble wrap and laid flat ready for delivery. The delivery address and contract number should be clearly marked on the wrapping. Remember that the top riser, the top nosing, the newels, the handrail, the balustrades, any winders and any shaped bottom steps are sent to site loose. These should also be well protected and labelled to ensure they are delivered to the correct address.

⑤ SHAPED ENTRY STEPS

Entry steps allow better access to the bottom of the flight and their use moves the newel at least one step further up the flight, giving a more open feel at the bottom of the stairs. There are several commonly used entry step shapes, including those described below.

Bull-nose

Probably the most common entry step shape, the bull-nose step has a 90° radiused end. When drawing this step, it is important to have at least 50 mm of straight riser before it comes into contact with the newel. A common mistake is to have the springing line on the face of the newel, which can look awkward.

Traditionally, the riser of this step would be reduced back to a veneer thickness around the shaped portion, and secured in place with a pair of folding wedges. In modern construction and in mass production, they are generally made of laminated ply. Once the former is made up, they can be produced very economically, saving much handwork.

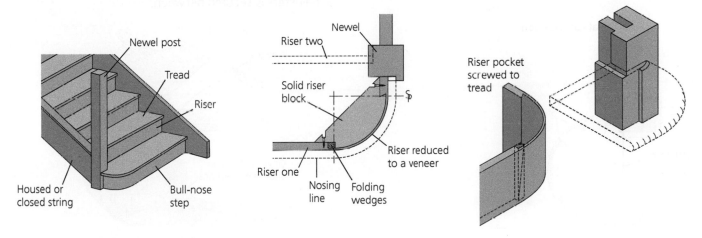

▲ Figure 7.29 Construction of a traditional bull-nose step

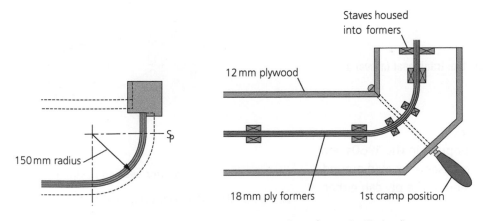

▲ Figure 7.30 Laminated bull-nose riser construction

Semi-circular ended step

Semi-circular ended steps are also known as 'round-ended' or 'D-ended' steps. This type of step turns through 180°. Again, there should be a short length of straight beyond the springing line before the tread meets the newel. Two methods of construction are shown in Figure 7.31.

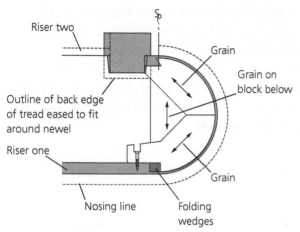

Traditional construction

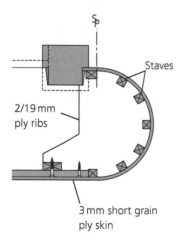

Modern construction

▲ Figure 7.31 Semi-circular ended step: traditional (above) and modern (below) construction

Splay-ended step

The splay-ended step became popular in the 1960s and is still occasionally specified. The riser is faceted around a block or pair of formers. The end of the ply can either be loose-tongued or biscuit-jointed.

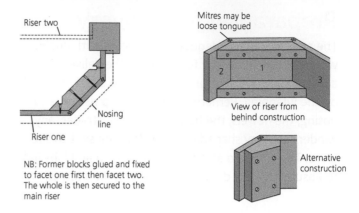

NB: Former blocks glued and fixed to facet one first then facet two. The whole is then secured to the main riser

▲ Figure 7.32 Splay-ended step construction

Curtail-ended step

Curtail-ended steps are also known as 'scroll-ended' steps. They are used only with cut strings and connect seamlessly with the well string of a geometrical stair. A handrail scroll will be formed over the scroll and a cage of balusters is secured between.

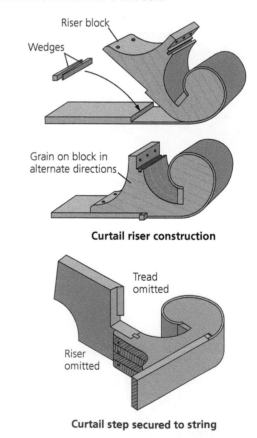

Curtail riser construction

Curtail step secured to string

▲ Figure 7.33 Curtail-ended step construction

Commode step

This is the name given to a shaped step where the riser face is continuously curved throughout its length. Any number of steps can have a commode shape. The shaped portion of the riser can be made in several ways. As with all entry steps, they are normally one-offs so a simple method of riser former construction is best.

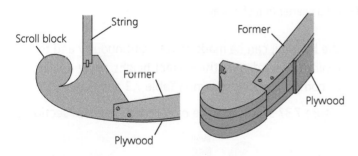

▲ Figure 7.34 Commode step construction

⑥ WINDER STEP CONSTRUCTION

As we saw earlier in this chapter, winders can take any tapered shape. Traditionally, winding steps turn through either 90° or 180°. In both cases, they are referred to as the square, kite and skew winder. Solid timber winders, whether hardwood or softwood, will require jointing in their width. Either a solid tongue-and-groove or a loose-tongued joint is used for strength. A biscuited joint should be avoided as this does not provide a continuous joint.

It is essential that the grain always runs parallel to the nosing, as this will reduce the effect of any shakes showing on the front of the nosing, and minimise the short grain present. It also looks better.

The shape of the winder can be obtained by making hardboard templets from the full-size rod or by making up skeleton frames using laths. These can determine the most economical amount of timber required to joint up the winders. Once jointed, the winders are ploughed (grooved) and nosed with the remainder of the treads and brought to shape from the plan.

Care must be taken when profiling the nosing due to its shape, and the power feed on the spindle should be used for safety.

Figure 7.36 shows how the winders and risers joint into the newel and the wall strings. Note how the nosing of the skew winder is squared off to joint into the wall string. This is to avoid an undercut scribe to the nosing. The narrow end of the kite winder is treated similarly. Note also how the newel is recessed for the risers.

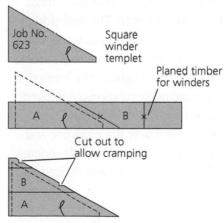

▲ Figure 7.35 Templet used to minimise waste when jointing winders

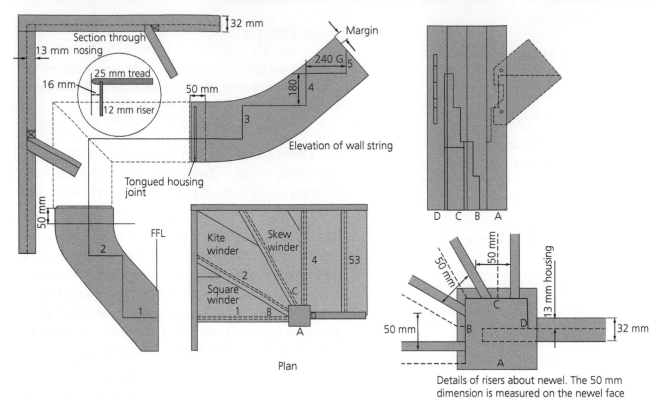

▲ Figure 7.36 Detailed setting out, marking out and construction details of a quarter turn of winders

7 HANDRAILS

Handrails must be comfortable to grip and strong enough to guard the open side of the stair and any landing. If the stair is wider than 1000 mm, Approved Document K (Regulation 1.34c) states that there must be a handrail on both sides of the stair. The wall-side handrail generally has a smaller section and is supported on handrail brackets. Sometimes a solid handrail section is specified for the wall side. The height of the handrail is also determined by the Building Regulations (Approved Document K Regulations 1.34–36). In domestic dwellings, it should be 900 mm over the pitch line and 1000 mm above the landing. In other types of building, it is higher, and the Regulations should be consulted for each staircase, to ensure conformity.

Continuous handrails

Continuous handrailing is required for geometrical or concrete stairs. With concrete stairs, the handrail is supported by iron balusters bonded into the concrete steps. At the top, an iron core rail connects the upper ends of the balusters. A wooden or plastic wrap handrail will finish this off. Where a wooden handrail is used,

the handrail can be made to run continuously and to maintain guarding at the correct height, using a range of components, as shown in Table 7.3 and Figure 7.38.

Figure 7.37 shows a range of common handrail sections.

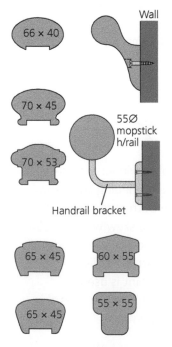

▲ Figure 7.37 Handrail sections (measurements in mm)

▼ Table 7.3 Common continuous handrailing components

Type	Purpose
Easings (ramps and knees)	These provide a transition between a raking handrail and a level handrail.
Swan neck	A combination of a knee and a bend, allowing the handrail to rise over the winders at the newel to ensure there is continuous support and guarding for the user.
Goose neck	A ramp combined with a mitred return to the newel, again allowing for support and guarding at a winding newel.
Level bend	This allows a handrail to turn around a corner where no newel is available.

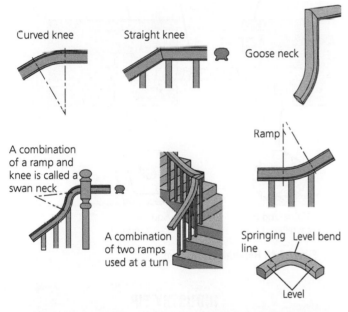

▲ Figure 7.38 Common single curvature handrailing components

Wreathed handrailing

Handrails need to rise and turn above the wreathed strings below them. A wreathed handrail will normally turn through 90° so two handrails will be required to turn through 180°. The two most common types of wreathed handrailing are **rake**-to-level and rake-to-rake. Figures 7.39 and 7.40 show examples.

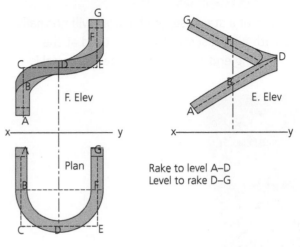

▲ Figure 7.39 Rake-to-level wreathed handrail

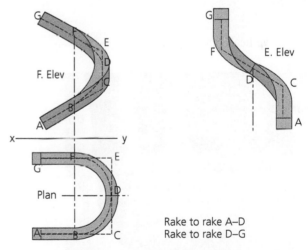

▲ Figure 7.40 Rake-to-rake wreathed handrail

In addition to wreaths, other handrail shapes may be required to maintain guarding at the correct height.

Wreathed handrailing is a very complex subject and the full details are beyond the scope of this qualification.

KEY TERM

Rake: the slope or pitch of the stairs.

349

Terminating scrolls

At the bottom of a staircase, the handrail will normally terminate with a level scroll. These are used at the lower terminating end of a handrail where a newel is not used.

▲ Figure 7.41 Vertical or monkey tail scroll

▲ Figure 7.42 Horizontal scroll

Jointing

Mortice and tenons are used to joint most handrails to newels. A handrail bolt is used where a handrail requires jointing in its length, such as at a scroll or handrail wreath. The handrail bolts are normally fitted in the joinery shop, then labelled up and disassembled. All the handrail components are sent loose to site for reassembly when the staircase has been fitted. Two dowels are normally introduced to the joint to prevent

the components from twisting if the glue joint fails. The bolt consists of three parts: a double-ended bolt, a captured nut (the square one) and a castellated nut (the tightening nut). Fitting these on site can be quite fiddly as the castellated nut has to be fitted from below and gravity is working against you.

While it is still possible to obtain these traditional bolts, these days a Zipbolt™ is more likely to be used instead. Connecting bolts are inserted and tightened on the underside of the handrail. Using an Allen key to tighten Zipbolts™ is much easier than trying to tighten the slotted nut of a traditional handrail bolt.

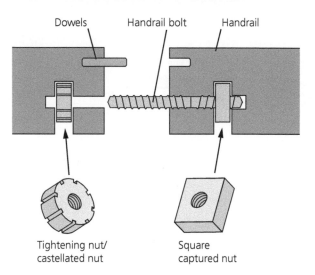

Dowels   Handrail bolt   Handrail

Tightening nut/ castellated nut   Square captured nut

▲ Figure 7.43 Handrail bolted joint

▲ Figure 7.44 Zipbolt™ used as a modern alternative to a handrail bolt

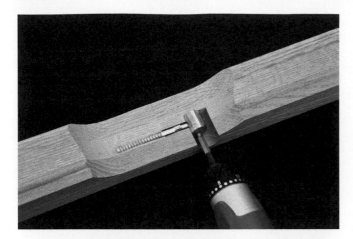

▲ Figure 7.45 Cutaway section showing how a Zipbolt™ works

ACTIVITY

Research alternative methods of jointing handrails in their length. State an advantage and disadvantage of two types and justify your preferred method.

⑧ NEWELS

As mentioned earlier in this chapter, newels are used to form a turn between straight flights of stairs. Newels in domestic flights will generally be left square in shape for reasons of economy. The tops of square newels can be finished with a **newel cap** or a solid square **turned finial**.

▲ Figure 7.46 Turned newels

▲ Figure 7.47 Square newel cap

KEY TERMS

Newel cap: a moulded solid timber cap recessed and applied to the top of a newel.

Turned finial: a turned decorative finish to the top of a post, the most common being 'acorn' shaped.

Turned newels are more expensive but will make the stair more visually appealing. Many companies now produce turned stair newels that can be inserted into a short newel. The advantage is that this can be jointed to the newel in the joiner's shop, without having to wait for it to be returned from the turner.

INDUSTRY TIP

In the best class of work, full-length newels will be sent away for turning. This ensures consistency of grain colour and characteristics.

Jointing strings and handrails to newels

Mortice and tenon joints are almost always used to joint strings or handrails to newels. On good-quality work, the tenon is centrally positioned and the shoulders housed into the face of the newel. This masks the effect of any shrinkage that may occur after fixing.

ACTIVITY

Carry out online research to see how manufacturers of mass-produced stairs connect the handrail to the newel. Explain the benefits over traditional mortice and tenon jointing.

Closed string newel jointing

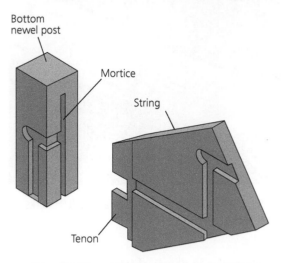

▲ Figure 7.49 Bottom newel joint

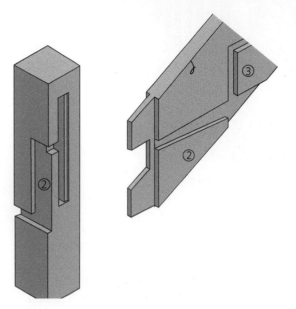

▲ Figure 7.48 Tenon shoulder housed into the newel face to hide string shrinkage

In cheaper work, the tenons on strings are generally **barefaced** on the outside face and the practice of housing in the shoulder is omitted.

KEY TERM

Barefaced: where there is a shoulder on only one side of a joint.

Figures 7.49–7.53 show common mortice and tenon arrangements for closed, cut and dog-leg stairs.

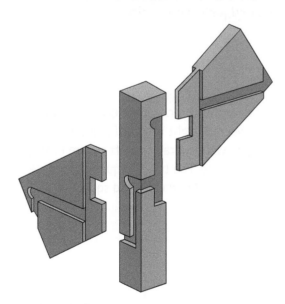

▲ Figure 7.50 Jointing detail of a common newel at a quarter-space landing

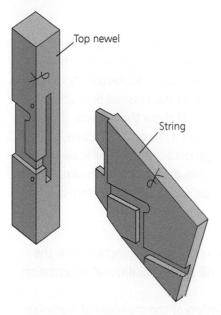

▲ Figure 7.51 Upper string to newel joint detail

Cut string stair newel jointing

Here the stair string is reduced in width as it is cut to the step profile. As such there is normally room for only a single tenon.

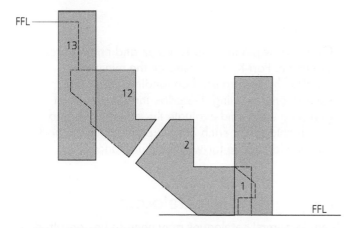

▲ Figure 7.52 Top and bottom newel to cut string jointing details

Dog-leg stair

As both strings intersect at the centre of the thickness of the newel, the handrail will be interrupted by the underside of the string of the flight above.

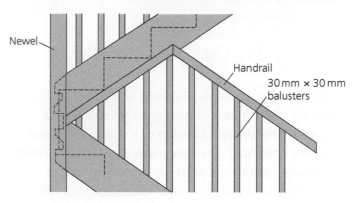

▲ Figure 7.53 String to newel jointing detail in a dog-leg stair

⑨ BALUSTERS

As with the newels, balusters could be square or turned depending on the quality of the work. Turned balusters are often called 'spindles'.

A variety of patterns is available, to fit any architectural style. The turned decoration is limited to the centre portion, leaving both ends square for fitting into the groove on the underside of the handrail and the jointing to the string, its capping or the treads (depending on the construction of the stair).

As has been mentioned earlier in this chapter, when fixing balusters the gap between them should not allow a 100 mm sphere to pass through (Approved Document K Regulation 1.39a).

Generally, all the balusters will be the same length, except where they are required for a dog-leg stair, where they are adjusted as shown in Figure 7.54.

▲ Figure 7.54 Graduated balusters

10 SETTING OUT STAIRS WITH TURNS

The main functional requirement of a staircase is to provide a safe means of passage from one floor to another. The **aesthetic** design of a staircase can also be very important: the staircase is often the dominant view when entering a prestigious building such as a city hall or mansion. In these cases, the stair will be constructed from a decorative hardwood rather than softwood, which would be used for an entirely functional residential stair.

KEY TERM

Aesthetic: relating to a visual sense of beauty.

IMPROVE YOUR MATHS

A customer requires 48 balusters for a staircase. If each baluster costs £9.75, what would be the total cost, including 20 per cent VAT?

Information sources required to set out stairs

Regulations

The manufacture of all staircases is regulated by Building Regulations Approved Document Part K ('Protection from falling, collision and impact'). Approved Document Part M ('Access to and use of buildings') may also need to be consulted in certain circumstances. In addition, British Standard **BS EN 15644**, the standard covering stair construction, provides additional information such as minimum component thicknesses, minimum length of tenons on a string and details of fixings required.

Drawings

All work shown on the architect's drawings should conform to the conventions of the above information sources. It is, however, the stair manufacturer's responsibility to ensure that the finished product conforms. The drawings will show the general arrangement of the stair at 1:10 or 1:20 scales, and will show any specific design details at a larger scale of 1:5 or 1:2. Some construction details will be shown, but the final construction will be decided by the setter-out.

Specification

The drawings should be read in conjunction with the specification, which will provide additional information, such as:

- the precise description of the standard of materials and workmanship that the architect requires
- the species of timber required
- the planned finish
- any British Standards that need to be met (relating to the construction of the product and **BS 1186**, 'Timber for and workmanship in joinery').

ACTIVITY

Go to www.planningportal.co.uk and find Approved Document Part K, then research the minimum handrail height required on landings for a stair serving one dwelling. Describe three methods of guarding stairs and evaluate the advantages and disadvantages of each one. Explain which method you would choose for which circumstances.

Manufacturers' catalogues

Manufacturers' catalogues may need to be consulted for any off-the-shelf products that are required or specified, such as standard turned balusters.

Job sheets

Joinery workshops use a variety of administration documents. One that you may come across is a job sheet. This outlines the main requirements and information for the job, such as:

- contract number and client
- architect's details and specification
- any schedules that are relevant (baluster types, etc.)
- finish required
- time allowed to complete, and delivery date.

If provided, this will generally be written by the joinery workshop manager.

Site survey

As covered in Chapter 6, before any setting out takes place a site survey is generally required to obtain accurate information. For stairs this could include:

- detailed measurements of the maximum available going and total rise from finished floor level (FFL) to the next FFL
- positions of trimming/trimmer joists
- details and positions of any door and window openings
- templets of curved walls (for geometrical stairs)
- notes on any access problems to the premises for delivery and parking, etc.

Armed with the information collected from the site survey we can start to 'set out' the stair. To do this we need to have comprehensive knowledge of Building Regulations Approved Document Part K, particularly Section 1 of the current version. Read these pages now, as they will be referred to throughout the rest of this chapter. You can find Part K at the Planning Portal website (www.planningportal.gov.uk) and download it free of charge. You will find a summary of this information in Chapter 8, pages 371–373.

Discrepancies in information

If the site survey highlights any differences between the architect's drawings and what has been built, the details should immediately be passed on to your supervisor. For example, it is not uncommon for the trimmer joists to be put in the incorrect position, leaving insufficient room for a stair to be constructed that meets the Building Regulations. Your supervisor will inform the architect, who needs to decide how these issues are to be resolved. It is best to make and keep a written record of these concerns and the architect's amended instructions email is ideal. The new instructions will now form part of the contract.

Setting out stairs with a quarter-space of winders

The data collected from the site survey can be used to produce a scale drawing of the opening and stairs. The following sections will take you through a worked example of how to set out a stair with a quarter-space of winders.

Figure 7.55 is an extract from an architect's drawing showing the requirements for a winding staircase in a residential property.

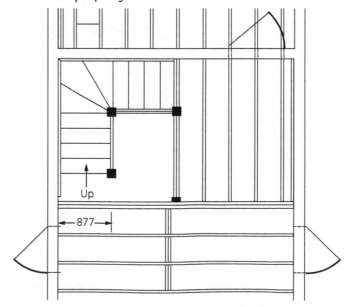

▲ Figure 7.55 Extract from architect's drawing showing the stair position and joisting arrangement

The **setter-out** will be the person whose responsibility it is to carry out the site survey and take accurate measurements. Two horizontal measurements are needed to establish the maximum going available in each direction. The vertical dimension is termed the 'total rise'. It is important that this dimension is measured from the point where the flight will start to the point where it will finish at the trimming joist. To achieve this, a line is levelled around the wall from the trimming joist to above the start of the stair and a height taken at this point. If a vertical height was taken directly below the trimming, it would not be accurate, as the ground floor may not be exactly level.

KEY TERM

Setter-out: an experienced joiner who has proved their competency, and is the person who produces the rods, cutting lists and orders for contracts.

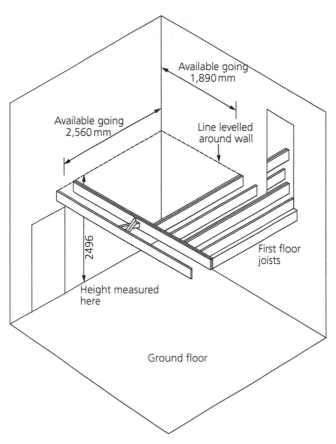

▲ Figure 7.56 Pictorial view showing which measurements need to be taken during the site survey

Stair calculations required to prove a stair will fit

Before the stair can be added to the scale drawing, calculations will have to be made to ensure it will fit and meet the requirements of the Regulations. The example below shows the calculations required to fit the stair to the opening in the survey (Figure 7.56).

Example

1 Calculate the number of risers required. To do this we divide the total rise by the maximum step rise allowed (220 mm – Regulation 1.4):

total rise ÷ maximum step rise = actual step risers required

$$2496 \div 220 = 11.35$$

2 You cannot have 11.35 risers – you need to have a whole number. If you reduce it to 11, the rise will exceed that allowed (2496 ÷ 11 = 226.9 step rise) so you need to round up to the next full number (12) and divide 2496 by this number:

$$2496 \div 12 = 208 \text{ mm}$$

This will be the step rise.

3 If there are 12 risers in the flight, there will be one less going, so there will be 11 treads. We now need to work out how much space the winders will occupy. To do this we need to set out the full-size winders on sheet material such as ply or MDF, with the minimum going of 50 mm at the newel face (Regulation 1.18) (see Figure 7.57).

From this we can determine the total going of the upper flight. The measurement is needed from the well string to the radiating riser point ('X' distance).

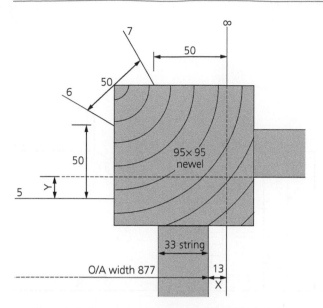

▲ Figure 7.57 Dimensions of minimum going allowed at the newel face

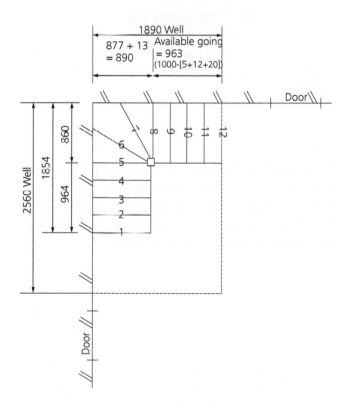

▲ Figure 7.58 Scale drawing showing riser arrangement

IMPROVE YOUR MATHS

Calculate the minimum number of risers required for a stair with a total rise of 2560 mm.

INDUSTRY TIPS

The term 'overall' can be abbreviated to 'O/A'.

'Flier' is the traditional term for 'tread'.

4 The total going (horizontal distance) between the riser face of the first tread in the upper flight and the trimming face is then calculated as follows:

total going for upper flight = total available going − (width of flight + 'X' distance)

total going for upper flight = 1890 − (877 + 13)

total going for upper flight = 1890 − 890 = 1000

total going for upper flight = 1000

5 Deduct fitting allowances of 5 mm at the wall, 12 mm for the riser thickness and 20 mm for clearance between the back of the riser and the trimming joist:

5 + 12 + 20 = 37 mm

We deduct this from 1000 to give the flight going:

1000 − 37 = 963 mm

Therefore, the total going for the upper flight is **963 mm**.

6 Next, we need to work out how many treads we have room for that will conform to the Building Regulations for a private stair. The minimum going allowed is 220 mm (Regulation 1.4), so divide the total going by the minimum going allowed:

963 ÷ 220 = 4.37 goings

You cannot have 4.37 goings – you need a whole number. If we round up to 5, the going will not be enough (963 ÷ 5 = 192.6) so we round down to 4 and divide 963 by this:

963 ÷ 4 = 240.75

This is rounded up to 241 mm.

7   When we have calculated a step rise and going for the upper flight, we need to confirm that it meets the **normal relationship** check (Regulation 1.5), which is that twice the riser plus the going should be between 550 and 700 mm:

$$(2 \times \text{rise}) + \text{going} = (2 \times 208) + 241$$
$$= 416 + 241$$
$$= 657 \text{ mm}$$

As 657 mm lies between 550 and 700 mm, the check is positive.

KEY TERM

Normal relationship: the average person's step length, according to Building Regulations Approved Document Part K. For stairs, it is twice the rise plus the going and should be between 550 and 700 mm.

ACTIVITY

Using Approved Document K, what is the maximum going allowed for a stair with a rise of 190 mm?

8   The normal relationship check being OK, we can complete the calculation. With four treads in the upper flight, there will be five risers (so, if there are 12 risers in total, in the upper flight there will be risers 8–12). There are three winder risers (5, 6 and 7), which leaves the lower flight containing the remaining four risers and four treads.

To calculate the total going of the lower flight we take the width of the flight, add the 'Y' distance (which will be the same as the 'X' distance, 13 mm in our example) and the four remaining goings.

$$\text{Total going of the lower flight} = 877 + 13 + (4 \times 241)$$
$$= 877 + 13 + 964$$
$$= \mathbf{1854 \text{ mm}}$$

This is well within our total going of 2560 mm so the stair will work for rise and going.

Headroom check

We need to check that there will be the required headroom (Regulation 1.10) of 2000 mm. We can do this by calculation but it is more usually carried out by scaling from the drawing. In our example, there are no clearance problems.

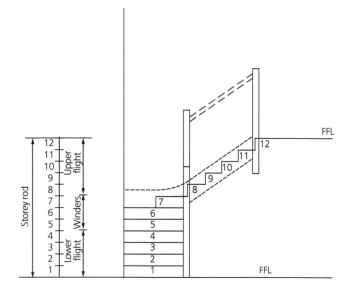

▲ Figure 7.59 Front elevation of stair including the storey rod

INDUSTRY TIP

A minimum of 11 risers' clearance between the bulkhead and the step below will normally satisfy the requirement of Part K.

Pitch

The quickest way to check that the stair pitch will be less than 42° (the maximum allowed for this type of stair according to Regulation 1.4) is to draw the full-size step dimensions and measure the angle with either a protractor or an adjustable set square.

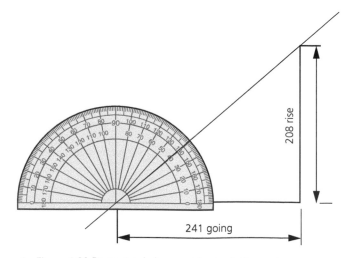

▲ Figure 7.60 Protractor being used to check the angle

If you prefer, you can use trigonometry to calculate the angle:

$$\tan (\text{pitch})° = \frac{\text{rise}}{\text{going}}$$

$$\tan (\text{pitch})° = \frac{208}{241}$$

$$\text{pitch} = \tan^{-1}\left(\frac{208}{241}\right)$$

$$\text{pitch} = 40.68°$$

As this is below the maximum allowed of 42°, we have a proved a solution that conforms to the Building Regulations.

IMPROVE YOUR MATHS

Calculate the stair pitch where the step going is 250 mm and the rise is 179 mm.

Drawing equipment required to set out and draw stair details

The equipment needed to set out and draw stair details is similar to that used and shown in Chapter 6, and includes:

- large set squares
- an adjustable set square
- trammel heads and beam
- dividers
- steel square
- large tee square
- line runner
- combination square
- 1 m rule
- a parallel straight edge.

Drawings required for a winding stair

With the proved calculations and the required drawing equipment, we can now produce a scaled line drawing (to 1:20 scale) showing the plan and the elevations of the wall strings. From these, the lengths of the strings and the additional width required for the winder housings can be obtained. The newels can also be added to this drawing to obtain their lengths.

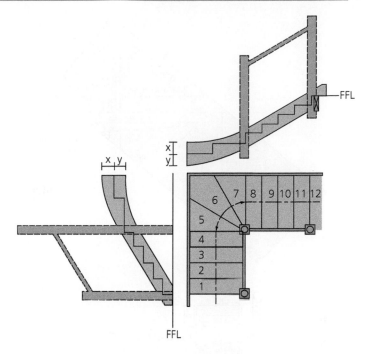

▲ Figure 7.61 1:20 scale drawing produced to obtain component lengths

The only full-size details required (shown in Figure 7.62) are:

- a section through one step – this will give the section sizes, profiles and the margin dimension
- sections showing handrail, string capping and finial detail (for moulding purposes)
- a plan of the winders – from this, the winder shapes and housings can be obtained; the corner joint between the two winder strings will be a tongue and trenched housing; because stairs are always fixed top down, the tongue must be on the end of the lower flight.

ACTIVITY

A prospective client requires a new hardwood cut string staircase. The total rise is 2470 mm and the available going is 3200 mm. Draw a full-size section through one step, showing:

- the joint you propose to use between the tread and riser
- the waist dimension suitable for the string
- an elevation of the decorative bracket
- the sizes of the components shown.

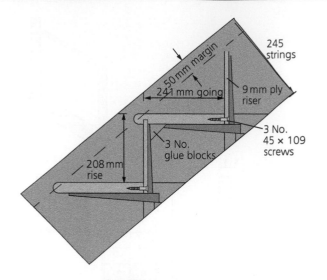

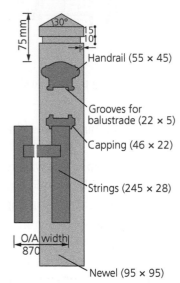

▲ Figure 7.62 Full-size details required when setting out stairs

Cutting sheet

The required sizes for all the stair components can now be obtained from both the scaled drawings and the full-size details. Item numbers are usually put on the scaled drawings and duplicated where necessary on the full-size details for clarity in a similar fashion to that shown on page 305 in Chapter 6.

page 305 in Chapter 6

ACTIVITY

Using Excel or another suitable program, produce a blank cutting list layout that you can use in the future. Explain the purpose of each column in your cutting list.

Jointing up

Some components may require jointing up to increase their width or thickness. In winding stairs made from solid timber, often the wall strings and winders and sometimes the newels will require jointing in their width. A loose-tongued and grooved joint would be used in preference to biscuit jointing as the **glue line** is stronger. Where the end of the joint will be visible, a stopped loose-tongue should be used to avoid seeing the end of the tongue.

KEY TERM

Glue line: the amount of glue contact area of a joint. This is particularly important when designing widening joints. The greater the glue line, the stronger the joint.

11 MARKING OUT COMPONENTS

Marking out strings

Once all the full-size details have been drawn and the timber has been prepared to size from the cutting sheet, you can start to mark out components. The first operation, as always, is to select the face side and edge and mark them. The sides that will be seen most should be as free from defects as possible. The strings should be sighted along their length for **camber**. If there is camber, both strings should be marked so that the camber is uppermost.

KEY TERM

Camber: a slight curve in the depth of the string. Remember, a string is a structural component, basically an inclined joist, so any camber should be positioned uppermost to resist deflection.

There are two main methods of marking out strings. The traditional method uses templets and the second uses a steel square and a fence.

Templets

Where strings are produced by hand, the following templets are needed:

- margin gauge
- pitch board
- riser and wedge templet
- tread and wedge templet.

The last two templets on the list are not required if the strings are being housed out using a stair jig and router, as only the tread and riser face lines are required. While the pitch board method is losing popularity, it offers the advantage of allowing you to test the angle of the newel to the string while fitting up, where this is not possible with the steel square.

Steel square

Many joiners will use a steel square and a fence to mark out the strings and newels. The steel square is quite heavy but is very efficient as it will save the time required to produce a pitch board. As the fence is fairly long, the marking out lines may not be as accurate if the string is not straight, as the fence will not be in positive contact with the edge of the string. A plywood square is a lightweight alternative to the heavy and cumbersome steel square, and has been used by the author for many years.

ACTIVITY

Produce a pitch board with a rise of 187 mm and a going of 245 mm. Calculate the hypotenuse length using Pythagoras' theorem and measure to check it for accuracy.

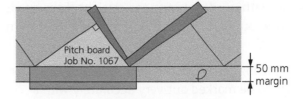

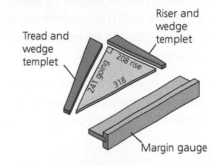

▲ Figure 7.63 The templets required for marking out strings and how these are applied

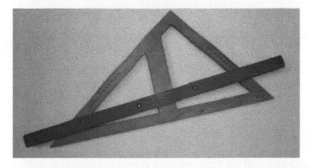

▲ Figure 7.64 Steel square with a fence as an alternative to a pitch board and margin gauge

▲ Figure 7.65 Lightweight plywood alternative to steel square

Apart from errors in the size of the pitch board or when setting up the steel square, errors can also easily snowball when marking out the tread positions on the strings. This can happen when moving the pitch board/steel square along the string from one position to the next. To avoid this, the hypotenuse dimension of the pitch board is stepped along the margin line of the first string marked. These positions are then squared up to the top edge of the string and transferred to its pair.

For stairs with winder strings, the straight flight treads are marked out first. The pair is then marked out from it as normal.

The wall strings of a winding flight require the winder housings to be marked out very accurately with a steel square. The measurements should be taken directly from the full-size rod. This informs you where additional material is required to increase the width of the string at this point.

Once these winder positions are marked, the tongue and trenched joint can be shown, and finally the easings to the top and bottom edges of the string are marked.

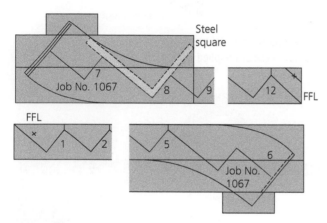

▲ Figure 7.66 Winding strings marked out

Marking out the handrails

The handrails will be marked out directly from the well string shoulders, as shown in Figure 7.67.

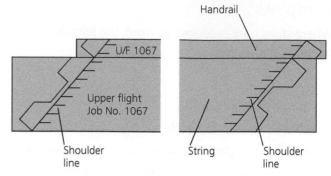

▲ Figure 7.67 Handrail shoulders marked out from string

Marking out the newels

The newels of winding stairs require quite a lot of careful marking out. If you are uncertain, the **stretchout** (as shown in Figure 7.68) of the faces can be drawn on the full-size rod to help you see what its final appearance will be. All the measurements are taken from the full-size plan of the newel.

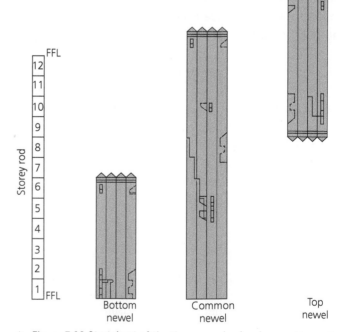

▲ Figure 7.68 Stretchout of the three newels showing marking out

Marking out the treads and risers

Marking out the treads and risers is a simple case of marking the face side and edge, indicating a length to be cut and marking the section details.

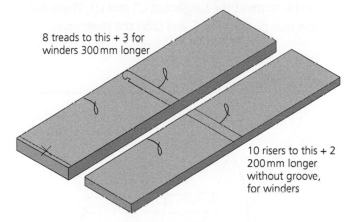

8 treads to this + 3 for winders 300 mm longer

10 risers to this + 2 200 mm longer without groove, for winders

▲ Figure 7.69 Treads and risers marked out

Marking out the winders

The winders should be cut to a templet. The final lines to be cut to are taken from the full-size plan of the quarter turn of the winders. Note that, in Figure 7.70, the direction of the grain is parallel to the nosing.

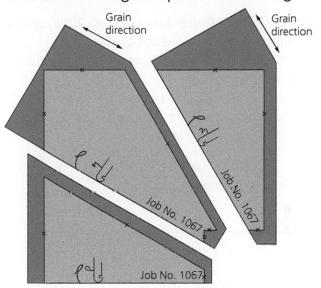

Grain direction

Grain direction

Job No. 1067

Job No. 1067

Job No. 1067

▲ Figure 7.70 Templets placed on jointed boards (or MDF) marking shape of winders

12 SETTING OUT, SURFACE DEVELOPMENTS AND CONSTRUCTING GEOMETRICAL STAIRS

As covered earlier in this chapter, a geometrical string runs continuously from the bottom of the stair to the top, generally with a curved portion along its length. The curved portion is known as a wreathed string. This is a short section of string jointed to the straight flights and takes the place of a newel, allowing the stair to turn (generally through 90° or 180°). The wreathed string rises and climbs at the same time. This type of string is normally cut and mitred, or cut, mitred and bracketed.

To mark out the shape of the step profiles on the wreathed face of the string, we have to geometrically develop its surface to obtain its actual shape by combining the step rises with their going around the curved face. The distance around the curved face is called the 'stretchout'.

The true length of a radius can be found geometrically as follows (see Figure 7.71):

1 The string radius 0–A is drawn through 90° to point B.
2 A tangent is drawn to the left of point B.
3 A 60° line is drawn through point A to touch the centre line at point X.
4 The 60° line is extended to meet the tangent at A1.
5 The length from A1–B is the true length of the radius A–B (the stretchout).

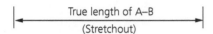

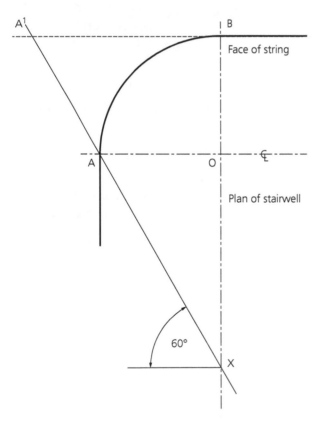

▲ Figure 7.71 Geometrically developing the true length of radius A–B

Any point along the radius line can be found as follows (see Figure 7.72):

1 Add any known points required, in this case points C and D (riser lines), onto the radius.
2 From point X, extend straight lines through points C and D to meet the tangent at C1 and D1. These are the true positions of C and D on the stretchout.

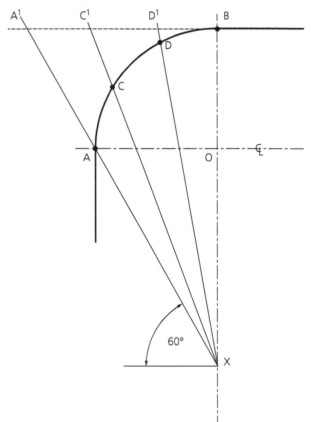

▲ Figure 7.72 Finding the riser positions on the stretchout

Setting out a stretchout for a wreathed string

Using the information above and incorporating a storey rod for the wreathed string area, a stretchout can be produced as shown in Figure 7.73. Note how one straight step is added at each end to allow for jointing to the straight string.

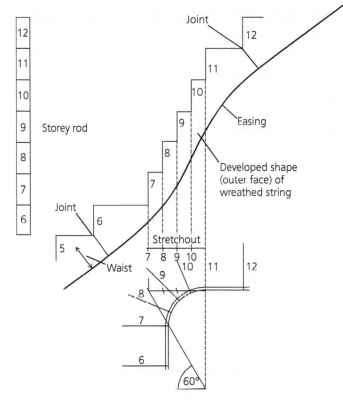

▲ Figure 7.73 Setting out the geometrical development (stretchout) of a wreathed string

Wreathed string construction

The shaped portion can be constructed in three ways:
1 built up (commonly termed 'staved')
2 laminated
3 solid.

Built-up construction

This is the most common method of constructing wreathed strings around a small radius turning through 90° or 180°.

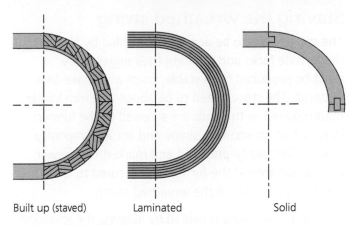

Built up (staved)   Laminated   Solid

▲ Figure 7.74 Methods of constructing wreathed strings

Former

The former (traditionally known as a 'caul' and commonly called a 'drum') needs to be accurate and robust in construction. It is made so that the outside shape matches the well of the string. The former must be strong enough to take temporary screws through the staves.

If the geometrical development is drawn on 1.5 mm birch ply, it may be used as the veneer for the wreathed portion. This will already have the marking-out lines to enable you to cut the string to the step profiles (the ply can be pre-veneered with a matching hardwood, if required).

> ### ACTIVITY
> Research the standard sizes available for 1.5 mm ply. Make a list of these.

Former constructed of 25 mm plywood ribs clad with 19 mm timber and plywood facings

Springing line

Setting out on face of veneer or ply

Vertical staves to be glued here

Springing line

Former

▲ Figure 7.75 Wreathed string section positioned to springing lines on the former

Staving the wreathed string

The staves have to be accurately profiled by hollowing their inside faces and tapering their edges. They must be produced from stable, knot- and shake-free material. The staves need to be about 200 mm longer than required as the ends are screwed to the former to maintain an accurate shape and act as temporary cramps. The ready-prepared and marked-out veneer can be laid around the former and secured to it by the short straight ends of the wreathed string.

Accurate positioning is helped by aligning the springing lines marked on the string with those marked on the former. The prepared staves are then thoroughly glued and screwed through the waste parts into the former. The solid end portion adds strength where the wreathed portion is connected to the straight strings.

> ### INDUSTRY TIP
>
> Having the face grain of the ply vertical makes it fold around the former more easily. (If veneered, the veneer should run in the direction of the pitch as far as possible.)

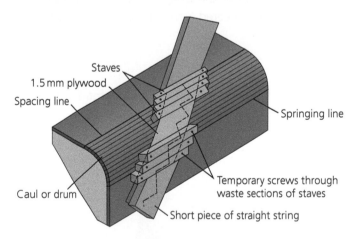

▲ Figure 7.76 String in staving process

Shaping and fitting up

When the glue is thoroughly dry, the screws can be removed and the wreathed string cut to the marked-out lines. Much of this can be carried out on the bandsaw and finished by hand. The joints can then be shot, loose-tongued, counter cramped and fitted individually to the straight flights at either end. The steps can now be fitted to the string, dismantled and sent to the site loose for fixing.

Counter cramp

The joint between wreathed and straight strings traditionally uses a counter cramp. This allows for assembling the string joint on site. The counter cramp consists of three morticed battens. The outside battens are secured to the straight string and the middle batten is screwed to the wreathed portion. The mortice in the centre batten is staggered by 5 mm. A pair of folding wedges is driven into the offset mortices, pulling the joint together. A loose tongue is fitted at the heading joint to increase the glue line of the joint. When the stair is fixed on site, the glued wedges are driven in and the remaining screws inserted.

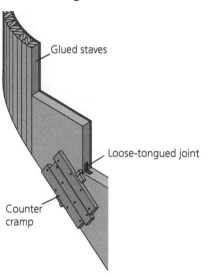

▲ Figure 7.77 Counter cramp used to joint wreathed string to straight flights on site

> ### ACTIVITY
>
> Make up a counter cramp and test just how efficient it is at tightening up the joints. Adjust as necessary.

> ### INDUSTRY TIP
>
> It is advisable to place clear polythene between the veneer and the former to prevent any excess adhesive from accidentally bonding the two together.

Laminated string construction

The laminated construction method is generally used to manufacture geometrical strings with a large radius. Circular (or, more correctly, helical) stair strings are normally constructed in this way.

▲ Figure 7.78 Helical stair

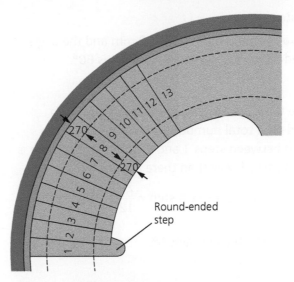

Round-ended step

▲ Figure 7.79 Showing 270 mm walking line dimensions

Setting out

When setting out this type of stair, make sure that the rise and particularly the going conform to the Building Regulations. In Figure 7.79, you will see that two radii are drawn 270 mm inside both strings. It is at both these points that the rise and going should conform (Regulations 1.25–28).

A full-size plan is required to obtain the tapered tread shapes. The development of the strings would normally be drawn to scale only to determine the length of the laminates required and the pitch of each string. The inside face of the laminate will be marked out prior to bending.

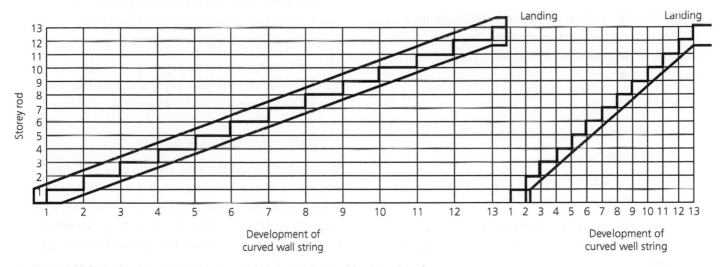

Storey rod

Development of curved wall string

Development of curved well string

▲ Figure 7.80 Scale drawing required to determine string pitches and laminate lengths

ACTIVITY

Next time you find a circular stair, climb it at various positions in its width. Note the position at which it is most comfortable to climb and whether it is between the two 270 mm margins.

A calculation can be made to double-check the going for both the wall and the well strings. The following data are required:

- the radius of the inside face of the wall string
- the angle contained between riser 1 and riser 13.

Example

In this example, the radius is 4000 mm and the angle contained between riser faces 1 and 13 is 60°.

Step 1

Calculate the angle contained by each step. To do this, divide the total number of steps into the angle contained between steps 1 and 13. (Remember there is always one step fewer than there are risers.)

$$\text{Angle contained by each step} = \frac{60}{12} = 5$$

Therefore, each step contains 5°.

Step 2

Next, calculate the distance between steps 1 and 13. To do this, calculate the circumference length of the inside face of the wall string:

Circumference = πd

Circumference = 3.142 × 8000 mm

Circumference = 25,136 mm

Step 3

Divide this circumference length by 360 (the number of degrees in a circle) then multiply the result by the number of degrees of one step, as calculated in Step 1 (5 in our example).

$$\frac{25,136 \text{ mm}}{360} \times 5$$

$$= 69.82 \times 5$$

$$= \mathbf{349.1 \text{ mm}}$$

Therefore, the going length on the development of the wall string should measure 349 mm.

This method is more accurate than the geometrical development as it is difficult to draw accurately at this size.

Former

A former will need to be made for both well and wall strings. The pitch line should be put around the former to position the string while it is being laminated.

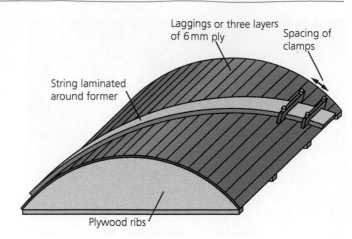

▲ Figure 7.81 Strings laminated around former

Manufacturing and assembly

The manufacturing method for this type of stair is similar to that for veneered and staved, in that a former is required and the laminates are held around it until the adhesive has set. A major consideration is the weight of the string and the ability to manoeuvre it while it is being worked on. The strings can be cut by hand or using jigs with a portable router. The tapered treads can be produced from a templet and each one will be shot to fit the plan, being adjusted where necessary when fitting to the strings.

Stairs of this type are fitted up in a vertical position, generally directly over the full-size setting out. They require robust support and safe working platforms.

Solid construction

This method is suitable only when a quarter turn is required at a quarter-space landing or at the top of the stairs to the final landing. This method offers very little strength and is difficult to shape. As the grain will run vertically, it will not follow through from the strings.

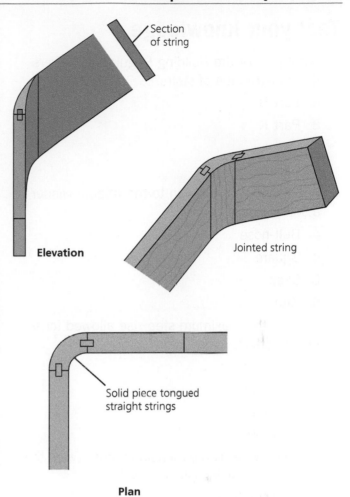

Section of string

Elevation

Jointed string

Solid piece tongued straight strings

Plan

▲ Figure 7.82 Assembling a helical stair

▲ Figure 7.83 Solid wreathed string construction

Test your knowledge

1 What part of the Building Regulations covers the construction of stairs?

   A Part B

   B Part K

   C Part L

   D Part M

2 What is the name given to the middle winder of a set of three?

   A Bull-nose

   B Square

   C Skew

   D Kite

3 What is the maximum step rise allowed for a private stair?

   A 200 mm

   B 220 mm

   C 240 mm

   D 260 mm

4 Between what two dimensions must twice the rise plus the going (2R + G) fall?

   A 500–600 mm

   B 550–600 mm

   C 550–700 mm

   D 600–700 mm

5 What is a suitable tread thickness?

   A 12 mm

   B 15 mm

   C 25 mm

   D 50 mm

6 Which two methods are both suitable for constructing a geometrical string?

   A Built up and housed

   B Housed and laminated

   C Laminated and cut

   D Built up and laminated

7 What joint is used between the lower and upper wall strings of a winding stair?

   A Barefaced tenon

   B Tongue and trenched

   C Tongue-and-groove

   D Barefaced dovetail

8 What is the name of a string shaped to the profile of the steps?

   A Cut

   B Well

   C Closed

   D Winding

9 On what document is the length, width and thickness of a string recorded?

   A Schedule

   B Bar chart

   C Cutting list

   D Specification

10 To whom would discrepancies found on a site survey be reported in the first instance?

   A Site fixer

   B Supervisor

   C Marker-out

   D Wood machinist

11 List four types of entry step.

12 Explain what measurements are required when carrying out a site survey for a quarter-turn winder stair.

13 A Calculate a suitable step rise where the total rise has a measurement of 2590 mm.

   B What is the maximum pitch allowed for a private stair?

   C What is the minimum height of a handrail on a private stair?

14 Explain how steps are jointed to:

   A a cut string

   B a closed string.

CHAPTER 8

FIXING STAIRS WITH TURNS

INTRODUCTION

Timber stairs are normally constructed off site in the workshop and then usually fitted at the first-fix stage when the building is watertight. Decorative staircases made from materials like oak are sometimes installed after the plastering has been completed. Installing turning staircases can involve the construction of landings between floors to incorporate the turn of the staircase. In the case of more complicated stair designs, winder treads, circular or helical stairs are used to form the turn. In most cases, the staircase's newel posts, handrailing, balustrade and any shaped first steps are usually installed and fitted on site.

LEARNING OBJECTIVES

By reading this chapter, you will learn how to:
1 understand stair types and component terminology
2 fix a flight of stairs with a turn.

1 TYPES AND ARRANGEMENTS OF TURNING STAIRCASES

Fitting and fixing turning stairs can be one of the most daunting and difficult tasks involved in first-fixing operations because of the number of components that make up a staircase. To remind yourself of stair types and stair component terminology, refer to Chapter 7. The information that follows covers additional considerations for fixing stairs with turns.

Stairs with turns vary considerably in their design. You may come across simple straight flights fixed between landings to form the turn, typically in domestic dwellings (either built between walls or with one side of the staircase up against a wall with newel posts and balustrade). You might also see more elaborate turning stairs that form a feature in more expensive domestic houses and public buildings.

The design, manufacture and installation of stairs is governed by the Building Regulations Approved Document Part K, which can be obtained through the planning portal.

Staircases that are required to turn achieve the turn in several ways:
- landings with either 90° or 180° turns
- tapered treads (often called winder treads)
- geometrical staircases
- circular or helical staircases.

See Chapter 7, page 332, for illustrations of staircases with turns.

The installation of these turning staircases requires the staircase to be installed either between walls, against a wall or free standing.

Staircases installed between walls

In these cases, the staircase is enclosed by walls on both sides throughout its rise, and the turn is achieved either by tapered treads or by intermediate landings. When intermediate landings are used, the bottom end of the top flight is sited on a landing (usually at the midpoint of the rise) while the top end of the top flight is fixed to the joist on the top landing. A second (lower) flight is then used from this turning landing to the lower floor level – again, the top of this second flight is fixed to the half landing and the bottom end sits on the lower level floor. These staircases require a handrail fixed to at least one side of the staircase.

▲ Figure 8.2 Quarter-turn staircase installed against a wall, incorporating tapered treads

▲ Figure 8.1 Tapered treads used to turn the staircase installed between walls; note the handrailing uses swept turns

Staircases installed against a wall

This type of staircase has one side fixed against a wall while the other side is open. Some form of guarding is required to the open side of the staircase, usually provided by at least two newel posts, handrailing and balusters. The newel posts are used as a means of joining the handrailing to the stair strings.

▲ Figure 8.3 Elaborate wreathed string staircase installed against a wall

Staircases that are free standing

Staircases that do not sit against a wall will have newel posts, handrailing and balusters fitted to each side of the staircase and tend to be grand, elaborate staircases with decorative features such as rounded first steps. Geometrical and circular (or helical) staircases have continuous handrailing usually terminating with a scroll, and are one of the most difficult types of joinery component to manufacture.

▲ Figure 8.4 Wreathed string free-standing staircase

2 FIXING A FLIGHT OF STAIRS WITH A TURN

Building Regulations

When designing, manufacturing and installing staircases, careful thought and interpretation of the Building Regulations are required. The construction, design and installation of stairs are controlled by the Building Regulations Approved Document Part K. The categories of staircase covered under the Regulations are as follows.

● **Private:** flights that are installed within one dwelling (for example, a two-storey house).

● **General access:** flights installed where a substantial number of people will gather (for example, a public building).

● **Utility:** flights in all other buildings. These are also known as 'common' stairs as they are used by more than one dwelling – for example, flats).

Different requirements are necessary for the different categories of stairs. Table 8.1 outlines some of the requirements that apply to the different types of stairs.

▼ Table 8.1 Different types of stairs and their requirements

	Private stairs	General access stairs	Utility stairs
Minimum rise	150 mm	150 mm	150 mm
Maximum rise	220 mm	170 mm	190 mm
Minimum going	220 mm	250 mm	250 mm
Maximum going	300 mm	400 mm	400 mm
Application of rise and going	For dwellings, external tapered steps and stairs that are part of the building, the going of each step should be a minimum of 280 mm. The maximum pitch for a private stair is 42°. For school buildings, the preferred going is 280 mm and the rise is 150 mm. For existing buildings, the dimensions shown above should be followed unless, due to the sizes on site, they are not possible. An alternative method should be discussed with the local authority's building control department, with particular attention paid to disabled access (Approved Document Part M). The Building Regulations Approved Document Part K also states that steps for straight flights should have the same rise and going, and that twice the rise plus the going (2R + G) should be between 550 mm and 700 mm.		

➡

▼ Table 8.1 Different types of stairs and their requirements (continued)

	Private stairs	General access stairs	Utility stairs
Handrail heights for flights and landings	Between 900 mm and 1100 mm, measured from pitch line or floor. Handrails can form the top of the guarding if the heights can be matched. All flights should have at least one set of handrails; if the stairs are 1000 mm or wider then a handrail must be provided on both sides. In buildings other than dwellings, continuous handrails must always be provided on both sides of flights and landings.		
Guarding of stairs – in the form of wall, screen or balustrade	When there is a drop of more than 600 mm. Guarding height for landings = between 900 mm and 1100 mm.	When there are more than two risers. Guarding height for landings = 1100 mm.	When there are more than two risers. Guarding height for landings = 1100 mm.
	Guarding should be provided in all buildings. In buildings likely to be used by children under 5 years old, the guarding should be designed so that a 100 mm sphere cannot pass through any opening in the guarding. Children should not readily be able to climb the guarding. 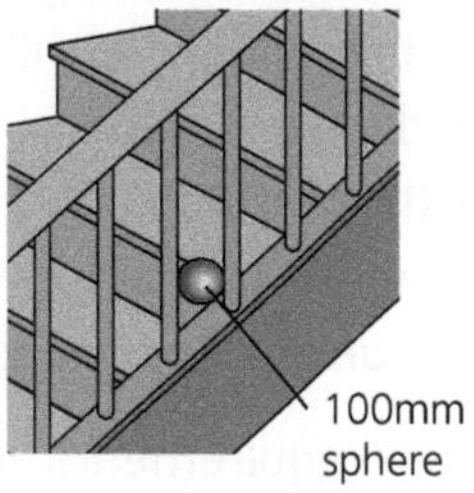 100mm sphere ▲ Figure 8.5 A 100 mm sphere should not be able to pass through any part of the staircase including the balustrade and open steps		
Minimum headroom height	2 m	2 m	2 m
	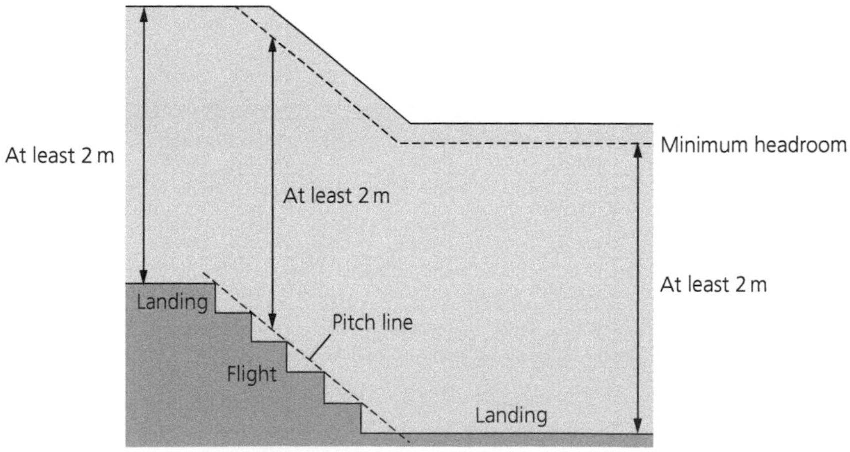 ▲ Figure 8.6 Headroom requirements The minimum headroom height can be relaxed in situations such as loft conversions, where a clear headroom height can be deemed to be satisfactory if it measures 1.9 m at the centre of the stair width, reducing to 1.8 m at the side of the stair.		
Widths of stairs	No minimum width unless used as a fire escape or for disabled access, when stairs must be at least 1 m wide. Although there are no recommendations given by the Building Regulations for minimum stair widths for dwellings, off-the-shelf flights are available at 860 mm widths for domestic installations.		

▼ Table 8.1 Different types of stairs and their requirements (continued)

	Private stairs	General access stairs	Utility stairs
Landings	Landings need to be provided at the top and bottom of every flight of stairs, with the width of the landing being at least as wide as the narrowest flight. Each landing should be clear of permanent obstructions. A door can open onto a landing at the bottom of a flight provided it will leave a clear space of at least 400 mm across the full width of the flight.		
	▲ Figure 8.7 Relationship between doors and stairs on landings		
Tapered treads	To comply with the required going for tapered staircases of less than 1 m wide, measure in the middle of the staircase. For staircases exceeding 1 m in width, measure 270 mm from the side. The going of tapered treads should measure at least 50 mm at the narrow end.		

ACTIVITY

Download the Building Regulations Approved Document Part K, and identify and highlight the following sections:

- Handrailing height requirements
- Tapered tread requirements
- Landing requirements
- Headroom clearance.

Using the information obtained from the Building Regulations, produce a simple step-by-step guide to the above sections of the regulations.

Fixing stairs with turns

The first consideration when fixing stairs with quarter turns is to work safely. You should therefore be able to prepare appropriate risk assessments and method statements, and to undertake initial checks to ensure the stairs will fit. This includes planning the sequence of installation and understanding the physical difficulties encountered when manoeuvring and fixing stairs into position as part of a team.

ACTIVITY

Refer to Chapters 1 and 7, and online sources, to identify how health and safety regulations apply to fixing stairs with turns. In particular, consider PPE, PUWER, Working at Height and Manual Handling regulations. Then draw up a risk assessment covering the process of installing stairs with turns. Check it as you carry out the fixing tasks identified in this chapter, and update it as you encounter and overcome new risks.

The installation and fixing methods required to install staircases with turns will be determined by several installation and finishing factors, and these factors will determine:

- Where is the staircase to be installed? (e.g. between walls, against a wall or free-standing)
- How is the staircase to be fixed? (e.g. screwed through the wall string)
- Where can the staircase be fixed? (e.g. underside of tread)
- Are any landings required? (e.g. quarter or half landings)
- How and where is the handrailing to be positioned and fixed?
- Are any balusters required for the staircase and landing?
- Are there any requirements for spandrel panelling or other under-stair framework and doorways?

Design principles associated with stairs that turn

To enable a staircase to turn, it can be built:

- using a landing known as a 'quarter 90° turn or half 180° landing
- with tapered treads, often referred to as a winder staircase
- as a geometrical staircase.

INDUSTRY TIP

One type of tapered tread is called a kite winder, which is a trapezium tread shaped like a kite.

Turning staircases using landings

This is possibly the simplest method to adopt when producing a turning staircase. When using landings, the total rise is achieved by using two straight flights of the correct individual rise. The total rise of each flight can differ, but the individual rise and going of each step must be the same in both cases. The landing should be constructed so its finished height, including the floor covering, meets the required specification for the staircase's total rise. The finished landing acts as the ground-floor level for the upper flight and the bulkhead for the lower flight.

Any intermediate landing used with turning staircases should be constructed in a similar manner as those used with stud partitions, with floor joists sitting on top of the framework to provide the landing area for the finished flooring to be laid on and the staircases to be fixed to. Alternatively, the landing framework can be built from suitably sized studwork. In either case, the studwork should consist of a set of double headers and a sole plate, as this doubling up helps to distribute the loading from the staircase or floor joists sitting on top of the studwork.

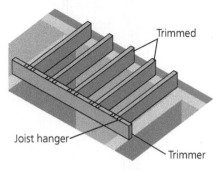

▲ Figure 8.8 Example of framework that could be used to form a landing used for either a quarter turn or half turn that is supported on two walls

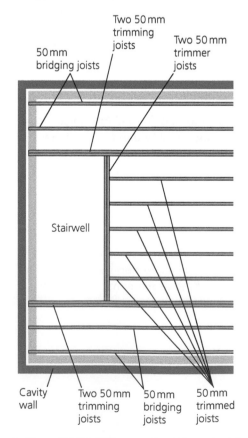

▲ Figure 8.9 Example upper floor trimmed out for a stairwell and landing

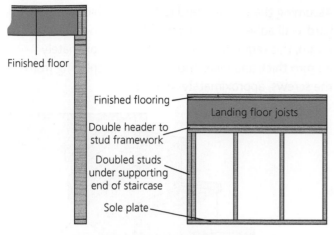

▲ Figure 8.10 Example landing construction

Any landing joists or stud framework should be positioned so that the bottom riser of the top flight sits over the top of the joist for maximum support.

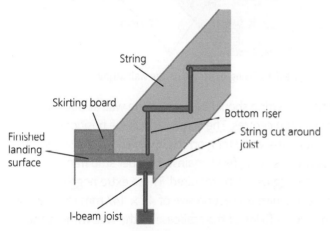

▲ Figure 8.11 Bottom step and string positioned over joist

INDUSTRY TIP

The staircase should always bear over the joist and not on any boarding used to form the finished surface of the landing. This allows for the surface boarding to be replaced if it becomes damaged.

INDUSTRY TIP

If a landing is to be used as a room under the staircase, the finished framework will require plasterboarding and skimming, as well as a door lining and door. The panelling and door frames associated with under-stair rooms are referred to as spandrel panelling. In some cases, the spandrel panelling incorporates the newel post, which runs from the landing through to ground level and is called a storey newel.

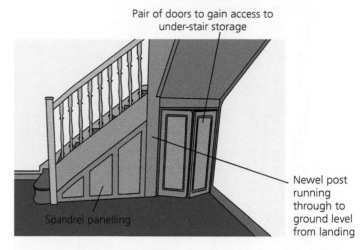

▲ Figure 8.12 Spandrel panelling and doors used to form a room below a quarter-turn staircase

ACTIVITY

Produce a set of sketches for the quarter landing staircase in Figure 8.12 that could be used as a guide to the construction of the framing arrangements for a spandrel panel incorporating a door.

The finished landing should be carefully positioned and levelled before being fixed back to the wall and floor with suitably sized screws or similar fixings. PU foam is a useful fixing addition to screws or coach bolts but should not be used as the sole fixing agent.

When constructing and positioning landings, always ensure you have at least 2000 mm headroom clearance above the finished staircase in order to comply with the Building Regulations. This should have been taken into account when the staircase was designed. Where a doorway opens onto a landing, there must be an unobstructed area measuring at least 400 mm square from the open edge of the door to the start of the staircase, to comply with the Building Regulations.

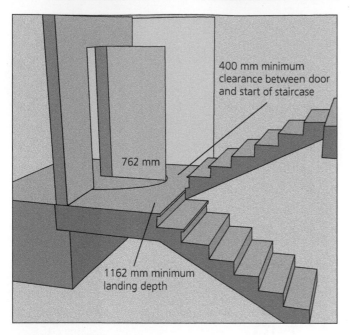

▲ Figure 8.13 Minimum landing size requirements when incorporating a doorway

It is usual to fix the landing and stairs just before plastering is to take place, allowing the staircase to be fixed to the bare wall. In these cases, the wall side of the string will require packing out to the same thickness as the hard-wall adhesive, plasterboard and skim. This is achieved by fixing timber strips to the outer edge of the wall string.

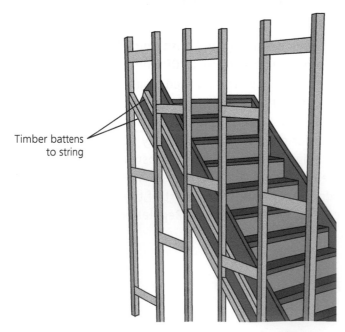

▲ Figure 8.14 Timber battens used to fix stair string to stud framework

Assuming the plasterboard is 12.5 mm thick with a hard-wall adhesive thickness of 12 mm and a skim of 2 mm, this requires a timber batten approximately 25 mm thick and wide enough to resist splitting from the screws, approximately 45 mm.

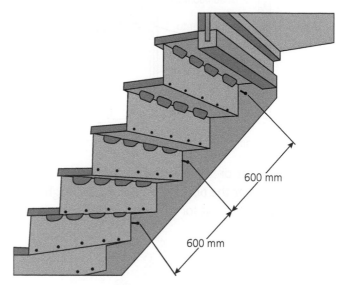

▲ Figure 8.15 Fixing positions on the wall string

The stair string should be fixed to the wall at 600 mm maximum centres. The screws should be positioned through the wall string on the underside of the steps. If the staircase is being fixed against a timber stud framework, extra noggins will be required. These extra noggins should have a minimum section size of 47 × 147 mm to allow for the secure fixing of the staircase to the wall without the risk of them splitting. These noggins should be positioned and fixed between the vertical studs at the required height to match the rise of the staircase.

After installation, the staircase will need to be protected from all the other trades that will use the flight during first- and second-fix stages. This can be achieved by covering each step with plywood or hardboard, covering the strings with cardboard, and covering the handrailing and balustrade with bubble wrap.

INDUSTRY TIP

In some cases, if the staircase is constructed from hardwood or is to receive a clear finish, it will be fixed after plastering and as late as possible, to reduce any likelihood of damage.

The fitting and positioning of the two straight flight staircases will be the same as for a single straight flight, as outlined in *Level 2 Site Carpentry and Architectural Joinery,* Chapter 4. The new intermediate landing just acts as another floor between the flights.

Newel posts

The design and construction of the landing will depend on whether a newel post is required, as in the case of an open-sided or free-standing staircase, or whether the staircase is to be installed between two enclosing walls with no newel post. The termination of the newel post usually falls into one of the following three types.

1 Terminating at landing/floor level – the newel post is notched to fit over the floor joist and finishes just short of the ceiling line.

2 Passing through the landing but terminating before floor level – this is known as a drop pendant newel. It is notched to fit the floor joist but passes through the landing and terminates just below the ceiling level, usually forming a decorative feature in the ceiling.

3 Passing through the landing and terminating at floor level – this is known as a structural newel. It is notched to fit the floor joist but continues through the landing, terminating at ground-floor level and acting as part of the structural support for the staircase.

Newel posts come as either full-height newels, which can be square or partly turned, or short newel post bases that are made up to full height with a pin newel insert. Pin newel inserts are fixed after the staircase has been fitted, by gluing the spigot into the socket in the newel base.

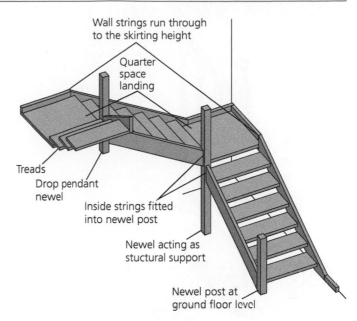

Wall strings run through to the skirting height

Quarter space landing

Treads

Drop pendant newel

Inside strings fitted into newel post

Newel acting as stuctural support

Newel post at ground floor level

▲ Figure 8.16 Newel post types and positions

Newel post notching

Newel posts will require notching or cutting out to fit over and onto the joist they are being fixed against. The newel post and string should sit on top of the joist and up against its face. For I-beams, the newel and string should sit on top of and against the top web, and against the face of the bottom web. The height of the notching for the newel post enables the newel to sit up against the joist with the top face of the loose tread nosing finishing level with the top edge of the landing's finished floor. This may require the loose top nosing to be rebated on its underside to sit on the top of the joist and finish level with the landing's floor.

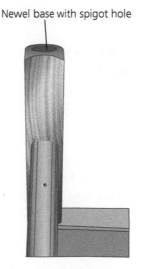

Newel base with spigot hole

▲ Figure 8.17 Newel base with spigot hole

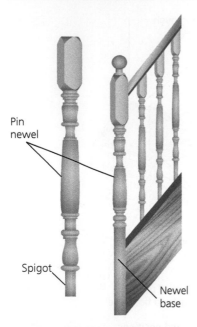

▲ Figure 8.18 Pin newel (left) and pin newel fitted into newel post base (right)

ACTIVITY

Draw to a suitable scale the relationship between the landing bulkhead, the newel post, the top of the stairs (including the nosing and loose riser) and the apron lining.

INDUSTRY TIP

Before fitting newel posts, ensure you can still manoeuvre the staircase and gain access to the stairwell from the assembly room or area. It is surprising how much room is required to move and position a staircase with its newel posts fitted.

For an image of the finished joist layout of a quarter-turn winder flight, see Chapter 7, page 355.

Fixing a turning staircase with winder treads

When the winder flight of stairs is delivered to site, it usually arrives in its completed form as far as possible. However, for ease of handling and access, its newels, winder treads, handrail, balustrade, shaped bottom step, nosing and top riser are usually supplied loose, ready for on-site assembly and completion.

The following step-by-step guide outlines the procedure for the final assembly and fixing of a quarter-turn winder tread staircase, which consists of:

- bull-nose bottom step
- three winder treads above the bull-nose step
- two short newel posts with pin newels
- handrailing and balusters.

1 Before installing a flight of stairs, make sure a suitable risk assessment has been carried out, and always ensure you comply with all its requirements. This should include making sure you meet relevant regulations, such as manual handling regulations, and wear appropriate PPE.

2 Check that the total rise between floor levels is correct, the available going is correct and the floor surfaces are level. Use the specification to check the height of the skirting that is to be used, and check that the upper flooring material is the same thickness as the thickness of the treads – it may be necessary to rebate the underside of the top tread to fit over the trimmer to ensure that the finished floor level and the level of the top tread are the same. Clear the stairwell and ensure that other trades do not try to access the upper floor via the stairwell until it is safe to do so.

3 Carefully lift the flight and position it on saw benches, with the wall string (remember to choose an assembly location that will allow for easy access to the stairwell). It is normal for the wall string to be left 'long' to protect it from damage and to allow for site conditions (e.g. incorrect levels or any other problem that might present itself during the fixing operations).

4 Check if the top nosing requires rebating in order to finish level with the landing floor covering, and rebate as required. In most cases the thickness of the treads is 22 mm – the same thickness as the floor sheeting – and therefore no rebating is required.

5 Starting with the top newel, measure the joist thickness and mark this measurement on your newel post. Start your measurement from the underside of the top nosing (remember, if the nosing was rebated, measure from the underside of the rebate), and allow an extra 5–10 mm clearance gap beneath the joist for fitting manoeuvrability.

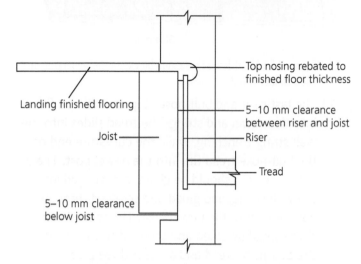

▲ Figure 8.19 Measure from the underside of the nosing for the cut-out of the newel and wall string

6 The wall string and the newel post will require cutting right back towards the riser, underneath the nosing. A clearance gap of approximately 5–10 mm can be left between the riser and the joist face to allow for adjustments and squaring up of the staircase in the opening. The top riser will be held in place with glue and screws.

7 The newel post will have been dry fitted in the workshop and should fit straight onto the string. Once you are happy that the newel fits correctly, apply glue to the tenon and dowel the newel into place. The newels will be draw bored onto the strings to pull the newel into place. Clean off the tops of your dowels.

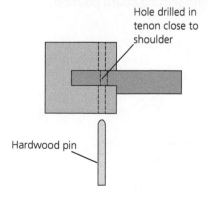

▲ Figure 8.20 Draw-bored joint

8 Cut the bottom newel post to match the rise of the step. This measurement should be taken from the top of the trenching for the bottom tread and measured down the riser trenching.

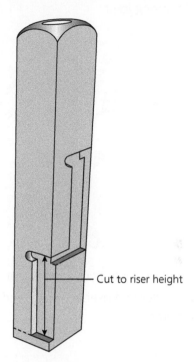

▲ Figure 8.21 Cut the bottom of the base newel so the riser height is correct

9 Once the newel post has been cut and fitted in the same way as the top newel, the winder treads can be slid into place, starting with the top one. Make sure the winders are fully fitted into the string with their nosing pushed down into the nosing recess of the string. You can screw through the wall string into the treads to pull them home. Then drive home the glued wedges and screw the riser through its back side into the tread in at least three locations.

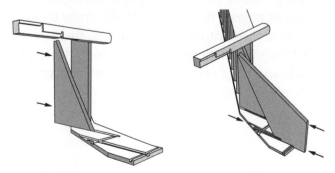

▲ Figure 8.22 Fitting the winder tread (right: drawing rotated for clarity)

381

10 The wall string, with its seat cut for the landing applied, can now be glued and screwed to the main stair string, ensuring they are correctly aligned and square. This joint is usually a half lap joint or a grooved housing joint.

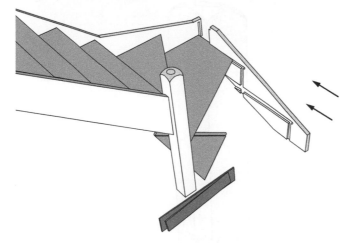

▲ Figure 8.23 Fit short wall string onto the main stair string

11 The last winder and riser are glued and wedged into place as before.

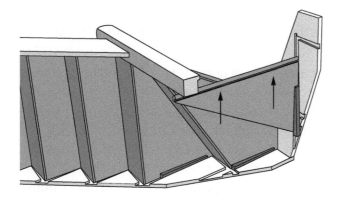

▲ Figure 8.24 Fit the last winder tread and riser

12 Apply generous amounts of glue to each glue block, fit at least three blocks to all winder treads and push them into place. You can pin these into the treads while the glue sets. Ensure all the winder treads and risers are screwed together through the back of each riser in at least four positions.

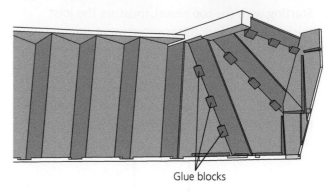

Glue blocks

▲ Figure 8.25 Fit at least three glue blocks to the underside of each winder tread

13 The last riser and bull-nose step can now be fitted into the newel and string. The tread slides into the wall string trenching, while the curvature end of the bull-nose tread fits into the newel post. The bull-nose step should be glued and wedged into the wall string, and glued and screwed through its back edge into the newel post. The second step riser should be glued and screwed to the tread of the bull-nose tread and at least three glue blocks should be used. Ensure the newel post is square to the tread and not twisted in relation to the string.

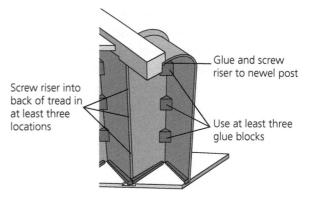

Glue and screw riser to newel post

Screw riser into back of tread in at least three locations

Use at least three glue blocks

▲ Figure 8.26 Bull-nose tread fitted

14 You can now mark and cut the wall string for the finished floor line and skirting board position.

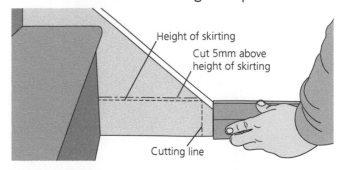

Height of skirting

Cut 5mm above height of skirting

Cutting line

▲ Figure 8.27 Marking the height of the skirting board on the wall string

15 The flight can now be lifted and offered to the trimmer (ensure the top riser and nosing are fitted). A temporary strut can be fitted to the staircase to help with its positioning and levelling, and to help keep the flight in place during the fixing process. Check the flight is fully located against the trimmer and landing (the floorboarding should have been cut back to allow the nosing and newel post to fit snugly against and onto the joist). Use a boat level to check that the treads are level in both directions, and pack and adjust the staircase as required to ensure it is level and plumb. Check the newel posts for plumb.

HEALTH AND SAFETY

The lifting process will require a team lift to ensure the staircase is adequately supported to reduce the risks of damaging the staircase and limit the risks of personal injury while lifting and positioning the staircase.

INDUSTRY TIP

Have a selection of wedges and packers available during the fitting process to help pack and level the staircase.

16 The staircase can now be drilled and screwed or bolted through the wall string into the wall; there should be one fixing through the wall string between the bottom and top steps, and at least another five or six fixings through the underside of the string – 600 mm is the maximum spacing between fixings to ensure the staircase is fixed securely to the wall. All fixing positions should be below the steps where possible. For staircases fixed between walls with no access available to the underside, the fixing holes should be counter-bored and filled with timber pellets after fixing. If there are uneven walls, packers should be used to close the gap between the wall and the stair string at the appropriate fixing points. The top newel post will require gluing and bolting or screwing to the trimmer at landing level to ensure a strong fix, with the top riser also glued and packed into position.

17 For plastered or solid walls, PU foam can be applied down the wall strings after fixing the staircase to fill any small gaps between the wall and the wall string. The PU foam also provides additional fixing points down the staircase.

18 The pin newels and handrails can now be fitted. In some cases, full-height newel is used; this can have mortices cut into it to receive the handrailing, which is pre-cut and tenoned to the required length and angle. If so, the handrailing should be fitted when the newel posts are fitted. Glue the pin newels and fox wedge them into place. (See Figure 8.28.)

19 The trimmer joist and stairwell that the staircase is fitted against can now be finished with an apron lining.

INDUSTRY TIP

A fox wedge is used in a blind or stumped mortice or socket, to force open the tenon or spigot, to help tighten the joint.

INDUSTRY TIP

Circular or helical staircases that are fixed against a curved wall are fitted and fixed in the same manner as a straight flight staircase fixed against a wall.

Apron linings

The outer faces and edges of the joists used to form the stairwell landing are typically covered either by plasterboard or a decorative timber facing, or by plywood that can be surfaced-finished to the required finish. Timber or plywood apron linings can be either pinned and glued or screwed into place. The top of the apron lining usually has a decorative timber nosing that butts up against the edge of the floor covering and overhangs the apron lining slightly. This nosing is typically finished with a bull-nose detail to its outer edge. The base rail used with the landing balustrade can sit over the joint between the nosing and the landing flooring.

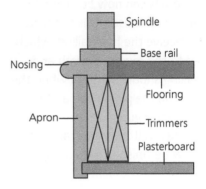

▲ Figure 8.28 Example of apron lining fitted to stairwell opening

ACTIVITY

Produce a 'quick guide' handout outlining the main fitting and fixing strategies for fitting a winder tread staircase against a wall. Then deliver a toolbox talk outlining the strategies in your 'quick guide' handout. Emphasis the importance of maintaining health and safety.

▲ Figure 8.29 A part freestanding curved staircase

INDUSTRY TIP

Toolbox talks should be given regularly when working with other people, to outline the working strategies of each person.

Fixing freestanding staircases

A freestanding staircase is fixed at its top to the landing in the same way as any other staircase. As there are no side-fixing points to a wall down the strings, the bottom of the staircase needs to be fixed to the floor.

Freestanding staircase flight designs can be:
- straight
- winder tread
- flights with quarter or half turns
- curved
- spiral.

Newel posts or any central spiral columns can be fixed to the floor in several ways. In all cases, accurate marking and positioning of the fixing is vital if the staircase is to remain level and correctly positioned.

▲ Figure 8.30 A part freestanding spiral staircase

▼ Table 8.2 Typical examples of fixing used for floor fixing newel posts and spiral columns

Steel fixing plate 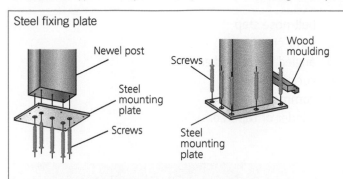	The fixing plate is screwed to the underside of the newel and then screwed down to the floor. Decorative trim can conceal the fixing to the floor, or floorboards or laminate flooring can be used to conceal the fixings.
Steel pins 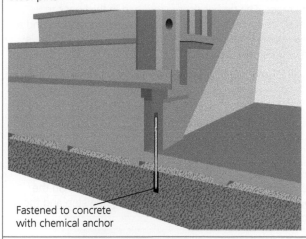 Fastened to concrete with chemical anchor	Holes are drilled into the concrete floor to receive steel pins and the pins are glued into the concrete with a special two-part rapid-setting glue. The newel post or spiral column is drilled to accept the pins, which are again glued into place. Always fully clean out any dust and debris from the drilled hole.
Steel base plate 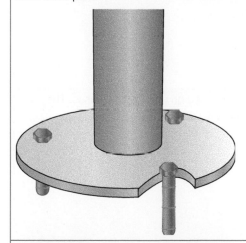	Spiral staircases tend to use a steel base plate that is screwed to the floor, with the central column of the staircase having a socket to receive the spigot on the base plate. A turn cover mould is used to hide the base plate fixings to the floor.
Angle brackets 	Steel angle brackets should be fixed to at least two sides of the newel post. The newel post can be finished with a decorative trim to complement the newel post and hide any fixings.

385

Shaped steps

Staircases are often finished with a decorative first (and sometimes second) step, particularly with more elegant, designed staircases. In these cases, the straight riser and tread are replaced with a rounded step and riser. The design and shape of these decorative steps fall into six main design classifications:

1 bull-nose step
2 semi-circular
3 angled or splayed end
4 curtail ended
5 commode
6 winder or tapered.

▼ Table 8.3 Design and shape of decorative steps

Type of step	Description
Bull-nose step 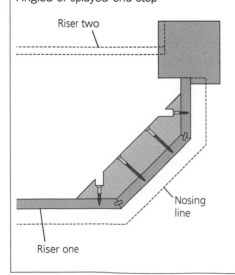	Usually at just one end of the step, but sometimes at both ends, the step is rounded using either flexi-ply or solid timber. The timber riser is reduced in thickness to approximately 2 mm where the bend is required, and is glued and wedged around a former. The tread is shaped to follow the riser, allowing for the same nosing, and is glued and screwed from the underside to the riser.
Semi-circular step 	Constructed in a similar manner to the bull-nose step except that the curtail step has a semi-circular end or ends. It can also have a deeper tread surface than the other treads, and extends out past the side of the staircase, giving the step and the staircase an increased effect of elegance.
Angled or splayed-end step 	The simplest design of shaped step is typically used on cheaper staircases. The angle on the riser has a mitred and loose-tongue construction.

▼ Table 8.3 Design and shape of decorative steps (continued)

Type of step	Description
Curtail-ended step	This is sometimes called a 'scroll-ended' step. It is used with cut string staircases and scroll handrailing, with a cage of balusters secured between.
Commode step 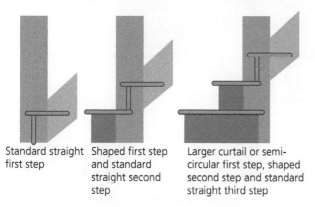	This type of step has a rise that is continuously curved throughout the length of the riser. Any number of steps can have a commode shape and, as with the curtail step, they are usually associated with grand, highly detailed staircases.

Fixing tapered (winder) treads is discussed below (page 389).

The bull-nose step is the most common of these designs in use. All the designs are usually fitted into the base newel post; as a result, they are supplied loose for on-site fixing. The newel post will have been recessed to receive the shaped step, but a small amount of adjustment, due to site conditions, may be required.

Standard straight first step

Shaped first step and standard straight second step

Larger curtail or semi-circular first step, shaped second step and standard straight third step

▲ Figure 8.31 Examples of bottom step and newel post configurations for typical shaped steps

Guarding stairs

The Building Regulations require all staircases to be guarded. This usually means some form of handrailing and balustrade fitted to the open side of the staircase.

Fitting handrails

Handrailing is usually attached to the newel posts on the open side of the staircase or, for staircases fixed between walls or against a wall, the handrail is fixed to the wall. Handrails for staircases should be between 900 and 1000 mm above the pitch line of the staircase, and between 1000 and 1100 mm on landings. Staircases that are more than 1000 mm wide must have handrailing fixed to both sides of the staircase.

> **INDUSTRY TIP**
>
> When fixing handrailing, ensure the handrail is fitted between 900 and 1000 mm above the pitch line on staircases, and between 1000 and 1100 mm for landings, in accordance with the Building Regulations.

In most cases, handrails will be straight lengths of timber that are either fixed between newel posts or fixed to a wall using brackets. The design of 'pig's ear' handrailing is intended to be fixed directly to the wall.

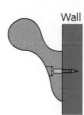

Wall

▲ Figure 8.32 Pig's ear handrailing fixed directly to the wall

With turning staircases, the handrail will be required to follow the turn of the staircase. This can be achieved by:
- continuous handrailing using straight lengths of material with mitred joints, or straight lengths of material with curved sections used to form the turn
- wreathed handrailing to match the curvature of the staircase.

Continuous handrailing

The easiest method of producing continuous handrailing running through the same vertical plane is to use straight lengths of handrail that are cut and mitred to the required angle. This can be an external or convex turn, called a 'knee', or an internal or concave turn, called a 'ramp'.

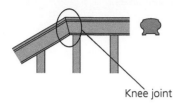

▲ Figure 8.33 Example of a straight knee jointed handrail

Other methods include using standard, mass-produced curved sections that are cut and fixed in the same way but eliminate any sharp angles.

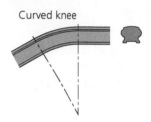

▲ Figure 8.34 Example of a curved knee joint in handrailing

Concave or internal curves used with handrailing are called ramps.

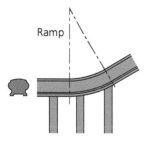

▲ Figure 8.35 Example of a ramp being used to form a vertical turn

In some cases, a combination of a knee and ramp is used to make a 'swan neck' turn.

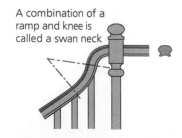

▲ Figure 8.36 Example of a swan neck turn

Easing is used to blend in straight lengths of handrail to curves and may be used in conjunction with gooseneck handrails.

▲ Figure 8.37 Example easing

Gooseneck handrailing incorporates a straight drop that blends into a turn.

▲ Figure 8.38 Example of a gooseneck handrail

Horizontal curves are also available and are often used straight after a newel post, particularly on landings, to form a turn.

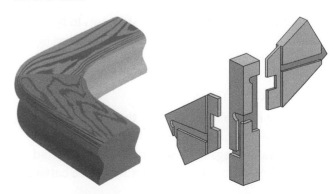

▲ Figure 8.39 Example of a horizontal curved turn being used straight after the newel post at landing height

Wreathed handrailing

Wreathed strings rise and turn at the same time, so the handrails above them must do the same. A wreathed handrail will normally turn through 90° so two will be required to turn through 180°. The two most common types of wreathed handrailing are rake-to-level and rake-to-rake.

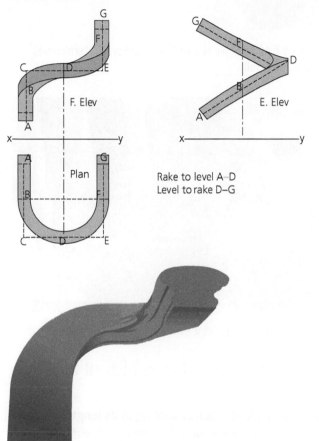

Rake to level A–D
Level to rake D–G

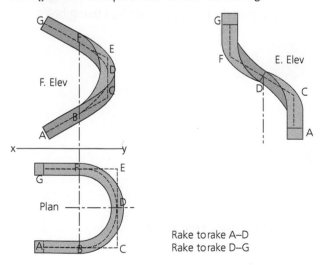

▲ Figure 8.40 Example of rake-to-level handrailing

Rake to rake A–D
Rake to rake D–G

▲ Figure 8.41 Example of rake-to-rake handrailing

Wall-mounted handrails

Where handrailing needs to be fixed to the wall of a turning staircase, either directly to the wall (as with pig's ear design) or via handrail brackets, careful consideration must be given to the joint positions that will form the turn.

Unless you are using a round handrail profile, the mitre profile edges of a handrail will not match at the corner mitre if both handrails are fixed at raking angles. This applies to both internal and external corners. To overcome this problem, the corners should be formed with a horizontal corner. The two horizontal corner lengths are then mitred to the required bisection angle of the raking handrail joint.

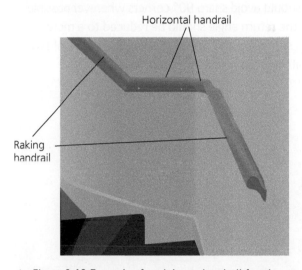

▲ Figure 8.42 Example of a pig's ear handrail forming a continuous handrail around a winder tread staircase

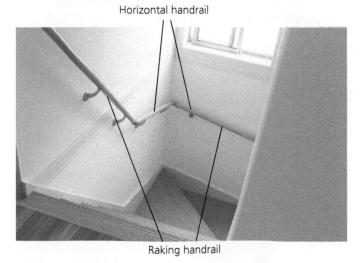

▲ Figure 8.43 Handrail fixed with brackets to form a continuous handrail around a winder tread staircase

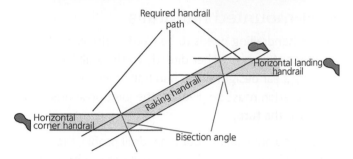

▲ Figure 8.44 Bisection angles for a handrail, transferring from horizontal to raking to horizontal

Handrails should return to the wall, and meet a newel post or decorative scroll at their end to alert people that the railing has ended. Wall returns can also act as a protection from bumping into a sharp edge of the handrail. Wall returns should avoid sharp 90° corners wherever possible; instead, the return angle should be reduced to a more pleasing, smoother and safer 45° angle, consisting of two 22.5° mitres (unless a curved return is used).

▲ Figure 8.45 45° wall return at the start of a raking wall-mounted handrail

Terminating the handrail

Where the handrail is not terminated at the bottom of the staircase by a newel post, the handrail will normally terminate with a scroll. There are numerous designs of terminating scroll, but they are generally classed as either vertical or horizontal scrolls.

▲ Figure 8.46 Example of a vertical (monkey tail) terminating handrail scroll

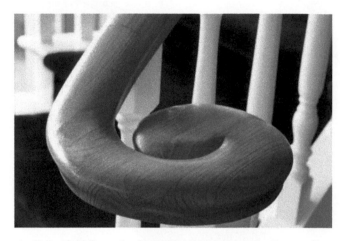

▲ Figure 8.47 Example of a horizontal volute terminating scroll

Jointing handrail to handrail and scrolls

Where a handrail requires jointing in its length, such as at a scroll or handrail wreath, a straight handrail bolt or Zipbolt™ can be used. For right-angled joints, there are 90° versions available. Care and accurate marking and drilling are required when using handrail fixing bolts.

▲ Figure 8.48 Traditional handrail bolt

▲ Figure 8.49 Modern Zipbolt™ alternative used to connect handrailing

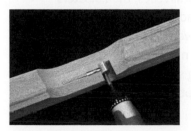

▲ Figure 8.50 Cutaway view of the Zipbolt™ fixing

ACTIVITY

Produce an illustrated onsite reference guide for the bisecting and cutting of angles associated with the fixing and terminating of handrails, including handrail scrolls.

Fitting balusters

When constructing a balustrade for either a landing or down the side of the staircase, the Building Regulations require that any gaps between the balusters making up the balustrade must not permit a sphere of 100 mm diameter to pass between them.

When positioning and fixing balusters on a staircase, usually two or three balusters will be fixed per tread. These are usually referred to as:
- 0 (level with the riser)
- ¼ (a quarter of the way across the tread)
- ½ (halfway along the tread)
- ¾ (three-quarters of the way across the tread).

The exact number will depend on the going of the tread, but remember not to exceed the 100 mm - sphere rule.

The positioning of any balusters on landing balustrades will require careful measuring and cutting of spacer blocks to maintain a constant gap between balusters.

It is easy to be a little out with the spacing for each baluster, so regularly check the positioning and vertical alignment of the balusters and adjust as required.

Setting out balusters

In most cases, two balusters will be required per tread and, depending on whether the staircase is of cut string design or closed string design, this will alter the way the individual balusters are marked out and cut.

Closed string staircase

The following steps outline a simple method of cutting and fixing balusters to a closed string staircase.

1. Set up a sliding bevel to the pitch of the staircase set from the face of the newel post, and set a sliding mitre saw or similar to this angle.
2. Carefully measure and cut the baluster base rail or string capping to the required length to fit between the newel posts.
3. Glue and pin the string capping onto the string.
4. Carefully measure the vertical height of the baluster from the bottom of the groove in the string capping to the bottom of the groove in the handrail, and transfer to a baluster ensuring both angles are facing the same way. Cut to length.

▲ Figure 8.51 Baluster cut to length

5. Test fit the cut baluster into the string capping and handrail, and ensure it is sitting vertically. Adjust length to suit.
6. Starting at the bottom of the flight, position the first baluster centrally between the newel and the first riser.
7. Measure the length of the gap between the face of the baluster and the newel post along the string capping and the handrail. Cut the fillets that were supplied with the string capping and handrail to these lengths. One end of the fillet can be left square; this will fit up against the face of the newel, while the angled end sits against the baluster. Check the baluster is still vertical and sitting in the correct position on the staircase, and adjust as required.

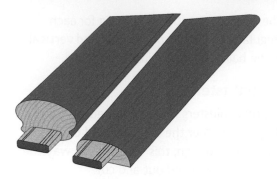

▲ Figure 8.52 String capping and handrail complete with fillets

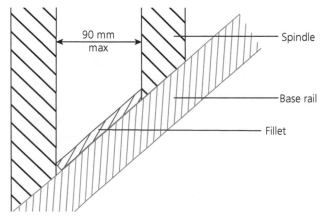

▲ Figure 8.53 Square end of fillet pushes up against the downward baluster, while the angled end butts up against the upper baluster

8   When you are satisfied the baluster and fillets are correct, the fillets can be glued in place. The baluster should be glued and pinned.

9   On the following treads, the balusters may be positioned: so the first baluster is level with the face of the riser and the second is an equal distance between the first and the next riser (known as 0 and ½); or so the first baluster is a quarter of the way across the tread and the second is three-quarters of the way across (known as ¼ and ¾).

10 Depending on the positioning used, measure and cut the balusters and fillets in the same manner as described in step 7 for the remaining treads except the top tread.

11 The top tread will have only one baluster and, as with the bottom step, this will be positioned centrally between the newel and the beginning of the riser. In the case of three balusters per tread (if the spacing between two balusters would exceed 100 mm), the same principle applies, but two balusters are used on the bottom and top treads, and three on all the others.

In the case of cut string staircases, the balusters sit on the face of the treads, either in a pocket cut into the treads or in a dovetail slot at the side of the tread. These housing methods will be carried out at the manufacturing stage of the staircase. For images of them, see Chapter 7, Figures 7.21 and 7.22.

▲ Figure 8.54 Glue and pin balusters in place

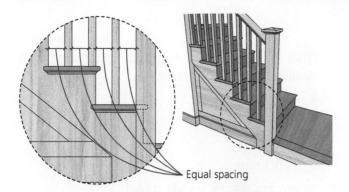

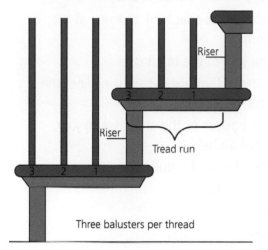

Equal spacing

Riser

Riser

Tread run

Three balusters per thread

▲ Figure 8.55 Baluster positions on cut string treads: two balusters (above) and three balusters (below)

IMPROVE YOUR MATHS

Calculate the minimum number of balusters and their spacing to comply with Part K of the Building Regulations. The landing specification is:

- clear gap between newel posts of 2135 mm
- each baluster is 32 mm square; you are required to comply with Part K of the Building Regulations.

Jointing when fixing balusters to stairs with quarter turns

Other methods of jointing to balusters are covered on pages 341–342 of Chapter 7. These are dovetailed and stub-tenoned. Baluster spacers fill the groove in the string capping and the underside of the handrail between the balusters to ensure consistent gaps.

Test your knowledge

1 What part of the Building Regulations covers staircases?

A Part L

B Part K

C Part A

D Part D

2 Where would an apron lining be found?

A Attached to the bull-nose step

B On top of the stair string

C Fitted between the handrail and the string capping

D Fitted around the face of the floor joists

3 What does the term 'balustrade' refer to?

A Joists forming the landing on which the strings sit

B A collection of balusters

C Combination of one tread and one riser

D Joint used to attach the newel post

4 What is the name of the continuous strings used to turn a staircase without the need for newel posts?

A Dog-leg strings

B Quarter landings

C Wreathed strings

D Winder strings

5 For domestic stairs, Building Regulations Approved Document Part K specifies that twice the rise plus the going should fall between what measurements?

A 550–750 mm

B 500–750 mm

C 550–700 mm

D 550–850 mm

6 According to the Building Regulations, domestic staircases should not exceed a maximum pitch of how many degrees?

A 42°

B 41°

C 40°

D 39°

7 What is the minimum height of handrailing on a landing for a private staircase?

A 800 mm

B 900 mm

C 1000 mm

D 1100 mm

8 Where are winder treads found?

A On staircases that turn and rise at the same time

B At the join to the landing

C On strings that are cut to fit against the trimmer joist

D At the bottom of the staircase to join the wall string

9 The shaped end of a bull-nose step is:

A quarter round

B tapered

C squared

D semi-circular.

10 In order to comply with Building Regulations, what is the largest sphere diameter allowed to pass between balusters?

A 70 mm

B 80 mm

C 90 mm

D 100 mm

11 What name is given to a newel post that passes through the landing but terminates before floor level?

A Pin newel

B Full-height newel

C Drop-pendant newel

D Short newel

12 What is the maximum distance between fixings down the wall string of a staircase?

A 400 mm

B 500 mm

C 600 mm

D 1000 mm

13 What is the minimum headroom clearance for private staircases?

A 1.8 m

B 2.0 m

C 2.2 m

D 2.4 m

14 What is the term for a convex angle on staircase handrailing?

A Ramp

B Swan neck

C Wreathed

D Knee

15 What do draw dowels join?

A The treads to the string

B The strings to the newel post

C The bull-nose step to the string

D The newel post to the trimmer joist

16 What is the pitch line on a staircase?

A An imaginary line that touches all the nosings in the flight

B An imaginary line along the handrail

C An imaginary line through the centre of winder treads

D An imaginary line used to fix the apron lining

17 What is the name given to the triangular framework beneath a staircase?

A Spindle

B Balustrade

C Baluster

D Spandrel

18 What is the maximum rise for a step in a domestic staircase?

A 200 mm

B 210 mm

C 220 mm

D 230 mm

19 Handrailing must be provided to both sides of a staircase if it is wider than what measurement?

A 1300 mm

B 1200 mm

C 1100 mm

D 1000 mm

20 Which of the following components would you normally expect to fit on site?

A Newels

B Wall string

C Risers

D Treads

CHAPTER 9

PRINCIPLES OF MAINTENANCE AND REPAIR

INTRODUCTION

This chapter provides information on the repair and maintenance of structural and non-structural components. Because timber is a natural material, it can deteriorate over time, particularly if proper precautions are not taken during the preparation of the component, its installation and while in use. Good component design, preparation and installation can reduce the need for repair or replacement.

On some occasions, replacement components may not be cost-effective or allowed due to planning restrictions so replacement parts will need to be formed and spliced in. Most timber can be repaired relatively easily, but it is essential when carrying out repairs to establish the reason why the damage occurred in the first place, and to address the cause as well as carrying out the repair. If this is not done, there is a risk that the damage could occur again.

Once a structure or component has been built, it is likely to need maintenance or repair at some point. Many individuals and even companies therefore specialise in the repair and maintenance sector, so it is possible to base your entire career on carrying out related tasks.

As always, it is vital to carry out all work to meet current health and safety legislation and Building Regulations.

LEARNING OBJECTIVES

By reading this chapter, you will learn:
1   the causes and prevention of deterioration and decay
2   how to repair and maintain carpentry.

1 CAUSES AND PREVENTION OF DETERIORATION AND DECAY

Causes of deterioration and decay

Timber components are continually open to attack from natural sources such as:
● biological decay
● insect infestation
● weathering (UV, water damage/ingress, storm damage).

Other forms of deterioration or damage can be a result of:
● wear and tear
● poor design
● poor workmanship
● poor or lack of maintenance.

Biological decay

There are several types of biological decay, which usually take the form of a fungus. These organisms get their food and energy from other organisms, such as timber – some from living material and others from dead matter. These types of fungi decompose, or break down, dead plants and animals. The most common types of decay caused by these fungi are:

- wet rot
- dry rot
- sap-staining fungi.

The fruiting body of a fungus is known as the **sporophore** and produces spores that are transported by wind, insects or animals to fertile wood, where germination takes place. Once the spores begin to grow, the fungus spreads out roots known as hyphae. These roots eventually become a mass called mycelium, which will in turn produce further fruiting bodies.

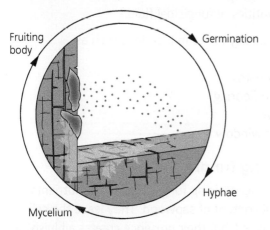

▲ Figure 9.1 Fungi life cycle

Wet rot

This is perhaps the most common form of rot and is associated with wet conditions. Wet rot is a term used to describe the effects of wood-destroying fungi that break down the cell walls of timber, causing it to decompose and collapse. This process can take place in hardwood or softwood timber with a moisture content of 20 per cent or higher, as this offers good conditions for the germination and nourishment of the fungi. Fungal spores germinate on wet timber and push out microscopic threads called hyphae. These hyphae penetrate the timber, releasing chemicals that cause the wood to gradually break down. Wet rot is typically confined to the area of dampness.

There are two types of wet rot: *Coniophora cerebella* (brown rot) and *Poria vaillantii* (white rot). Brown rot is the more common type. Its mycelium is dark brown or black and has a sheet-like growth pattern, with delicate brown threads sprouting from the rotting timber. White sheets of mycelium can be observed during the early stages of an attack.

The mycelium of wet rot may grow over adjacent masonry, but not extensively. Its brown, typically fanlike threads thrive in environments such as cellars. Spores of wet rot germinate in the pores of the timber, and the resulting mycelium feeds upon the **cellulose** of the timber and spreads out. Eventually the mycelium produces a sporophore that releases more spores which lodge in the pores of the timber.

The structure of the affected timber is undermined so that it shrinks and cracks, becomes brittle and takes on a darker colour, and the shrinkage creates cube-like sections. This type of attack will continue to spread unless action is taken to control outbreaks.

▲ Figure 9.2 Example of wet rot in floor joists

Wet rot can be caused by:

- unprotected timber in direct contact with moisture
- insufficient ventilation below suspended floors
- leaks from gutters, pipes, sinks and baths, as well as domestic appliances.

Wet rot can affect any unprotected timber and joinery.

In order to prevent and combat wet rot, it is important to ensure that all timber – including window and door frames – is well protected from rain and other sources of water by coating the timber in a suitable surface finish and preservative. All gutters and downpipes must be free from defects or leaks, and adequate subfloor ventilation is essential; poor ventilation will make any timber decay problems worse. Ensure that any new timbers introduced to a building are isolated from masonry with a physical membrane.

Wet rot is typically found in locations such as:
- the bottom of timber doors and window frames
- the bottom rail of wooden window sashes
- timber door and window cills
- joist ends.

Dry rot

The fungus commonly known as dry rot is called *Serpula lacrymans*. Dry rot is a far more destructive fungus than wet rot and is difficult to remove once established. The fungus attacks the cellulose in timber, hardwood or softwood and, once the spores germinate, fine strands of fungal growth sprout and spread quickly. Dry rot devours wood and will even penetrate mortar, plaster and concrete. It can cause widespread and often unseen damage over the entire height of a building. Timber that has a moisture content of 20 per cent or above combined with a lack of natural ventilation is open to dry rot attack.

If any of the following are seen, it is likely that dry rot is present:
- timber warping, shrinking and cracking both along and across the grain, and becoming easy to crush
- orange-red spore dust
- large fruiting bodies
- a musty smell
- hyphae that look like cotton wool.

▲ Figure 9.3 Example of dry rot to suspended floor joists

Dry rot can be caused by:
- unprotected timber in direct contact with moisture
- insufficient ventilation
- leaks from gutters, pipes or domestic appliances.

Dry rot can be prevented by ensuring that:
- the moisture content of timber remains below 20 per cent
- there is a good source of ventilation to roof and floor spaces
- all materials are sourced from reputable suppliers.

When treating dry rot, infected plaster should also be removed so that masonry can be scrubbed with a wire brush and treated. This approach is very effective but involves a considerable amount of time and effort. It also disrupts the fabric of the building, as it involves the removal of large quantities of sound timber and the use of large quantities of fungicidal fluid.

Dry rot is typically found in locations such as:
- rafters
- skirting boards
- joists and flooring
- architrave
- door and window frames.

Sap-staining fungi

Blue stain or sap-staining fungi obtain their nutrients from the cell content of sapwood. They leave the cell walls undamaged, but their presence creates a bluish stain or discoloration that cannot be removed. These fungi, like wet rot, attack wood with a high moisture content, so to reduce the effects of blue stain fungi, it is important to carry out conversion and drying of timber as soon as possible after a tree has been felled.

▲ Figure 9.4 Example of blue stain in floorboards

ACTIVITY

Design and produce an inspection checklist that could be used by maintenance operatives to help with the inspection and identification of wet rot and dry rot fungal attacks.

Eradicating timber decay

Before treatment of fungal decay begins, the reason why the decay set in should be established and eliminated, or controlled with proper maintenance. This could range from more frequent protection of the timber to a rigorous and regular painting or staining schedule of the timber, or ensuring, for example, that any leaking pipes and gutters are repaired so that the moisture content of the timber is not allowed to rise above 20 per cent.

Fungal decay treatment must not only kill off any existing infestation but also protect the affected and surrounding timber from further attack. To treat it effectively, all decayed timber must be cut out, the damp conditions that allow spores to germinate cured, and any other timber that is at risk of attack identified.

Disposing of timber affected by rot

In the case of wet rot, affected timber is unlikely to affect other wood once it has been removed. It can be disposed of easily in landfill via a local authority-approved disposal site or burned on site (though you must seek local permission to do this).

Timber affected by dry rot can still affect the integrity of a building even after it has been removed. All affected timber, and any discarded material in subfloor voids, should be taken away from the premises to a local authority-approved disposal site, or burned on site (again, local permission is required for burning). Any plaster that may be contaminated with strands or affected in any other way should also be removed and put into plastic bags, then removed from the building and disposed of with care at a specialist disposal site.

ACTIVITY

Design and produce a poster that could be used as part of an information drive to combat the potential problems of wet rot in older properties belonging to a housing association.

Insect infestation

A variety of wood-boring insects can attack timber, but it is the larvae that usually do the damage. In the UK, the most common forms of insect attack are from the:

- common furniture beetle
- powderpost beetle
- deathwatch beetle
- weevil.

Some moths can damage the foliage of trees, and the larvae of the wood wasp can damage trees that have been felled and left uncovered for too long.

To understand how beetle larvae attack wood, it is important to investigate the beetles' life cycle. Knowledge of how they live, what they eat and how they breed will help with the eradication of an infestation.

Life cycle of a wood-destroying beetle

The typical life cycle of a wood-destroying beetle is as follows:

1 A beetle lays eggs in cracks or crevices on the surface of the wood.
2 A few weeks later, the eggs hatch and the larva enter the wood. They eat the wood and tunnel further into it, leaving frass (sawdust) behind in the tunnel.
3 After one or more years of tunnelling (the time taken depends on the species of beetle), a larvae makes a small chamber just below the surface of the wood, where it pupates (turns into a chrysalis).
4 The pupa forms into a beetle and emerges from the chamber, leaving a flight hole on the surface. These holes are a first visible sign of insect attack.
5 The beetle then travels or flies to a new area, mates and lays eggs, beginning the cycle again.

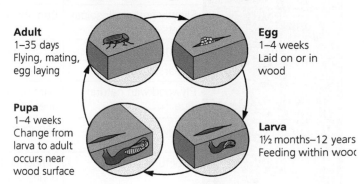

Adult
1–35 days
Flying, mating, egg laying

Egg
1–4 weeks
Laid on or in wood

Pupa
1–4 weeks
Change from larva to adult occurs near wood surface

Larva
1½ months–12 years
Feeding within wood

▲ Figure 9.5 Life cycle of a wood-destroying beetle

Common furniture beetle

This is a serious indoor pest, threatening wooden furniture and structures with the tunnelling action of its larvae.

During spring and early summer, adult beetles emerge from timber ready to mate. Immediately after mating (which can occur on the same wood from which the female emerges), eggs are laid, and the beetles push them into cracks and crevices in the wood. Two to four weeks later, the eggs hatch and the larvae burrow into the timber. Indoors, the life cycle takes between two and four years: the larvae pupate in chambers just beneath the surface and emerge leaving holes around 2 mm in diameter. They may also leave behind frass.

The powderpost beetle

The powderpost beetle is typically not found in homes, but in timber yards and furniture warehouses where there is a lot of fresh wood. This does not mean that it can be ruled out as a cause of domestic woodworm because, if you acquire timber or furniture from a location where the beetle is present, there is a chance your wood will be affected.

Powderpost beetle larvae spend months or years developing inside timber, feeding on its starch content.

Their presence becomes apparent only when they emerge as adults, leaving behind shot holes. They may also leave piles of powdery frass below. Female beetles may re-infest the same piece of wood by laying eggs, continuing the cycle for generations.

The deathwatch beetle

The deathwatch beetle is more common in the south of England, and adult deathwatch beetles can be found on flowers during the spring. The gut of this type of beetle contains micro-organisms that help it to break down wood cellulose. The adult female deathwatch beetle lays eggs in crevices on timber and the larvae tunnel in after hatching. They can remain there unnoticed until they emerge as adults, leaving distinctive exit holes.

Weevils

Wood-boring weevils attack only timber that has already started to show signs of decay, or timber that is damp. Both the adult and larvae bore into the timber. Attacks from weevils are a secondary problem – the source of decay or damp must be identified and removed as well as targeting the weevil.

▼ Table 9.1 Comparison of wood-destroying beetles

Name	Common furniture beetle	Powderpost beetle	Deathwatch beetle	Weevil
Size/colour	3–5 mm; reddish to blackish brown	5 mm; reddish-brown to black	6–8 mm; chocolate brown	3–5 mm; dark brown to black
Flight	May to August	May to September	March to June	June to October
Where eggs are laid	Crevices or cracks	Holes in the end grain of large-pored hardwoods	Crevices in decayed wood and flight holes	Just below the surface of the timber in holes and crevices
Size of larvae	Up to 6 mm	Up to 6 mm	Up to 8 mm	Up to 6 mm
Bore dust	Slightly gritty, ellipsoidal pellets	Bun-shaped pellets	Bun-shaped pellets	Coarse oval-shaped pellets with a gritty feel
Diameter of exit hole	2 mm	0.75–1.5 mm	3–4 mm	10 mm
Life cycle	2 years or more	10 months to 2 years	3–15 years	6–12 months
Type of wood attacked	Plywood with natural glue	Hardwood: oak, elm, obeche	Hardwood, especially oak	Any wet or rotting timber
Area of attack	Mainly sapwood	Sapwood	Wood previously decayed by fungus	Wood previously decayed by fungus
Location	Structural timbers	Timber and plywood drying or in storage	Roofs of old buildings	Any wet or rotting timber
Other comments	Accounts for 75% of beetle damage in the UK	Softwood and heartwood immune	Attacks that start in hardwood can spread to softwood	Best to remove source of damp and decay

Eradication of wood-destroying beetles

If you suspect an attack from wood-destroying beetles, all affected timber should be examined carefully to determine:

- what is being attacked (for example, furniture or structural timbers)
- the extent of the attack
- the characteristics of the flight holes
- the quantity and nature of the frass
- the moisture content of the timber
- the species of wood being attacked.

Once you have established the above information, it should be possible to work out which species of beetle is attacking the timber. Take the following measures to find out whether an attack is still in progress.

- Where possible, the affected area must be opened up. Cut away any wood that is badly affected, and carefully remove and dispose of it.
- If structural timbers are under attack, you must assess the stability of the structure. Any repair and replacement work on the affected timbers must involve appropriate shoring or support of the area.
- Remove all debris from the affected area to stop the spread of the attack.
- Treat with an insecticide if the attack is in a localised area. This type of treatment requires a reputable specialist contractor.

Disposal of insect-infested wood

As insects can fly, infested timber can still affect other wood even after its removal. It can be disposed of easily by taking it to a local authority-approved disposal site or burning on site (with local permission).

ACTIVITY

Design and produce an inspection checklist that could be used as part of the maintenance inspection process. The checklist should identify the signs to look out for with a possible infestation as well as areas susceptible to infestation.

Types of maintenance

Maintenance of a building's external and internal components will be required at some point to prolong the life of the individual components and the building.

Examples of typical components that require inspection and maintenance are:

- windows
- doors
- stairs
- fascias
- soffits
- cladding
- roof coverings
- guttering and downpipes
- rafters
- joist coverings
- joists
- ironmongery
- water and waste pipes.

Maintenance of the building's components should be part of a planned inspection and maintenance schedule, to ensure that components do not fail or break and lead to unnecessary remedial action to other components due to their failure. For example, if blocked or dripping guttering and downpipes are not cleared or repaired, decay to the fascia and soffit will certainly result, as will possible damp and rot to structural components such as joists and rafters. If no form of inspection and planned maintenance is undertaken, costly repairs are very likely later in the component's life.

The intervals between inspection and maintenance activities will vary depending on the component being inspected. For example, roof rafters do not need to be inspected as often as windows, and joist coverings will not need to be inspected as often as guttering and downpipes.

▼ Table 9.2 Example of typical inspection periods and components they could cover

Monthly	Quarterly	Annually	Biannually	Every five years
Water and waste pipes for drips	Door and window operation, ironmongery	Gutter and downpipes	Cladding, joist coverings, fascia and soffit	Rafter, joists, roof coverings

Planned maintenance

The best way to prevent damage and decay is with a planned maintenance scheme. In the long term, regular inspection and maintenance will save time, money and a great deal of effort.

For example, the best way to preserve the life of external doors and windows is to regularly protect them against the elements (such as rain and ultraviolet light from the sun) by applying a suitable wood stain or paint finish to the timber surface.

Another example is preventing rot in structural or non-structural timbers by repairing leaking pipes and gutters, and keeping the moisture content of the timber below 20 per cent.

INDUSTRY TIP

Ultraviolet (UV) light is the light from the sun and is responsible for most damage to exposed wood because it helps to destroy lignin, the component of wood that hardens and strengthens the cell walls.

In most modern housing, the structural timbers are pre-treated with a preservative of some sort. In some instances, if a particular type of insect attack is common in a given area, the local authority may insist that all structural timbers in the houses it owns are given appropriate treatment. The objective of any such treatment is to reduce the population of wood-boring insects to an insignificant level. The most effective way of doing this is to interfere with the insects' life cycle so that the infestation dies out.

ACTIVITY

Design and produce a maintenance plan or schedule for housing in your local area. It should identify typical components associated with local buildings and suggest a maintenance routine for each component. Give details of typical preventative measures to preserve each component and reduce the risks of decay or damage.

Unplanned (condition-based) maintenance

Although regular maintenance can significantly reduce the likelihood of components failing, some situations cause unexpected damage. Examples include severe weather, such as gale-force winds, water ingress from flooding, or direct damage, such as from a fallen tree branch. Under such circumstances, repair work should be undertaken as soon as possible to return the component or structure to its original condition – although sometimes it may be delayed while insurance claims are cleared, for example. In some instances, temporary repairs may be needed to prevent further damage while waiting for full repairs to be undertaken. The priority is to make the damaged area safe, secure and watertight. Of course, the damage may be reduced if the component or structure has been regularly maintained before the unplanned maintenance is required.

Costs and implications of maintenance and repair

Whether to repair or replace a component is not always a simple decision. There are usually several factors that must be taken into consideration before a final decision can be made, and these will ultimately affect your choice of repair and how it is carried out. You may need to weigh up:
- the cost of replacement against cost of repair
- the time required for replacing against time required for repair
- the time frame available to complete the task
- any local authority planning restrictions, such as listed building consent, heritage work or work being carried out in a conservation area.

In many instances, whether to replace or repair will be quite obvious – for example, a broken door handle is usually much cheaper to replace than to repair, in terms of time and the cost of a new door handle compared with the cost of trying to get it repaired. In other cases, it may be more difficult to make a judgement – for example, if the bottom rail and bottom of the door stile on a standard softwood door have started to rot, the time spent cutting and fixing a splice in the rail and door stile may be costlier than purchasing and fitting a new door. But if it was a bespoke handmade door, a repair may prove to be much more cost-effective.

In the case of listed buildings, heritage work or work being carried out in a conservation area, you will usually have a limited choice as to what type of repair or replacement you are allowed to do. The local planning authority will dictate which type of action can be taken.

Refer to Chapter 2 for more information about working with local authority requirements.

② UNDERSTANDING HOW TO REPAIR AND MAINTAIN CARPENTRY

Repairs to timber components are broadly referred to as either structural or non-structural.

Structural repairs tend to be difficult and time-consuming, requiring a means of access and possibly temporary support of the component being repaired. These types of repair are usually more expensive and involve items that form an integral part of the building, such as:

- floor joists
- pitch-roof rafters and purlins
- flat-roof joists and coverings.

Non-structural repairs are usually easier, quicker and cheaper, and include items such as:

- soffits, including guttering
- door and window frames
- doors and opening window lights
- staircases
- mouldings
- ironmongery associated with doors and windows, etc.

Structural repairs

Repairs to structural timber such as rafters, purlins and floor joists will almost certainly require some form of temporary support to the timber being repaired, and possibly to the timbers surrounding the area being repaired. A suitable risk assessment should be carried out on the repairs required and, in some cases, this will require the advice and guidance of a structural engineer, who will help to identify and devise suitable work practices incorporating temporary structural support methods and possible permanent additional supports for the repaired area.

Most structural repairs involve cutting out the affected section and splicing or piecing in a new section of timber. It is important to ensure that the timber being used for the repair is the same species as the timber requiring repair, as well as having the same moisture content to within 1 per cent. If the replacement section has too great a difference in moisture content, any further drying of the timbers may cause problems with the repaired joint due to different shrinkage or expansion rates of the two sections of timber.

When working on a building that falls under heritage or listed-building control, strict guidelines will be issued as to what timber can be replaced and what should be repaired. As a general rule, most of the original timberwork should be left in place as far as possible, with new sections being spliced in as necessary. In some instances, the original timbers may be beyond repair, requiring the sourcing of completely new replacement timber.

Where it is possible to repair structural timbers by inserting a new piece of timber into the original, there are three ways to achieve a satisfactory joint. In all cases, the repair requires cutting out the decayed or damaged section. The replacement section is usually joined to the existing section using one of the following methods:

- scarfing joint – this type of joint is used when the original length of timber is fully cut through and a new section inserted
- piecing in – with this type of repair, the original length of timber has only a part depth of the timber cut away and not its full section thickness
- splint repair – this type of repair is generally used when there are no restrictions placed on the joint used, as is usually the case with listed building and heritage work repairs. In a split repair, none of the timber is cut away. Instead, an additional length of timber or steel is fixed to its side and used as a stiffening piece.

Temporary support systems

Cutting out or altering structural timbers in any way is, by the nature of the task, altering or adjusting a structural component of the building and should not be attempted without:

- conducting a risk assessment and producing a safe system of work
- consulting a structural engineer, and getting professional advice on both the type of repair required and temporary support methods to use while the work is being carried out

- consulting the local authority building control department and establishing if building consent is required to carry out the repairs
- establishing whether the building falls under listed building or heritage control, and seeking advice and permission for the repairs In a split repair, none of the timber is cut away. Instead, an additional length of timber or steel is fixed to its side and used as a stiffening piece.

The advice and guidance from all these sources will establish the type and quantity of support methods employed. Never second-guess what and where support is required, and always follow expert advice.

▼ Table 9.3 Examples of temporary support methods when altering structural timbers

Type of support	Where used
Acrow props	Acrow props provide a vertical support system to support overhead loads such as floors and roofs. Once positioned, the screw collar of the prop is turned to raise the inner section and tighten the prop against the load. Acrow props are among the quickest, most efficient, cost-effective and safe options when supporting heavy or load-bearing objects. Acrow props are widely available and used throughout the construction industry. They should always be positioned on firm, sound ground that is capable of supporting the addition of loads.

▼ Table 9.3 Examples of temporary support methods when altering structural timbers (continued)

Type of support	Where used
Strongboy  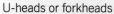	A Strongboy is an attachment that fits on top of an Acrow prop, with the blade of the Strongboy used to support the load. One advantage is the ability to offset the Acrow prop, allowing better access to the working area. The blade can also be inserted into a mortar joint in masonry to support the wall while repairs are carried out to the wall or any timber lintels. Once inserted, the screw collar of the prop is tightened until both the prop and Strongboy are taking weight – but do not over-tighten the collar as this can damage the Strongboy or cause the masonry to split apart.
U-heads or forkheads  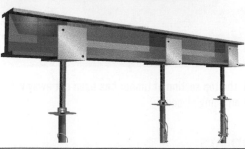	This type of attachment is used mainly to hold timber beams of steel RSJs while they are inserted and fixed in place. The 'U' or 'fork' stops the beam from slipping from the prop.

INDUSTRY TIP

Fix the top plate of an Acrow prop in place with screws to prevent accidental movement of the prop.

ACTIVITY

Produce a toolbox poster outlining the dos and don'ts when using Acrow props and their accessories.

Scarfing joint

This is among the most difficult types of joint to use with a repair, but also one of the most secure methods when accurately cut and fixed. The completed scarfing joint has a smooth or flush appearance on all its faces. In most cases, the scarfing joint should have a finished length between two and a half and four times the depth of the component being jointed.

Scarfing joints can vary slightly in design but fundamentally all do the same job of providing a strong, neat lengthening joint to the timber. In some cases, the scarfing joint might use timber dowels or stainless-steel bolts through it to help hold the joint together while the glue sets.

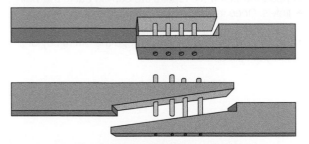

▲ Figure 9.6 Exploded view of a scarfing joint

▲ Figure 9.7 Scarfing joint used to replace rotten end of floor joist

Source: 'Repairing Historic Roof Timbers' by John Hoath on www.buildingconservation.com

In other cases, the scarfing joint might incorporate hardwood taper wedges that are driven into the joint. These wedges force the joint together and lock all of the surfaces while the glue sets.

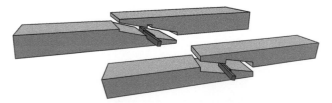

▲ Figure 9.8 Exploded view of a scarfing joint incorporating hardwood taper wedges

▲ Figure 9.9 Example of a completed scarfing joint with hardwood wedges

Sometimes, additional support may be recommended for the scarfing joint, usually in the form of steel 'fishplates' fixed to both sides of the joint and bolted together. The fishplates provide a considerable amount of extra support and are typically used when repairing timber that is prone to bending or bearing a considerable amount of weight, such as a purlin.

Fish plates

▲ Figure 9.10 Scarfing joint incorporating hardwood wedges and two bolted fishplates

Piecing in

Only a small section of the timber may require replacing. This could be on any side of the timber and could be a result of damage or decay. In these cases, only the damaged section is cut away and a suitable length of timber is 'pieced in' or set/let into the original piece. The new section can then be planed and sanded down to resemble the original section.

▲ Figure 9.11 The top section of timber has been cut away with a new section being pieced in

Splint repairs

If the method of repairing the structural timber is not defined by how it will look upon completion or by other restrictions dictated by a local authority, one of the simplest methods is to use a splint. The decayed section of the timber is cut away from the original timber with a simple 90° cut, and a replacement section is then cut from two lengths of timber with their ends butting up against each other. The two lengths are held together with timber splints that are fitted either side of the joint, with all three being bolted together.

▲ Figure 9.12 Timber splints used to join and strengthen floor joist

Source: 'Structural Timber Repairs' by Robin Russell on www.buildingconservation.com

In some cases, you will be advised to replace, repair or reinforce lengths of timber and/or their joints with metal plates. These can either be fitted between two lengths of timber (called a flitch plate) or fitted over and around the timber (referred to as metal strapping or shoes).

Flitch plates and beams

This type of repair or replacement requires a steel flat bar (flitch plate) to be inserted between two lengths of timber. In some cases, a whole new beam is formed containing a flitch plate. This form of construction has the advantage that smaller-sectioned beams or joists can be used to carry the same loads as a larger solid length of timber. This type of repair or replacement is often used when repairing or replacing floor joists and purlins.

▲ Figure 9.13 Single-flitch and double-flitch plate beams

Metal strapping and shoes

In some cases, support and strengthening can take the form of steel straps that are bolted or screwed to the face of the timber requiring the additional support. This additional strapping can also include steel shoes or branches to help support any additional timberwork that joins the side of the timber beam being repaired or strengthened. This type of repair is often used with load-bearing beams, where the branching side joists

join the main beam. Steel plates are often seen as a cost-effective way of repairing timber joists and beams while providing an invisible repair when viewed from the underside.

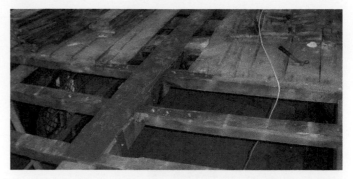

▲ Figure 9.14 Steel strapping incorporating branches used to support floor joists

Source: 'Repairing Historic Roof Timbers' by John Hoath on www.buildingconservation.com

Non-structural repairs

The repair of non-structural timbers will involve similar considerations to those for structural repairs in deciding whether a repair should take place or if a replacement component is required instead. Usually, any repair will be made by splicing in a new section of material; as with structural timbers, the type of timber and its moisture content should match those of the component being repaired. For small repairs, wood filler could be an option, particularly if the finished surface is to be surface finished with paint.

Splicing

When splicing in a new section of timber to a component that has a moulded finish, such as a window frame or window sash, the splice or replacement section should be made from timber that has been pre-moulded. This will enable the splice to better match the existing profile without requiring an over-long and difficult repair.

There are a number of ways to cut timber to accept a splice, and the location of the splice may determine precisely how this is done. Most positions likely to require a splice can be accessed and cut easily with an **oscillating multi-tool**.

▲ Figure 9.15 Oscillating multi-tool

Depending on the repair required, either a stepped splice or partial splice can be used. The stepped splice is used when a full sectional dimension piece of timber needs to be inserted. This should always be used when performing a splice on windows, rather than a 45% splice, so that the pieces of wood won't move over time. A high-strength glue should be used rather than screws to avoid rust.

The angle can be marked onto the section requiring the cut, while the replacement section can be set and cut easily on a chop saw. Any splice should be glued and, where possible, screwed in place and, when the glue has fully set, cleaned down with abrasive paper ready for surface protection and finishing.

When splicing in new sections, it is important to establish how far the rot or damage has penetrated. Remove all damaged timber until sound timber is found. In some cases, particularly with a partial-depth splice, it can be difficult to remove all the waste with a saw, so a bevel-edged chisel should be used for a good, flat finish ready to fix the splice in place. On occasion, the splice can protrude slightly above the existing timber. This allows for a plane and abrasive paper to finish the splice flush with the existing timber, ready to accept the surface finish. The fixings used to hold the splice in place must be finished below the surface and filled with external filler.

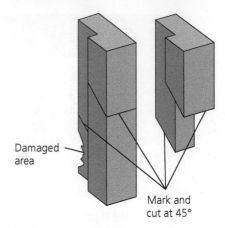

Damaged area

Mark and cut at 45°

▲ Figure 9.16 Stepped splice used with a full sectional repair

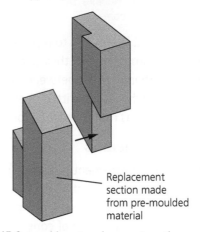

Replacement section made from pre-moulded material

▲ Figure 9.17 Cut and insert replacement section

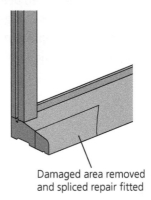

Damaged area removed and spliced repair fitted

▲ Figure 9.18 Partial sectional repair to a window cill using a splice repair

▲ Figure 9.19 Forming a splice using an oscillating multi-tool

▲ Figure 9.20 Bottom portion of jambs and mullions in a window cut away ready to accept the replacement splices

Staircases are often prone to damage and require repairs, the most common areas requiring attention being treads, handrailing and the balustrade. Treads often suffer from broken or damaged nosings, while the balustrade suffers from loose balusters and newel posts.

Repairs carried out on damaged and broken stair nosings usually involve splicing or fitting in a replacement section to the stair tread, very much as would be done in any other splicing task. Because the tread nosing is such an important safety feature of a staircase, additional support to the underside of the nosing repair may be required to prevent the nosing repair breaking away when trodden on, particularly if a large section of the nosing has been removed and replaced. In such circumstances, an additional length of beading can be fixed to the face of the riser and the underside of the nosing.

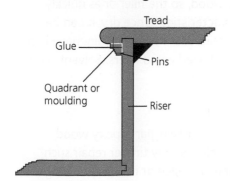

▲ Figure 9.21 Additional moulding used to provide support to a nosing repair on staircases

In some cases, it may be a better option to completely cover the treads and risers with new material. If so, all components must be suitably glued and fixed to the existing treads and risers.

▲ Figure 9.22 Overlaying existing treads and risers

Squeaky treads are loose, and are best cured from the underside of the tread, if access is available, by fitting additional glue blocks or by removing the existing tread wedge and replacing it with a new one, which is glued and fully knocked home. Where the underside of the step is inaccessible, an extra piece of moulding or beading can be used at both the top and bottom of the step.

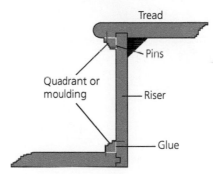

▲ Figure 9.23 Beading used at the top and bottom of a step to help prevent movement of squeaky treads

Handrailing can become loose where it is fixed to the newel or in its length, such as at the point of turns and scrolls. In some cases, access to the fixing can be gained by removing the timber plug or filler used to hide the fixing, so it is only a matter of tightening up the fixing and refilling the fixing entrance.

Where there is no access to the means used to join the handrail, such as with mortice and tenon joints used with newel posts, or dowel and glue joints used in lengthening handrails, additional fixings such as screws and glue can be used to pull the joint together. Additional fixings should be positioned so that there is sufficient access to fully tighten them, but at the same time they are unobtrusive and easily hidden.

Loose balusters can have a small amount of glue applied to the fixing point and pinned back into place with panel pins, which are punched below the surface. Newel posts will usually require further fixings through the newel into the trimmer joist or the stair string, with the fixing heads being filled with either wood filler or timber pellets.

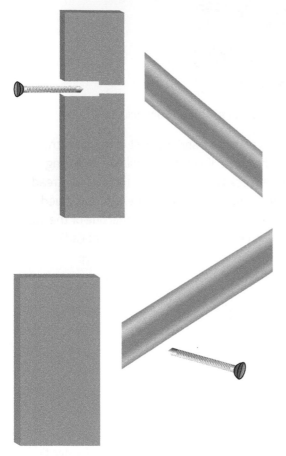

▲ Figure 9.24 Additional fixing used to secure handrailing to newel posts

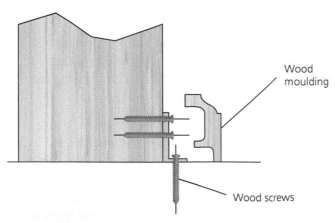

Wood moulding

Wood screws

▲ Figure 9.25 Angle bracket fixed to newel and floor providing additional support to the bottom of the newel

With the majority of non-structural repairs, the finished repair is likely to be on show, so it will usually require a smooth, high-quality finish. The repair needs not only to be structurally sound and securely fitted, but also to be finished to a suitable standard to accept the required surface finish. It is likely that the finished splice or repair will have small imperfections, such as:

- inaccuracies in the joints
- inaccuracies in the finished levels of the repair
- any fixing holes used to secure the splice may need filling and finishing.

These imperfections can be filled with a suitable wood filler, and then sanded and finished to accept the surface coverings.

Wood fillers

There are several types and makes of wood filler, but in general terms they all fall under one of the following types:
- water-based filler
- solvent-based filler
- epoxy-based filler.

Water-based wood filler

Water-based wood filler is used for its easy cleaning properties, and it does not dry out easily in the container. It can shrink when it dries after application, so it is best to overfill the repair. It sands easily when dry. This is a good type of filler for getting unfinished wood ready to paint. It is ideal for making light repairs to woodwork and furniture, but should not be used for external repairs.

Solvent-based wood filler

Solvent-based wood fillers are more heavy duty. They contain actual wood, so the filler dries quickly and with similar characteristics. Once dry, it can be sanded, drilled and stained. Shrinkage is minimal when compared with water-based wood fillers. Solvent-based filler is ideal for external use.

Epoxy-based filler

Epoxy-based wood filler or two-part epoxy wood fillers can be used on almost any timber repair such as repairing and filling damaged or rotten mouldings, wooden window cills, and door and window jambs, window sashes and staircases. Epoxy-based fillers are prepared by mixing together two parts (a hardener and a putty) to form a putty that sets very hard. When the putty has hardened, it can be sanded back to produce a fine surface finish that does not shrink. This type of filler is generally considered the best for external use.

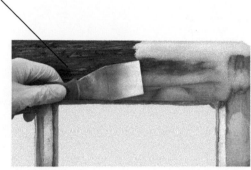

Area of rot removed and filled with two-part epoxy filler

▲ Figure 9.26 Window sash repaired with two-part epoxy wood filler

Whichever type of filler is being used, it is generally best practice to use several thin applications to fill the defect rather than one deep fill. This means each layer dries out quickly, allowing for a faster finish to the repair. If one deep fill is used, the surface of the filled repair will dry quickly but, as you work this top surface, the deep part of the fill is still unstable and pliable.

When the filler has set fully, it can be finished smooth with abrasive paper. It is usual to start with a coarse-grit abrasive paper and progressively move up to a smaller-grit paper to achieve a fine finish suitable for painting or staining. Using a sanding block or power sander helps to keep the surface flat. Once you have sanded back to the surface, cover the filler with the appropriate finish.

ACTIVITY

In your centre, practise cutting and splicing a door or window frame and a piece of existing architrave. Ensure that you use the correct tooling and protect the surrounding area from damage. Practise using filler and sanding the work to an acceptable finish.

Ironmongery repairs and maintenance

In most cases, the ironmongery used on timber components such as doors and windows requires very little maintenance, and only a limited amount of adjustment or repair can be carried out (see Table 9.4).

▼ Table 9.4 Examples of types of repair and maintenance that can be carried out on ironmongery

Type of ironmongery	Type of maintenance or repair
General-purpose hinges	Light lubrication with oil to moving parts, or replace if damaged
Friction hinges Tension	Light lubrication to moving parts using oil Adjust tension block screw for ease of use

▼ Table 9.4 Examples of types of repair and maintenance that can be carried out on ironmongery (continued)

Type of ironmongery	Type of maintenance or repair
Locks and latches 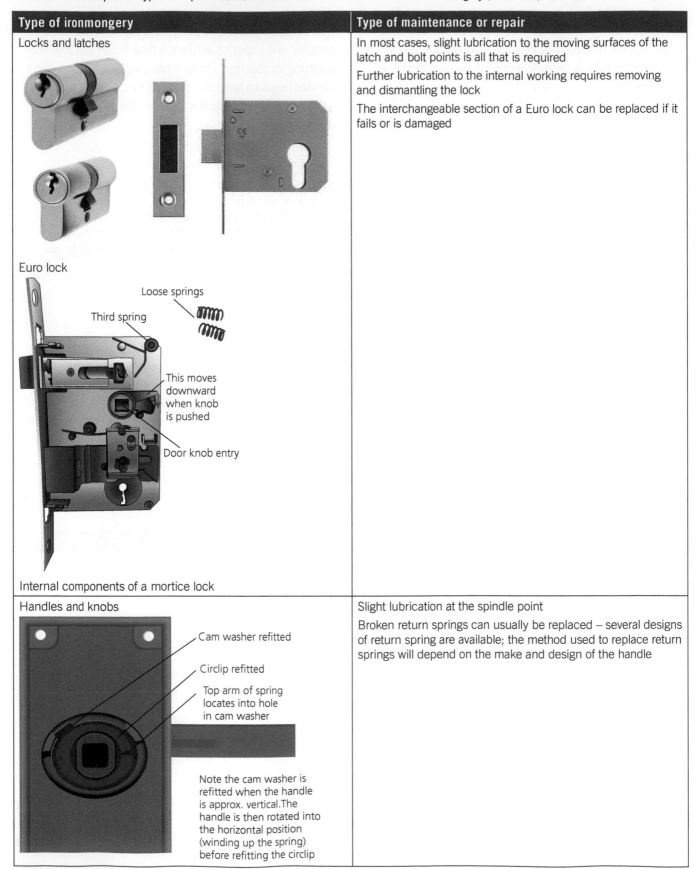 Euro lock Internal components of a mortice lock	In most cases, slight lubrication to the moving surfaces of the latch and bolt points is all that is required Further lubrication to the internal working requires removing and dismantling the lock The interchangeable section of a Euro lock can be replaced if it fails or is damaged
Handles and knobs	Slight lubrication at the spindle point Broken return springs can usually be replaced – several designs of return spring are available; the method used to replace return springs will depend on the make and design of the handle

Loose springs

Third spring

This moves downward when knob is pushed

Door knob entry

Cam washer refitted

Circlip refitted

Top arm of spring locates into hole in cam washer

Note the cam washer is refitted when the handle is approx. vertical. The handle is then rotated into the horizontal position (winding up the spring) before refitting the circlip

Repairing guttering and downpipes

Gutters and downpipes can be manufactured from a range of materials, including wood, cast iron, plastic and composite materials. Plastic is most common today, as it has the advantage of being very light and easy to fit, but it is also brittle and easily damaged. When fitting plastic guttering, always ensure the rubber sealing strip used at the joints is fully located in its slot, otherwise the guttering is likely to drip at the joint.

There are several standard profiles available. The most common are:

- ogee (or OG, for 'ornamental gutter') – a modern redesign of a Victorian gutter profile; it is used where a period look is important, on new-builds as well as Victorian refurbishments
- square section – very popular in the 1980s and 1990s, it is used with square-section rainwater pipes; square section has a very good rainwater capacity
- round – the standard gutter profile, used on many domestic properties throughout the UK
- deep flow – a deeper version of the round profile. It is slightly elliptical, and is generally used on larger or steeper angled roofs where the velocity and volume of the water entering the gutter is high.

The guttering of a typical house consists of the components shown in Figure 9.27.

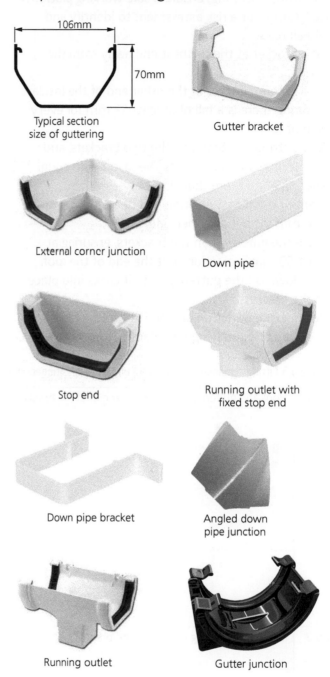

106mm

70mm

Typical section size of guttering

Gutter bracket

External corner junction

Down pipe

Stop end

Running outlet with fixed stop end

Down pipe bracket

Angled down pipe junction

Running outlet

Gutter junction

▲ Figure 9.27 Common components used in rainwater guttering

Fitting guttering

When fitting guttering, ensure a safe working platform is used. Carry out a risk assessment to identify and rectify all risks.

1 Fit a bracket at the furthest end away from the outlet.
2 Fit a second bracket at the other end of the fascia, ensuring there is a fall of at least 12 mm for every 3 m of run.
3 Run a string line between the two brackets, and fit intermediate brackets to the fascia with equal spacing of no more than 1 m between them.
4 Identify the outlet position and fit a bracket 150 mm from it at either side.
5 Place the guttering on the brackets, ensuring at least 50 mm projection past the end of the roof. Pull down on the guttering until it clicks into place in the brackets.
6 Position the outlet union and fix it to the **fascia**, then clip the guttering into place.

> **KEY TERM**
>
> **Fascia:** the flat panel that sits vertically beneath the eaves of a roof.

Table 9.5 outlines the process to use when fitting replacement guttering into pre-positioned brackets.

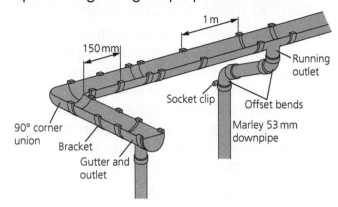

▲ Figure 9.28 Components used in a typical run of rainwater guttering

▼ Table 9.5 Fitting replacement guttering into a pre-positioned bracket: step by step

Step 1	Hook the guttering under the rear clip of the bracket.	
Step 2	Push the guttering down into the bracket.	

▼ Table 9.5 Fitting replacement guttering into a pre-positioned bracket: step by step (continued)

Step 3	Hook the guttering under the front clip of the bracket and pull the guttering down into it until it clicks into place.	
Step 4	Make sure you leave a 5 mm gap between the bracket and the end of the gutter, to allow for the plastic expanding and contracting as the temperature changes.	

Downpipes

Many components are used in fitting downpipes. Figure 9.29 shows the position of these components, along with the guttering. The text that follows the diagram describes each part.

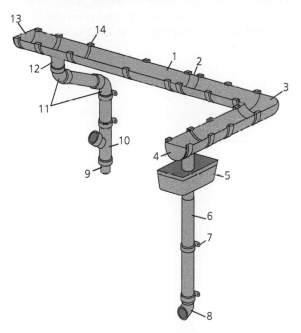

▲ Figure 9.29 Parts of a rainwater guttering and downpipe system

1 Gutter: available in various lengths, the guttering must slope gently downwards (by at least 12 mm every 3 m) towards the running outlet in order to provide effective drainage. It should be fixed at 1 m intervals with gutter support brackets.

2 Gutter union bracket: connects two gutter pieces.

3 90° gutter angle: allows a run of guttering to continue around a corner. A gutter support bracket must be fixed within 150 mm on either side of the angle.

4 Stop end outlet: sits at the end of a run of guttering to close the pipeline, and releases water by connecting to the downpipe. Usually supplied as a 'right end' or 'left end', depending on the layout.

5 Hopper: funnel-shaped rainwater collector that diverts to a downpipe.

6 Downpipe: available in various lengths. Releases water down to the shoe and is fixed to the wall with downpipe brackets.

7 Downpipe bracket: secures the downpipe to the wall. A bracket should be positioned near the top of the first pipe, and then around every 1 m after that.

8 Shoe: fitted at the base of the downpipe to change the direction of the flow of water, discharging it clear of the wall into a drain or hard standing. It should be positioned around 40 mm above the ground to allow water to run away from the building.

9 Downpipe connector: allows more than one downpipe to be connected in series. Plastic downpipes sit with the upper pipe inside the lower, which creates a watertight seal.

10 Branch: single branch for joining two downpipes together, to divert the water from another roof section into the same drain.

11 Offset bends: at either a 112.5° or 92.5° angle, these bends bring the downpipe close to the wall, ensuring water runs vertically.

12 Running outlet: provides an outlet to the downpipe for rainwater along the length of guttering. Unlike a stop end outlet, it connects to a guttering run at both ends.

13 External stop end: closes off a run of guttering.

14 Gutter support bracket: attaches the guttering to the fascia at 1 m intervals.

Fitting downpipes

As with fitting guttering, ensure a safe working platform is used when fitting downpipes. Carry out a risk assessment to identify and rectify all risks.

1 From the running outlet, drop a string line and plumb bob.

2 Fit a downpipe bracket close to the outlet, leaving one screw out.

3 Measure equal spaces of no more than 1 m, down to where the shoe will be fitted.

4 Fit the remainder of the brackets, leaving one screw out.

5 Position the downpipe into the brackets and fit the remaining screws, using connectors as required.

6 Finish by positioning the shoe at the bottom of the guttering.

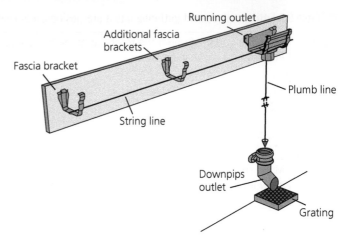

▲ Figure 9.30 Lining through guttering and downpipe with a string line and plumb bob

ACTIVITY

Imagine a potential customer has asked you to produce a quote for the materials required to replace their rainwater system. Undertake an inspection of the rainwater guttering and downpipe system used in a building near you. Identify the type of profile being used and all the components required to replace the full rainwater system.

Using the information you have obtained, find a price for each of the components and produce a quote identifying each component, and its replacement cost.

Test your knowledge

1 What are the roots of fungi called?

  A Mycelium

  B Hyphae

  C Sporophore

  D Dry rot

2 The fruiting body of fungi is known as:

  A mycelium

  B wet rot

  C sporophore

  D hyphae.

3 At what percentage of moisture content will timber start to rot?

  A 17 per cent

  B 15 per cent

  C 10 per cent

  D 20 per cent

4 Which of the following is a cause of wet rot?

  A Leaking gutter

  B Painted timber

  C Varnished timber

  D Insect infestation

5 What type of rot is shown below?

  A Wet rot

  B Dry rot

  C Black rot

  D Grey rot

6 What type of rot is shown below?

  A Wet rot

  B Grey rot

  C Black rot

  D Dry rot

7 What is the recommended way to dispose of timber infected with dry rot?

  A Burn on site (with permission)

  B Bag and seal it prior to disposal

  C Leave on site

  D Place in a skip

8 What is the life cycle of a furniture beetle?

  A Larva, pupa, egg, larva

  B Beetle, larva, pupa, egg

  C Egg, larva, pupa, beetle

  D Egg, pupa, larva, beetle

9 What should be looked for if an insect infestation is ongoing?

  A Flight holes and frass

  B Mycelium

  C Hyphae

  D Blue stain

10 What joing is usually used for spliced repairs in structural timbers?

  A Halving joint

  B Housing joint

  C Mitred splice joint

  D Scarfing joint

11 What is the common name given to this type of prop?

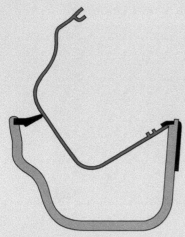

A Pressure prop

B Awning prop

C Acrow prop

D Protection prop

12 What name is given to an attachment commonly used with steel props?

A Tallboy

B Bigboy

C Toughboy

D Strongboy

13 What is the name given to the steel bracing used either side of a structural splice?

A Flitch plate   C Fishplate

B Side plate   D Dogplate

14 What is the name given to the steel plates fixed between structural timbers?

A Flitch plate

B Centre plate

C Fishplate

D Dogplate

15 What is the normal splicing angle used for spliced repairs?

A 25°

B 35°

C 45°

D 55°

16 What is the recommended fall of guttering over a length of 3 m?

A 15 mm

B 10 mm

C 18 mm

D 12 mm

17 Which of the following is generally considered the best type of wood filler for external repairs?

A Water-based filler

B Epoxy filler

C Solvent-based filler

D Putty filler

18 What is the purpose of a string line and plumb bob when erecting guttering and downpipes?

A It helps keep the guttering and downpipe horizontal

B It helps keep the guttering and downpipes vertical

C It helps with the location of the rubber gaskets in all joints

D It helps with the positioning of guttering and downpipe brackets

19 What does a stop end do when erecting guttering and downpipes?

A It directs the water into the drain

B It connects two lengths of downpipe

C It stops water running out of the end of the guttering

D It provides support to the downpipe at its lowest point

20 What is the purpose of a shoe when erecting guttering and downpipes?

A It directs the water into the drain

B It connects two lengths of downpipe

C It stops water running out of the end of the guttering

D It provides support to the downpipe at its lowest point

Answers

CHAPTER 1

Answers to activities
Improve Your Maths, page 26
75°

Test your knowledge

1	C	6	D
2	A	7	B
3	B	8	A
4	C	9	B
5	A	10	

11 • Do not remove the guardrail: you are not a trained scaffolder and you have not been authorised to do so.
   • Leave the scaffold: it may be unsafe to continue to work at height without the correct collective protective measures in place to prevent a fall.
   • Report the incident to your supervisor.
12 • It is a practical test designed to assess the fit of half-mask respirators and the level of protection they may offer. The test may involve you wearing other items of PPE at the same time to ensure they are compatible and still protect the user.
   • You have a legal obligation to co-operate with your employer, to enable it to meet its duties under the Health and Safety at Work Act.
13 96 people.
14 At 80 dB hearing protection should be made available, at 85 dB it is mandatory to wear it.

CHAPTER 2

Answers to activities
Improve Your Maths, page 60
20.2 m

Improve Your Maths, page 61
Area of rectangle: $5 \times 3 = 15$ m²
Area of circle: $3.142 \times 3^2 = 28.28$ m²

Improve Your Maths, page 61
Volume: $0.6 \times 0.3 \times 29.2 = 5.256$ m³

Cost: $5.256 \times 90 = £473.04$

Improve Your Maths, page 61
Volume: $240 \times 0.225 \times 0.05 = 2.7$ m³

Cost: $2.7 \times 450 = £1215$

Improve Your Maths, page 62
a 96
b 90
c 649.8

Activity, page 68
a 4.5 weeks
b 8 weeks

Test your knowledge

1	B	6	D
2	C	7	C
3	B	8	B
4	B	9	C
5	D	10	B

11 Refer back to the text on page 56, under the heading 'The tender process'.
12 Example hourly rate £20.02 × 2 hours = £40.04
   £18.74 for the hire of a cordless plane
   £40.04 + £18.74 = £58.78
   6% of £58.78 (overheads) = £3.53
   £58.78 + £3.53 = £62.31
   21% of £62.31 (profit) = £13.09
   £62.31 + £13.09 = £75.40
   20% of £75.40 (VAT) = £15.08
   £75.40 + £15.08 = **£90.48**

13 Here are some examples:
- Posters: advantages – clear, everyone receives the same message; disadvantages – can be overlooked, can become outdated
- Verbal: advantages – quick, the recipient can ask questions for clarity; disadvantages – no written record, can be misunderstood
- Safety signs: advantages – clear, simple, quick; disadvantages – their meaning may not be understood

14 Speak to your tutor or supervisor about the Toolbox talk you have prepared. They will be able to identify the strong points of the presentation and the areas that could be developed further.

15
- Air and ground source
- Solar/photovoltaic panels
- Wind turbines
- Tidal (water)

CHAPTER 3

Answers to activities
Activity, page 131

0.83 mm

Test your knowledge

1	A	11	A
2	A	12	D
3	D	13	C
4	D	14	C
5	A	15	A
6	B	16	A
7	B	17	B
8	A	18	B
9	C	19	B
10	A	20	C

CHAPTER 4

Answers to activities
Improve Your Maths, page 185

1.25 m

Improve Your Maths, page 198
Pitch 26.6°

Test your knowledge

1	C	11	C
2	F	12	D
3	A	13	B
4	C	14	A
5	D	15	B
6	A	16	B
7	D	17	A
8	A	18	C
9	C	19	D
10	D	20	B

CHAPTER 5

Test your knowledge

1	A	6	C
2	C	7	C
3	D	8	B
4	C	9	D
5	A	10	B

11 Refer to pages 280–282 for information about installing and fixing window boards.

12 Refer to the section headed 'Sealing window frames' on page 277 for information about the use of smoothing tools.

13 Speak to your supervisor about methods used at your place of work.

14 Refer to the section headed 'Fixing foam' on pages 276–277 for information about precautions to take when using expanding foam.

15 Refer to the section headed 'Meeting stiles' on pages 234–235 for the answer to this question.

CHAPTER 6

Answers to activities
Improve Your Maths, page 293

1500 mm

Test your knowledge

1	C	6	D
2	B	7	A
3	C	8	B
4	B	9	B
5	D	10	C

Ask your supervisor or trainer to check your answers to questions 11–15.

CHAPTER 7

Answers to activities

Improve Your Maths, page 342

20

Improve Your Maths, page 349

£47.25

Improve Your Maths, page 354

£561.60

Improve Your Maths, page 357

12

Activity, page 358

320 mm

Improve Your Maths, page 359

35.6°

Activity, page 359

14 mm

Test your knowledge

1	B	6	D
2	D	7	B
3	B	8	A
4	C	9	C
5	C	10	B

11 Bullnose, 'D' end, commode, curtail

12 The total rise is taken from finished floor level to finished floor level. In addition, the total available going is required in each direction in which a flight of stairs is required.

13 a 2590 ÷ max rise (220) = 11.77. You can't have 11.77 so divide by the next whole number: 2590 ÷ 12 = 215.83, say 216 mm.

b 42°

c 900 mm

14 a The riser of the step is mitred to the step profile of the string, then glued and pinned or screwed and blocked from behind.

b The step is wedged into a housing on the inside face of the string.

CHAPTER 8

Answers to activities

Improve Your Maths, page 393

16 balusters, with a spacing of 95.4 mm

Test your knowledge

1	B	11	C
2	D	12	C
3	B	13	B
4	C	14	D
5	C	15	B
6	A	16	A
7	B	17	D
8	A	18	C
9	A	19	D
10	D	20	A

CHAPTER 9

Test your knowledge

1	B	11	C
2	C	12	D
3	D	13	C
4	A	14	A
5	A	15	C
6	D	16	D
7	A	17	B
8	C	18	D
9	A	19	C
10	D	20	A

Glossary

Abrasive timbers: timbers that will quickly blunt or dull the cutting edge of tooling.

Acute angle: an angle between 0° and 90°.

Aesthetic: relating to a visual sense of beauty.

Angle of hook or rake: the angle at which the face of the saw tooth slopes from the tooth tip, either down and forward from the tip, as in the case of negative tooth profiles for crosscutting, or down and backwards from the tooth tip, as in the case of positive tooth profiles for ripping.

Anti-friction rollers: adjustable, free-running rollers set into the machine bed of a thicknesser, enabling the material to move through the machine without sticking.

Approved Code of Practice (ACoP): a document, usually produced by the HSE, that provides practical advice on how to interpret regulations, for example COSHH, The Provision and Use of Work Equipment Regulations or Safe Use of Woodworking Machinery.

Architectural technician: a technician who is employed by an architectural practice to produce construction drawings.

Backset: the distance from the face of the latch to the centre of the spindle.

Barefaced: where there is a shoulder on only one side of a joint.

Barefaced tenon: a tenon that is shouldered on only one face of the joint.

Bisection: the division in two geometrically of an angle or line.

Black japanned: finished with a black enamel lacquer originally associated with products from Japan.

Bolection mouldings: small timber sections moulded with a rebate, usually a decorative profile. The rebate on the moulding provides a secure fixing, and hides any potential gaps created as a result of shrinkage in the timber at a later stage. Bolection moulds are commonly used to secure glazing and panels into doors, windows and screens.

Borrowed light: light that enters a room or corridor from an adjoining room through a glazed opening, usually above a door.

Box frame: the name traditionally given to vertically sliding windows, due to the nature of the jamb and head construction.

BREEAM: Building Research Establishment Environmental Assessment Method, a certification of the assessment of a building's environmental, social and economic sustainability performance.

Brick-on-edge: where bricks are laid on edge with the stretcher face uppermost.

Camber: a slight curve in the depth of the string. Remember, a string is a structural component, basically an inclined joist, so any camber should be positioned uppermost to resist deflection.

Case hardening: a defect caused by the timber being dried too rapidly, leaving the outside dry but the centre still wet. It typically causes the material to bend and twist during cutting, resulting in binding on the saw blade and kickback.

Casting: the curling and movement of timber.

Cellulose: a substance found in the structure of plant cells. Cellulose is a major component of timber and helps to make it strong.

Centre: (when making curved joinery) a temporary structure made by a carpenter or joiner to allow a bricklayer to turn a brickwork arch around it. When the mortar has set, the centre may be 'struck' (removed).

Centres: the distance from the centre of one rafter to the centre of the next.

Centrifugal force: a force created by a rotating body, such as by a rotating cutter block, where objects are forced away from the centre of rotation.

Collet: a means of centralising the auger in a chisel headstock.

Combination frame: a door frame that also extends to incorporate glazed and solid panelled areas.

Compass plane: a plane with a flexible sole, which can be adjusted to both concave (inside of a curve) and convex (outside of a curve) shapes.

Compound cut: a cut that consists of two angles – for example, the bevelled angle from a canted saw blade and the mitre angle (or crosscut angle) from the fence.

Conservation area: an area, often in a town or city, that has been designated to be of special architectural or historical interest, which is protected by law from changes that might reduce its character and importance.

Contract documents: these comprise the working drawings, schedules, specifications, bill of quantities and contracts.

Convex: in timber, a face that bulges out in the middle.

Crown: the uppermost part of a shaped headed door or frame.

Cut roof: a construction technique that builds up the roof without the use of trusses.

Cutter block: the machine block that holds the cutting knives or cutters.

Decibel (dB(A)): the unit that measures sound levels. The A weighting is applied to instrument-measured sound levels to try to account for the relative loudness perceived by the human ear.

Demountable power feed unit: an automatic feed system that can be moved out of the way. It is used mainly for continuous feeding of material during cutting or profiling operations.

Dihedral angle: the angle given to the top edge of the hip rafter running up its length from the eaves to the ridge.

Dimension paper: paper with vertically ruled columns, on which building work is described, measured and costed.

Dishing: where the face or edge of the timber has been scooped out by the base of a sander as a result of not maintaining an even pressure, or sanding in one place for too long.

Door leaf: one half of a pair of double doors.

Dormer: an opening in the roof surface where a secondary roof projects away from the plane of the existing roof, typically used to give greater head room within the primary roof – for example, in a loft extension.

Draw-bored joint: a method of creating joints that allows them to be pulled together without the need for cramps.

Dropping on: on a cutting machine, the pushing of the material into the revolving cutters part way along.

Dry fit: the process of fitting joints one at a time to ensure the shoulder fits on both sides with the shoulder square or at the required angle, that the joint is not twisted, and it is flush or parallel as required with the face.

Dubbing off: where the face or edge of the timber has been bevelled as a result of not maintaining direct contact with the surface while it was planed or sanded.

Duty holder: anyone with a legal responsibility under health and safety legislation is referred to as a duty holder. For example, employers, employees and the self-employed all have legal responsibilities to comply with, therefore they are 'duty holders'.

Egress: a means of exiting a room, building or space. Also a term used in construction law to describe a way out or exit. Employers have a legal responsibility to provide safe access to and egress from your place of work.

Equilibrium moisture content: the moisture content at which the wood is neither gaining nor losing moisture.

Espagnolette bolt: a multi-point locking device that is normally on the locking side of the door.

Façade: the exterior face of a building; also referred to as the 'front face' of the building.

Face side and face edge: the two sides that have been planed square to each other on the surface planer.

Facet: a section forming a flat face – for example, the panes of glass within a segmental bay window.

Fair ending: trimming the end of a board to remove defects such as splits and to square it up.

Fascia: the flat panel that sits vertically beneath the eaves of a roof.

Ferrous: containing iron.

Flier: the traditional term for parallel treads/steps.

Float glass: the standard type of glass readily available.

Folding wedges: a pair of wedges used 'back to back' to create a pair of parallel faces. By sliding one wedge against the other, the total parallel thickness of the folding wedges can be adjusted to pack out the required gap.

Forend face plate: the part of the lock that is seen when the lock is housed in the door.

Forest Stewardship Council (FSC®): an international not-for-profit organisation that promotes responsible management of the world's forests.

General operative: a person working on a construction site without a specific set of trade skills. They are usually employed by a contractor to carry out labouring tasks, such as mixing plaster and mortar, keeping the site clean, and stacking and storing materials.

Geometrical staircase: a staircase that turns and rises without the use of landings. The stair strings are formed to rise and turn and are also known as 'wreathed strings'.

Geometry: a branch of mathematics that deals with points, lines, angles, surfaces and solids.

Glue line: the amount of glue contact area of a joint. This is particularly important when designing widening joints. The greater the glue line, the stronger the joint.

Glulam: a common abbreviation for 'glue laminated'.

Going off: the part-drying of an adhesive.

GRP: glass reinforced plastic, a composite material. For example, garage doors are often manufactured with a combination of insulating foam, timber, PVC and GRP, making them resilient to the weather, low-maintenance and well insulated.

Heading joint: a lengthening joint between two components.

Health and Safety at Work etc. Act 1974 (HASAWA, HSWA or HASWA): the primary piece of health and safety legislation that employers, the self-employed, employees, people in control of premises, manufacturers, designers, importers and suppliers of equipment have a duty to comply with.

Health and Safety Executive (HSE): the main UK body responsible for the encouragement, regulation and enforcement of health, safety and welfare in the workplace. The HSE is also responsible for research into occupational risks in Great Britain.

Health and safety policy: a legal document written by employers with five or more employees outlining how health and safety will be managed on site.

Heat recovery units: units installed in buildings to extract the stale air from the inside and replace it with fresh air from the outside. The heat from the stale air is trapped and used to heat the colder fresh air being introduced into the building.

High-build exterior stain: a microporous multi-layer timber coating that resists surface mould/algae and protects against UV light, resulting in a humidity-controlling timber finish.

High-performance: modified to give superior performance in terms of thermal and sound insulation.

Hygroscopic: able to absorb or lose moisture.

Hypotenuse: the side opposite the right angle; the longest side of a right-angled triangle.

Imposed loads: the live loads and any additional loads a roof may incur, such as the weight of snow (so it is sometimes called the snow load).

In stick: where the timber is stacked with thin laths ('sticks') separating the boards, which allows air to circulate and the moisture content to acclimatise.

Interlocked doors: on a narrow bandsaw, doors that are impossible to open while the machine is in motion and which prevent the machine from starting if the doors are open.

Interlocking grain: also known as 'refractory grain'; grain that spirals around the axis of a tree but reverses its direction regularly, causing a poor finish whichever way it is planed.

Joint line: the fit of a joint where two pieces of material are bonded to each other.

Kickback: when the material is thrown forcefully back towards the operator, potentially causing serious injury.

Kinetic lifting: a controlled method of lifting that ensures that the risk of injury is reduced to an acceptable level.

Knee: the convex or external joint in handrailing.

Knives: the part of the cutting block that does the cutting; they can also be known as 'blades' or 'cutters'.

Laminated glass: two pieces of standard glass with a thin plastic interlayer that acts as an adhesive holding the glass together in the event of breakage.

Lead time: the delay between the initiation and execution of a process. For example, the lead time between placing an order for a staircase from a joinery manufacturer and its delivery may be anywhere from two to eight weeks.

Lead-in and lead-out: where a templet can come into positive contact with a ring fence before the cutter comes into contact with the timber. This prevents snatching and allows the work to be controlled as it is fed in and out.

LEV: local exhaust ventilation system – a collective term referring to extraction systems, because not all airborne hazards are dust – they could also be mist, vapour, gas or fumes.

Lip cut: the part of the purlin that sits under the hip or valley rafter and touches the other purlin.

Listed building: a building that is protected by law and has restrictions on what you can and can't do to develop it. Listed buildings are categorised as Grade I (the most significant), Grade II* or Grade II. Before carrying out any building work on an old property, it is important to establish whether it is 'listed', by contacting your local authority.

Live loads: the weight of the structure – for example, the timber and tiles or slates on the roof.

Matchboarded doors: a series of timber boards jointed together (usually by a tongue and groove arrangement) in their width.

Mechanical joints: joints that, due to their design, hold or pull themselves together.

Meeting rail: the name given to both the top rail of a bottom sash and the bottom rail of a top sash as they meet when closed.

Method statement (a): a document that provides a logical sequence of how to complete a task safely. Method statements are not a legal requirement, but are often used to manage hazards identified in risk assessments.

Method statement (b): a document written at the planning stage of a construction project to manage any hazards identified in the risk assessments; it will identify resources that need to be costed and organised to ensure they are available when required.

Near miss: an incident that has happened, which had the potential to cause injury, illness or damage but did not (e.g. bricks falling from a scaffold platform and landing close to a worker).

Newel cap: a moulded solid timber cap recessed and applied to the top of a newel.

Noggins: timber used as a brace between rafters to help stiffen and provide a fixing point for restraint straps.

Nominated first aider: a person who has been recognised by their employer as having the necessary knowledge and skills to perform the role of a first aider at their place of work.

Non-ferrous: not containing iron.

Normal relationship: the average person's step length, according to Building Regulations Approved Document Part K. For stairs, it is twice the rise plus the going and should be between 550 and 700 mm.

Obscured glass: 'frosted' glass that provides a level of privacy by obscuring the view through the glass. Different patterns are available depending on the level of privacy required, graded from 1 (least) to 5 (greatest). This type of glass is commonly used in bathroom windows and front entrances to residential properties.

Obtuse angle: an angle between 90° and 180°.

Oscillating multi-tool: a specialist power saw that can perform accurate cuts in tight spaces that traditional saws find difficult to access. Multi-tools work by vibrating the saw head thousands of times per minute to create a sawing motion, and produce a fine, controlled cut.

PAT (portable appliance testing): there is a legal requirement for all portable electrical appliances to be tested on a regular basis (dependent on use). PAT testing itself isn't a legal requirement, but is one method of managing the process. Most companies test appliances annually.

Pelmet: boxing-in that is fixed in front of running gear to hide the track of a sliding door and provide a decorative finish.

Peripheral speed: on a circular saw, the speed at the periphery of the blade – how fast the outer edge of the blade is travelling.

Permitted development: the right to construct without obtaining planning permission. This is sometimes used to convert a loft or garage, or to add a single-storey extension or porch to an existing building.

Pick-up: where timber is fed into a cutter against the grain.

Piecing in: applying diamond-shaped pieces to the timber to repair a damaged area. Good piecing will be hardly visible.

Pitch line: a line two-thirds of the way down from the top of the common rafter, used to set out the lengths of the rafters.

Pitch marks: small circular cuts produced by all rotary planing machines.

Purlin: a large horizontal structural timber used to support rafters.

Quantity surveyor: a job role that involves producing the bill of quantities, and working with a client to manage costs and contracts.

Rack: a stair or other manufactured component that has been pushed out of shape, allowing the glue line to fracture and the joints to weaken and break.

Radiused corner: any corner whose sharp point has been softened by a radius.

Rake: the slope or pitch of the stairs.

Reamer: a type of rotary conical cutter used for cutting metal.

Reporting of Injuries, Diseases and Dangerous Occurrences Regulations (RIDDOR): legislation that puts duties on employers, self-employed people and those in control of work premises to report certain serious accidents in the workplace, occupational diseases and specified dangerous occurrences (near misses).

Revolutions per minute (rpm): the number of times the block goes round in one minute.

Risk assessment: a written document used to assess the level of risk in the workplace and the measures that should be taken to manage this risk. Risk assessments are a legal requirement under the Management of Health and Safety at Work Regulations.

Safety data sheet: a document produced by a product manufacturer to fulfil their legal duty under the HASAWA to provide information on products or substances that are hazardous to health.

Scale: a reduction in size by a given ratio, e.g. 1:5 on a location plan means that the measurement on the scale is five times smaller than the actual structure and area shown. In this case, if the full-size object is 500 mm long, the scale-size object shown on the drawing will be 100 mm long.

Second seasoning: where timber is sawn to the sizes required on the cutting list and left 'in stick' for as many days as time will allow. The freshly sawn faces will acclimatise to the atmospheric conditions in the joiners' shop, minimising the amount the timber will distort after planing.

Secondary machining: jointing and profiling of planed components.

Setter-out: an experienced joiner who has proved their competency, and is the person who produces the rods, cutting lists and orders for contracts.

Setting-up costs: costs involved in preparing the site for construction work, such as the costs of hoarding, temporary services and temporary site accommodation.

Sheathing: timber-based material used to cover the roof surface.

Shoot in: the process of fitting a door or sash to an opening with a parallel gap that allows for fitting and finish clearance. Allow 2–3 mm for a painted varnish and 1–2 mm for a polished finish.

Single curvature: (when manufacturing curved joinery) where work is shaped in one view only, either elevation or plan.

Site induction: a meeting or training session providing information to everyone entering a construction site on the risks that they are likely to face.

Snatching: in a machining context, this is where the rotating cutter hooks into the grain, causing it to break or the work to be thrown back at the operative. The operative could lose control or have an accident.

Snib: a small knob on bathroom lever handles used to throw the bolt on a sash lock.

Soakers: lead installed between each course of tile or slate to prevent water ingress.

Spandrel frame: the panelling used to enclose the space under a staircase. Spandrel frames often have a door built into them to allow access to the void under the stairs, turning it into a usable storage space or a small room where a toilet and sink can be installed.

Specified diseases: diseases that are reportable under the Reporting of Injuries, Diseases and Dangerous Occurrences Regulations (RIDDOR) 2013.

Spelching: when, during cutting on saw machinery, part of the material breaks away in an uncontrolled manner, damaging the components being cut.

Splinter guard: a part of the striking plate that extends over the front edge of a door frame or lining.

Spokeshave: a tool used to shape curved surfaces, consisting of a blade fastened between two handles. The sole can be flat or round bottomed, for planing concave or convex shapes respectively.

Sporophore: the fruiting part of the fungus.

Spreading: where the roof structure starts to move and pull outwards from the wall plate, causing the ridge to sink and possibly destabilising the whole structure.

Spring set: the bending of the top third of alternating teeth in opposite directions around the saw blade.

Spring-back: where the tension in the timber wants to pull the curve flat and it loses its intended shape.

Springing line: where a curved section starts to 'spring away' from a straight line.

Statutory undertakings: the services that are brought to the site, such as water and electricity.

Stretchout: a drawing showing the unwrapped face of a newel or geometrical string.

Stuck: a traditional term for the working of the moulding, grooves and rebates onto solid timber using hand tools and machines.

Take off: taking the length, width and thickness of components and recording them on the cutting list.

Taken off: materials measured from the contract drawings.

Templet: also known as a 'template', this is a thin piece of hardboard, MDF or plywood that is cut to the shape required and is then used to help reproduce that shape.

Tender: a process of formally estimating the costs of potential building work, presented by a number of contractors to the client.

Throat: the horizontal distance from the cutting edge of the blade to the machine casing.

Thrust wheel: the small wheel behind the blade of a narrow bandsaw, which prevents the blade from being pushed backwards during cutting operations.

Timber mouthpiece: used to prevent material becoming trapped between the machine bed and the blade; this should be replaced as often as required.

Tinnitus: a permanent ringing or hissing in the ears, often caused by exposure to loud noise such as that from woodworking machinery.

Toggle clamp: an adjustable quick-acting clamp often used on cutting machines.

Total going: the horizontal distance between the first and last riser in a straight flight.

Toughened glass: standard float glass that is heated and rapidly cooled to give added strength. To conform to BS 6262–4:2005, all toughened glass must be permanently marked (usually etched in the corner) to demonstrate it has achieved product safety standards. It must also display the manufacturer's name or trademark.

Tracking: tilting the top pulley wheel of a narrow bandsaw to make the blade run in the centre of the pulley.

Tungsten carbide tips (TCTs): the cutting edges of the saw blade. They are very hard wearing but are brittle and can be easily damaged when changing the blade.

Turned finial: a turned decorative finish to the top of a post, the most common being 'acorn' shaped.

U-value: also known as thermal transmittance, this is the rate of heat transfer through a structure (in watts). It is expressed as W/m²K (watts per metre squared Kelvin).

Unit rate: the labour rate + the material rate, resulting in a figure that can be used in quotations and tender documents.

VOC: the volatile organic compounds measure shows how much pollution a product will emit into the air when in use.

Walkthrough illustration: an animated virtual-reality sequence as seen at the eye level of a person walking through a proposed structure.

Warranty: an assurance of performance and reliability given to the purchaser of a product or material. If the item fails within the warranty period, the purchaser is entitled to a replacement or to have it repaired; depending on the terms of the warranty.

Weathering: a means of preventing water from gaining entry. This can sometimes be a simple profiled angle on the top edge of the timber exposed to the elements, e.g. on a door frame cill.

Well: the gap between two flights of stairs; also the space that the staircase fits into.

Winders: tapered steps used to save space by allowing extra risers to be incorporated. They generally turn through 90° or 180°.

Index

Notes